Bob Reuben.

UNDERWATER WELDING

SOUDAGE SOUS L'EAU

UNDERWATER WELDING

Proceedings of the International Conference held at Trondheim, Norway,
27-28 June 1983 under the auspices of the International Institute of Welding

SOUDAGE SOUS L'EAU

Communications présentées à la Conférence Internationale tenue à
Trondheim, Norvège, les 27 et 28 Juin 1983 sous les auspices de
l'Institut International de la Soudure

Published on behalf of the

INTERNATIONAL INSTITUTE OF WELDING

by

PERGAMON PRESS

OXFORD · NEW YORK · TORONTO · SYDNEY · PARIS · FRANKFURT

U.K.	Pergamon Press Ltd., Headington Hill Hall, Oxford OX3 0BW, England
U.S.A.	Pergamon Press Inc., Maxwell House, Fairview Park, Elmsford, New York 10523, U.S.A.
CANADA	Pergamon Press Canada Ltd., Suite 104, 150 Consumers Road, Willowdale, Ontario M2J 1P9, Canada
AUSTRALIA	Pergamon Press (Aust.) Pty. Ltd., P.O. Box 544, Potts Point, N.S.W. 2011, Australia
FRANCE	Pergamon Press SARL, 24 rue des Ecoles, 75240 Paris, Cedex 05, France
FEDERAL REPUBLIC OF GERMANY	Pergamon Press GmbH, Hammerweg 6, D-6242 Kronberg-Taunus, Federal Republic of Germany

First edition 1983

Library of Congress Cataloguing in Publication Data
Main entry under title:
Underwater welding.
English and French.
Papers presented at the first International Conference
on Underwater Welding, Conference internationale sur le
soudage sous l'eau.
1. Underwater welding and cutting–Congresses.
I. International Institute of Welding. II. International
Conference on Underwater Welding (1st: 1983: Trondheim,
Norway) III. Title: Soudage sous l'eau.
TS227.2.U53 1983   671.5'2   83-6312

British Library Cataloguing in Publication Data
Underwater welding.
1. Underwater welding and cutting–Congresses
I. International Institute of Welding
671.5'2   VM965
ISBN 0-08-030537-7

In order to make this volume available as economically and as rapidly as possible the authors' typescripts have been reproduced in their original forms. This method unfortunately has its typographical limitations but it is hoped that they in no way distract the reader.

Printed in Great Britain by A. Wheaton & Co. Ltd., Exeter

Other Pergamon Titles of Interest

ALLSOP & KENNEDY	Pressure Diecasting, Part 2
ASHWORTH	Ion Implantation into Metals
ATTERAS	Underwater Technology: Offshore Petroleum
BARRETT & MASSALSKI	Structure of Metals, 3rd Edition
COMINS & CLARK	Specialty Steels and Hard Materials
GIFKINS	Strength of Metals and Alloys (ICSMA 6)
HENLEY	Anodic Oxidation of Aluminium and its Alloys
IIW	The Physics of Welding
MANN	Bibliography on the Fatigue of Materials, Components and Structures, Volume 3
MASUBUCHI	Analysis of Welded Structures
OSGOOD	Fatigue Design, 2nd Edition
PARKIN & FLOOD	Welding Craft Practice
UPTON	Pressure Diecasting, Part 1

Pergamon Journals of Related Interest

Acta Metallurgica

Canadian Metallurgical Quarterly

International Journal of Engineering Science

International Journal of Mechanical Sciences

Materials Research Bulletin

Metals Forum

Ocean Engineering

Scripta Metallurgica

Welding in the World

PREFACE

The present volume contains the collection of papers presented on the occasion of the first International Conference organised by the International Institute of Welding on 27 and 28 June 1983 at the start of the 1983 Annual Assembly in Trondheim (Norway). This is the first time that the usually distinct events which mark the first two days of each Assembly (Houdremont or Portevin Lecture, Public session and Colloquium) are devoted to the same theme, that is to say, this year, underwater welding.

This theme is introduced by the Portevin Lecture of Professor Cotton (United Kingdom), developed by three lectures presented by invited authors: Messrs. Amiot (France), Christensen (Norway) and Masubuchi (USA) and studied in more detail in 33 papers prepared by authors who have replied to the call circulated by the Organising Committee.

This Committee, consisting on the one hand of members of the Norwegian Organising Committee and, on the other, of members of the Select Committee "Underwater welding" of the IIW, which had proposed the subject and programme of the Conference. The aim of the Select Committee was to assemble the maximum amount of scientific and technical information on underwater welding, to place this information at the disposal of interested parties and to promote, within the IIW, the studies required to ensure progress in the relevant subjects, whether associated with welding technology or the testing and inspection of assemblies carried out under water.

The holding in Norway, a country where the subject of underwater welding is of prime importance, of this first International Conference, shows the concern of the International Institute of Welding to concentrate its activities on contemporary problems whose urgency justifies an appeal for international cooperation.

H. Granjon
Scientific and Technical Secretary
Chairman of the Select Committee
"Underwater welding"

PREFACE

Le présent volume constitue le recueil des contributions présentées à l'occasion de la première Conférence Internationale de l'IIS qui ouvre, les 27 et 28 juin 1983, l'Assemblée annuelle de Trondheim (Norvège). C'est en effet la première fois que les manifestations habituellement distinctes qui marquaient les deux premiers jours de chaque Assemblée (Conférence Houdremont ou Portevin, séance publique et colloque) sont consacrées au même thème, à savoir, pour cette année, le soudage sous l'eau.

Ce thème est introduit par la Conférence Portevin du Professeur Cotton (Royaume-Uni), développé par trois conférences présentées par des auteurs invités: MM. Amiot (France), Christensen (Norvège) et Masubuchi (USA) et étudié plus en détail à la faveur de 33 communications préparées par les auteurs qui ont répondu à l'appel qui a été diffusé par le Comité d'Organisation.

Ce Comité était constitué d'une part de membres du Comité Norvégien d'Organisation de l'Assemblée, d'autre part de membres du Comité Restreint "Soudage sous l'eau" de l'IIS, qui avait proposé le sujet et le programme de la Conférence. L'objectif du Comité Restreint était de rassembler le maximum d'informations scientifique et techniques sur le soudage sous l'eau, pour les mettre à la disposition de tous les milieux intéressés et pour promouvoir, au sein de l'IIS, les études nécessaires aux progrès des techniques mises en jeu, qu'il s'agisse de la technologie du soudage ou du contrôle et de l'inspection des assemblages réalisés sous l'eau.

La tenue en Norvège, pays où le thème du soudage sous l'eau est d'une importance essentielle, de cette première Conférence Internationale, constitue un témoignage du souci de l'Institut International de la Soudure de centrer ses activités sur les problèmes dont l'actualité et l'urgence justifient les efforts de coopération internationale.

H. Granjon
Secrétaire Scientifique et Technique
Président du Comité Restreint "Soudage
sous l'eau"

INTERNATIONAL CONFERENCE ON UNDERWATER WELDING
CONFERENCE INTERNATIONALE SUR LE SOUDAGE SOUS L'EAU

LIST OF PAPERS
LISTE DES COMMUNICATIONS

SECTION I

WET AND DRY WELDING/METHODES DE SOUDAGE DANS L'EAU ET A L'ABRI DE L'EAU

SECTION II

INSPECTION AND PERFORMANCE / CONTROLE ET COMPORTEMENT

SECTION III

PHYSICAL, METALLURGICAL AND MECHANICAL PROBLEMS / PROBLEMES PHYSIQUES, METALLURGIQUES ET MECANIQUES

SECTION IV

REPAIR AND OTHER APPLICATIONS / REPARATIONS ET AUTRES APPLICATIONS

PORTEVIN LECTURE   CONFERENCE PORTEVIN

Welding Under Water and in the Splash Zone
– a Review

H. C. Cotton

Summary

The author reviews the parameters controlling underwater welding especially
arc welding with covered electrodes, MIG, TIG and with cored wire. It
describes wet welding and hyperbaric welding, with the use of nitrogen and
noble gases for shielding and in the welding chamber. The specific problems
of underwater welding are described in detail: hardness of the HAZ, cold
cracking by hydrogen, method of calculating the thermal input, consequences
of pressure on fusion characteristics.

KEY-WORDS

Underwater welding
FCA welding
MIG welding
MMA welding
GTA welding
Process parameters

Le soudage sous l'eau et dans la zone de marnage — vue d'ensemble

H. C. Cotton

RESUME

'auteur passe en revue les paramètres gouvernant le soudage sous l'eau,
ɔtamment le soudage à l'arc avec électrode enrobée, MIG, TIG et avec
il fourré. Il décrit le soudage en pleine eau et le soudage en condition
yperbare, avec utilisation de l'azote et de gaz nobles comme gaz protec-
eurs ou comme atmosphère de caisson. Les problèmes spécifiques du soudage
ous l'eau sont décrits en détail : dureté de la zone thermiquement
ffectée, fissuration à froid par l'hydrogène, calcul de l'apport de cha-
eur, effets de la pression sur les caractéristiques de fusion.

MOTS-CLES

oudage sous l'eau ; soudage avec fil fourré ; soudage MIG ; soudage
anuel avec électrode enrobée ; soudage TIG ; paramètre

Welding Under Water and in the Splash Zone
– a Review

H. C. Cotton

Introduction

It is a matter of great personal satisfaction to me to have been invited to deliver to the IIW the bi-annual lecture which honours the memory of Professor Portevin. I am proud to recall that his eminence as a metallurgist - perhaps I may say a welding metallurgist - was recognised in the United Kingdom by his election to the Royal Society; he was, I believe, the first, but happily not the last, welding specialist to be accorded that distinction.

As a visiting Professor in welding at the Cranfield Institute of Technology I would also like to pay tribute to a pioneer in welding education. Professor Portevin was closely associated with the Ecole Supérieure de Soudure Autogène which, with more than 50 years behind it, occupies a senior position in the world among welding schools. Though at Cranfield we cannot claim such a long history in welding education we follow the same principles in treating welding technology as a post-graduate specialisation.

As a welding metallurgist, Professor Portevin would have been fascinated by the problems of underwater welding which we are to discuss today and to morrow. During his working life, before electric arc welding was thought to be safe enough to use for the welding of ships'hulls, attempts were made to use the process to salvage wrecks. For the purpose of emergency repair, metal arc welding under the surface of the water was then and is now an invaluable tool. The simplicity of the process and the ability to use plant and equipment generally available for normal atmospheric welding are an added bonus. Wet welding using stick electrodes is attractive. If the fundamental rules widely accepted as controlling the successful welding of structural steel seem to be violated, this is often ignored provided that the requirement for speedy repair is accomplished. Excellence may be sacrificed for expediency. Temporary works in particular are usually not required to be perfect.

INTRODUCTION

C'est pour moi un grand honneur et une grande joie d'avoir été invité à prononcer la conférence biennale de l'IIS dédiée à la mémoire du Professeur Portevin. Je suis fier de rappeler que sa position éminente en tant que métallurgiste - peut-être puis-je dire de métallurgiste du soudage - a été reconnue au Royaume-Uni puisqu'il fut élu à la Royal Society ; il fut, je crois, le premier spécialiste en soudage à recevoir cette distinction.

En tant que chargé de conférences sur le soudage à l'Institut de Technologie de Cranfield, j'aimerais également rendre hommage au pionnier de l'enseignement du soudage que fut le Professeur Portevin, étroitement associé aux activités de l'Ecole Supérieure de Soudure Autogène qui, avec plus de 50 ans d'existence, occupe une des première places parmi les écoles de soudage du monde entier. Bien que, à Cranfield, nous ne puissions pas prétendre à une si longue expérience dans le domaine de l'enseignement du soudage, nous suivons les mêmes principes en considérant la technologie du soudage comme une spécialisation après un premier diplôme d'ingénieur.

En tant que métallurgiste du soudage, le Professeur Portevin aurait été passionné par les problèmes du soudage sous l'eau que nous allons discuter aujourd'hui et demain. Au cours de sa vie active, avant que le soudage à l'arc électrique n'ait été considéré comme suffisamment sûr pour le soudage des coques de navires, des essais ont été faits pour l'application de ce procédé au renflouage de navires. Le soudage à l'arc était alors et est encore un moyen inestimable pour l'exécution de réparations d'urgence sous l'eau. La simplicité du procédé et la possibilité d'utiliser le matériel, généralement employé pour le soudage dans une atmosphère normale, constituent un avantage supplémentaire. Le soudage dans l'eau avec électrode enrobée présente des attraits. Si les règles fondamentales, largement reconnues comme assurant la qualité du soudage de l'acier de construction, semblent être violées, on ne tient souvent pas compte de ce fait, à condition que les exigences d'une réparation rapide soient effectivement remplies. La perfection peut être sacrifiée au profit de l'opportunité. Les travaux provisoires, en particulier, n'ont normalement pas besoin d'être parfaits.

Avant même que le soudage à l'arc n'ait été jugé assez fiable pour pouvoir servir à la construction de coques de navires, des tentatives avaient été faites pour l'utiliser lors de fins de sauvetage d'épaves. Pour les répara- tions d'urgence, le soudage à l'arc en dessous de la surface de l'eau a été et reste encore un outil inestimable. La simplicité du procédé et la possibilité de mettre en oeuvre du matériel généralement disponible pour le soudage en atmosphère normale constituent des atouts supplémentaires. Le soudage en pleine eau avec électrodes enrobées est un procédé intéressant Certes, les règles fondamentales qui sont généralement admises comme étant nécessaires à l'obtention de bonnes soudures semblent être violées, mais ce fait est souvent ignoré, l'essentiel étant que la réparation soit exécutée rapidement. La qualité est parfois sacrifiée au bénéfice de la rapidité mais il est juste de rappeler qu'en général on n'exige pas que les travaux provisoires soient parfaits.

5

Wet welding allows cheap and simple repairs to be effected in a variety of situations and the results achieved are a convincing argument for the tenet of 'fitness for purpose'. (23). During the sixty years or so after its innovation, wet welding still finds many adherents, but the number of applications for which the process is suitable is limited. As far as is known, there is no reason to suppose that welds made under water, whether in the wet or dry environment, are in any way exonerated from compliance with those vital parameters which control welding in air at atmospheric pressure.

Controlling parameters for welding

Freedom for underbead and weld metal cold cracking depends mainly upon the composition of the steel to be welded, the quench rate, the hydrogen content of the weld metal and heat affected zones (HAZ) and the restraint stress. The inter-relationship of these and other factors was explained by Suzuki recently. (16). Except for the welding of very low carbon steels, the result of the application of these principles is that it has become normal practice, for the welding of structural steels of significant thickness, to use low hydrogen electrodes, to keep them dry and often to preheat so as to facilitate the diffusion of hydrogen out of the joint while it is still warm. The extent to which these precautions need to be applied depends, of course, upon several factors; suffice it to say that preheating is widely applied, even for the welding of relatively thin pipelines, and has become mandatory for offshore platform welding. When welding in the wet, compliance with these requirements cannot be achieved. (Fig. 1, Table 1). The hydrogen content is increased enormously and the quench rate becomes so rapid that the effect of thickness is almost obscured. (Table 2).

Masubuchi claimed some degree of success even in the wet welding of a very strong steel - HY80 - which is rather difficult to weld even in dry atmospheric conditions but the application related to emergency conditions and perfection was by no means achieved. Later, when he chose to employ the Tekken test to evaluate cold cracking tendency in wet welding a number of steels using a variety of consumables, he effectively demonstrated the crucial importance of restraint stress in its contribution to cracking. (1)

His investigation (3) confirmed some previous work by the Welding Institute (TWI) (2) which indicated also that the importance of composition, thermal input and quench rate etc. applies as much to wet welding as to normal atmospheric welding. For the avoidance of cold cracking at the level of restraint offered by the Tekken test configuration at a thermal input of about 1.5 kj/mm a carbon equivalent (IIW) of 0.3% seemed to be the maximum tolerable when the welding arc is submerged in water. Nor were attempts to prevent cracking under these conditions by the use of austenitic electrodes always successful. (3). Such a technique has often been proposed for wet welding steels with a carbon equivalent (CEq.) over 0.4%. (4). Because of the restraint stress and the presence of a brittle bond line, which is difficult to avoid in welding dissimilar joints of this type even in more favourable conditions, cracking was experienced. Hydrogen also possibly had some contributary effect.

From these results it seems reasonable to conclude that the SMAW process, when submerged in water, displays deficiencies which might only be tolerable when making welds in steels of very low CEq. and then only at low levels of restraint. In practice such benevolent conditions may often be encountered, particularly in the welding of sheet piling and some of the low strength steels used in older platforms if connections are fairly simple. These are

Le soudage en pleine eau permet dans de nombreux cas d'effectuer à peu de frais des réparations simples, et les résultats obtenus plaident en faveur du principe de "l'aptitude à l'emploi" (21). Depuis son apparition, il y a une soixantaine d'années, le soudage en pleine eau a conservé de nombreux partisans mais ses possibilités d'application sont limitées. Autant que nous sachions, il n'y a aucune raison de supposer que des soudures exécutées sous l'eau, que ce soit en caisson ou en pleine eau, échappent aux principes essentiels qui gouvernent le soudage à la pression atmosphérique.

Paramètres gouvernant le soudage

L'absence de fissuration à froid sous cordon ou dans le métal fondu dépend principalement de la composition de l'acier à souder, de la vitesse de trempe, de la teneur en hydrogène dans le métal fondu et dans la zone thermiquement affectée, ainsi que du bridage. L'interaction de ces facteurs avec d'autres facteurs a été décrite récemment par Suzuki (16). Sauf pour le soudage d'aciers à très bas carbone, l'application de ces principes a conduit au fait qu'il était de pratique courante, lors du soudage d'aciers de construction de forte épaisseur, d'utiliser des électrodes à bas hydrogène et bien sèches, et, souvent, de préchauffer les pièces afin de favoriser la diffusion de l'hydrogène hors du joint pendant qu'il est encore chaud. La rigueur avec laquelle ces précautions doivent être observées dépend évidemment de plusieurs facteurs ; il suffit de dire que le préchauffage est largement utilisé même pour le soudage de pipelines à paroi relativement mince, et qu'il est devenu obligatoire pour le soudage des plates-formes offshore. En soudage en pleine eau, il n'est pas possible de satisfaire à ces exigences (Fig. 1, Tableau 1). La teneur en hydrogène croît considérablement et la vitesse de trempe devient si élevée que l'effet de l'épaisseur en est presque éclipsé (Tableau 2).

Masubuchi a signalé un certain succès dans le soudage en pleine eau d'un acier à très haute résistance (HY 80) qui est plutôt difficile à souder même à la pression atmosphérique, mais l'opération de soudage décrite était liée à une réparation d'urgence et les résultats obtenus étaient loin d'être parfaits. Plus tard, lorsqu'il choisit d'utiliser l'essai Tekken pour évaluer la tendance à la fissuration à froid lors du soudage en pleine eau de quelques aciers avec emploi de divers métaux d'apport, il a clairement démontré le rôle prépondérant du bridage dans le phénomène de fissuration (1).

Ses recherches (3) ont corroboré des travaux antérieurs du Welding Institute (2) qui indiquaient également que la composition chimique, l'apport de chaleur et la vitesse de trempe, etc., étaient des facteurs aussi importants en soudage en pleine eau qu'en soudage à la pression atmosphérique. Pour éviter la fissuration à froid au niveau de bridage correspondant à l'éprouvette Tekken soudée avec un apport de chaleur d'environ 1,5 kJ/mm, la valeur maximale admissible pour le carbone équivalent (IIS) semblait être de 0,3 % lorsque l'arc est immergé dans l'eau. D'autre part, le recours à des électrodes austénitiques pour essayer d'éviter la fissuration dans ces conditions n'a pas toujours été couronné de succès (3). Une telle solution a souvent été proposée pour le soudage en pleine eau d'aciers ayant un carbone équivalent supérieur à 0,4 % (4). Du fait de la contrainte de bridage et de la présence d'une zone de liaison fragile difficile à éviter dans les joints hétérogènes de ce type, des fissures se sont produites même dans des conditions plus favorables. Il est également possible que

7

the conditions under which wet welding has often been reported as successful. Tackling the wet welding of stronger steels with the shielded metal arc welding consumables presently available, especially when the restraint is high, presents serious cold cracking risk. There is some evidence that flux coatings could be developed which could greatly improve the cracking resistance of wet welds and this possibility deserves further study. (5).

HAZ HARDNESS

The avoidance of hydrogen cracking alone may not always provide an acceptable solution to the problem. The hardness of the weld HAZ may still be unacceptably high.

Silva in a generalisation commented that, in wet welding, the width of the heat affected zone ranged between 25% and 50% of that zone in otherwise comparable welds made in air. The width of the HAZ is a useful measure of quench rate so it can be confidently assumed that wet weld HAZs are significantly harder than their counterparts in normal welding. Pellini and Wells both drew attention to the danger of local brittleness as a potent source for brittle fracture initiation. Hard HAZs are often also brittle and tend to occur mostly at the toes of fillet and lap welds. These locations are often 'hot spots' where the stress is intensified, so it is recommended that wet welds are not used where such risk applies.

Nevertheless, wet welding has its place in the scheme of things and much ingenuity has been displayed in attempts to ameliorate its deficiencies. When the thermal input is sufficiently high, thermal insulation of the surface of the area to be welded has been shown to be of some advantage in suppressing the cooling rate. The formation of bubbles on the surface of the cooling weld has beneficial effects also. (10). The control of weld sequence in order to take some small advantage of tempering effects is hopefully practised. To be effective this latter technique relies upon almost perfect positioning of the tempering bead. The probability of placing the tempering bead in the exact location required is hampered by the poor visibility caused by the generation of steam and other gases by the welding process itself and by the buoyancy of the bubbles formed. Touch welding is relied upon heavily for guidance purposes. Evidently notable skill is required for success in wet welding except in very simple applications subject to low restraint. (4) (6).

DRY HYPERBARIC WELDING

During the period 1960-1970 the almost exponential demand for crude oil and gas lead to fears that traditional reserves might soon prove to be insufficient or even become exhausted. New fields were sought and found in locations hitherto thought to be uneconomic but worth exploring on the assumption that a world shortage might raise prices. Alaska is an example of such a location on shore but the value of known offshore fields also began to be of interest. At first only sources close to land in fairly shallow water were sufficiently attractive but, as the price of crude oil began to rise, exploration began to stray into deeper water and large fields such as Forties in the North Sea were discovered. Existing technology in the design and fabrication of the large offshore structures and pipelines required for the exploitation of these finds became strained and fears of possible structural failure began to be voiced. The cost of these investments was so great and the risk to life so appalling that top quality repair technique were demanded and these included under-water welding. (7). Under-water welds of a quality quite equal to the best that could be achieved on land in

'hydrogène ait contribué à cette fissuration.

'après ces résultats, on peut raisonnablement conclure que le soudage anuel avec électrodes enrobées, employé en pleine eau, présente des inuffisances qui ne pourraient être admises que pour le soudage d'aciers yant un carbone équivalent très bas, et à la condition que les contraintes e bridage soient faibles. Dans la pratique, des conditions aussi favorables e rencontrent assez souvent, notamment lors du soudage de palplanches ou de ertains aciers à faible résistance utilisés sur d'anciennes plates-formes, ans la mesure où les assemblages sont relativement simples. Voilà les pplications pour lesquelles le soudage en pleine eau a pu être utilisé vec succès. Par contre, le soudage d'aciers à haute résistance avec les lectrodes actuellement disponibles présente de sérieux risques de fissuation à froid, surtout lorsque le bridage est important. Certains faits emblent indiquer qu'il serait possible de mettre au point des enrobages apables d'améliorer sensiblement la résistance à la fissuration des souures exécutées en pleine eau ; ce problème devrait faire l'objet de echerches plus approfondies (5).

DURETE DE LA ZONE THERMIQUEMENT AFFECTEE (ZTA)

e seul fait d'éviter la fissuration par l'hydrogène ne permet pas toujours e résoudre complètement le problème. Il arrive, en effet, que dureté de a ZTA atteigne des valeurs inacceptables.

ilva, en généralisant, a indiqué qu'en soudage en pleine eau la largeur e la zone thermiquement affectée représentait entre 25 et 50 % de la argeur de la même zone de soudures comparables exécutées à l'air libre. a largeur de la ZTA constitue une indication utile de la vitesse de trempe : n peut supposer en toute confiance que la ZTA de soudures en pleine eau st beaucoup plus dure que celle de soudures exécutées à la pression tmosphérique. Pellini et Wells ont attiré l'attention sur les dangers que eprésente une fragilité locale en tant que source potentielle d'une amorce e rupture fragile. Les ZTA dures sont souvent également fragiles et ont endance à se trouver au raccordement des soudures d'angle et à recouvreent. Ces endroits constituent souvent des "points chauds" où les contraintes ont plus élevées ; il est donc recommandé de ne pas exécuter de soudures n pleine eau lorsqu'un tel risque existe.

outefois, le soudage en pleine eau a sa place dans l'ordre de la nature et es chercheurs ont déployé beaucoup d'ingéniosité pour essayer de corriger es imperfections. Par exemple, lorsque l'apport de chaleur est suffisamment levé, un isolement thermique de la surface de la zone à souder s'est révélé fficace pour modifier la vitesse de refroidissement. La formation de bulles la surface de la soudure pendant son refroidissement exerce également un ffet favorable (10). D'autre part, en contrôlant les séquences de soudage, n espère tirer un léger profit de l'effet de revenu mais, pour être effiace, cette technique exige une localisation presque parfaite de chaque asse. Malheureusement, la probabilité de déposer un cordon de soudure à un ndroit précis est encore réduite par la mauvaise visibilité due à la prouction de vapeurs d'eau et d'autres gaz, par le soudage lui-même et 'évacuation des bulles formées. Afin de mieux guider l'électrode le long u joint, on compte beaucoup sur les électrodes à contact. Pour être atisfaisante, une opération de soudage en pleine eau exige évidemment ne grande habileté de la part du soudeur, sauf pour des assemblages très imples et peu bridés (4, 6).

9

normal conditions were demanded but this could not be achieved at that time
Serious development work in both diving techniques and welding to supplement
the excellent work already performed by the US Navy, which had other but
similar interests, began about 1970. (8).

The desirability of excluding water from the weld area was soon evident.
Various techniques for the provision of localised dry spots were examined
including the portable dry spot and water curtain techniques. Such devices
which are of particular interest for some specialised applications, were
found to be too complicated in use for the geometry of much of the repair
work envisaged. The difficulties in providing relatively large dry areas,
even in the very deep sea, were also found to have been considerably over-
stated.

Sufficient strength to cater for a differential pressure of at most one bar
is required. Dry hyperbaric welding began to be practised.

The shielded metal arc welding process in the dry hyperbaric environment

Simply excluding the water from the location of the joint seemed to be all
that was required. The convenience and ready availability of the SMAW
process and the familiarity with it of diverse already professing skill in
wet welding lead to its wide spread adoption. The practice of coating the
electrode flux with a thin layer of varnish or equivalent material to keep
the water out and to prevent the flux disingrating in the sea continued to
be used. (Table 1).

In still water the achievement of sufficient dryness in a chamber to allow
for 'dry' welding was easily accomplished but there are difficulties even
to this day in using hyperbaric chambers under stormy conditions in the
splash zone or at shallow depth when subjected to wave forces. However,
the design of such equipment is another matter; so far as the welding
process is concerned, it was found that the mere exclusion of the water is
not in itself enough to ensure quality welding of the standard required.
Simple code requirements could easily be met, but these were thought to be
insufficient to define the quality of weld required for parts of platforms
or pipelines of primary structural importance.

Welding in air at hyperbaric pressure (SMAW)

The danger of fire when welding in compressed air is ever present.
Robinson (9) reported that, at a depth of only 46m, clothing burns in air
at six times the rate of burning at atmospheric pressure. He indicated
that, in respect of catastrophic burning, compressed air is not a safe
working environment even when the depth exceeds only a few feet. Yet in a
heliox atmosphere, sufficient to sustain life at that same depth of 46m,
he found it difficult even to ignite his test strips of cotton cloth and
sometimes impossible to maintain flame propagation even of those few strips
which were successfully set alight. The relative safety of the heliox
mixture is notable.

Apart from the danger to the diver of fire, contamination by air can have
a devastating effect upon the soundness and mechanical properties of weld
metal. To reduce this effect to bearable limits the shielded metal arc
process was invented. Carbon dioxide and monoxide shields are favoured
because these useful protective gases can easily be produced by the cal-
cining of chalk, crushed marble or limestone included in the flux. Alternatively, or in addition to the lime, cellulosic materials may be added

SOUDAGE HYPERBARE EN CAISSON

Au cours des années 60, les besoins quasi-exponentiels en pétrole et en gaz naturel ont fait craindre que les réserves traditionnelles ne deviennent bientôt insuffisantes ou même s'épuisent totalement. De nouveaux gisements ont été recherchés et trouvés, dans des lieux jugés jusqu'alors non rentables mais intéressants à explorer si l'on suppose qu'une pénurie mondiale ferait monter les prix. L'Alaska est un exemple de ces gisements à terre, mais les gisements en mer, également, commençaient à se révéler intéressants. Au début, seuls les gisements situés près des côtes, dans des eaux peu profondes, étaient suffisamment rentables mais, à mesure que le prix du pétrole augmentait, les explorations commençaient à s'étendre vers des eaux plus profondes, et c'est ainsi que furent découverts les gisements des Forties dans la mer du Nord. Les méthodes classiques de calcul et de construction des canalisations et structures marines de grandes dimensions nécessaires à l'exploitation de ces gisements ont été peu à peu transgressées et des mises en garde contre d'éventuels accidents ont été faites. Le coût de ces investissements était si élevé et les risques pour les vies humaines si préoccupants que l'on exigea des techniques de réparation de la plus haute qualité. Parmi ces techniques figurait le soudage sous l'eau (7). On a exigé des soudures sous l'eau ayant une qualité égale à celle des meilleures soudures réalisées à terre dans des conditions normales ; mais cette exigence n'a pas pu être satisfaite en ce temps-là. C'est vers 1970 que commencèrent sérieusement les travaux de développement, aussi bien des techniques de plongée que de soudage, destinés à compléter les excellents travaux déjà effectués par l'U.S. Navy, qui avait des objectifs similaires, par certains aspects (8).

Une évidence apparut bientôt : il était souhaitable d'éliminer l'eau de la zone à souder. Diverses solutions ont été étudiées pour disposer d'un espace sec pour le soudage, l'arc étant isolé de l'eau soit par une enceinte portative, soit par un rideau d'eau. De telles solutions, qui sont particulièrement intéressantes pour certaines applications spéciales se sont révélées trop compliquées à l'usage, compte tenu de la géométrie de la plupart des structures à réparer. Par contre, on s'est aperçu que les difficultés liées à la mise au point d'enceintes relativement grandes, même pour les très grandes profondeurs, avaient été considérablement exagérées.

Les enceintes doivent être assez résistantes pour supporter une pression différentielle d'un bar au maximum. Grâce à ces enceintes (ou caissons), il a été possible d'exécuter les premières soudures en conditions hyperbares.

Soudage avec électrodes enrobées en condition hyperbare

Au début, il semblait que la seule condition à remplir fût d'éliminer l'eau de la zone à souder. La facilité de mise en oeuvre du soudage manuel avec électrodes enrobées et la bonne connaissance qu'en avaient les plongeurs déjà habitués au soudage en pleine eau ont favorisé l'extension du procédé. On a continué à pratiquer la méthode consistant à revêtir l'enrobage d'une couche de vernis ou d'un produit similaire pour l'isoler de l'eau et l'empêcher de se désagréger dans la mer (Tableau 1). En eau calme, il est facile d'obtenir un caisson suffisamment sec pour souder "en atmosphère sèche" mais même de nos jours, il est difficile en cas de tempête d'utiliser des caissons hyperbares dans la zone de marnage ou à de faibles profondeurs où l'action des vagues est ressentie. Toutefois,

which produce similar gas as they burn at the electrode tip. These shields may seem to be sparse but they are usually very copious. It has been estimated that the volume of shield generated by the melting and burning flux of a typical SMAW electrode is about equivalent to a CO_2 gas flow of 12 litres per minute. (10). At ambient pressure, provided that the arc is kept reasonably short, this is about sufficient to protect the molten pool from the atmosphere. This is essential to ensure an absence of porosity and to achieve the required mechanical strength. That there is little to spare is evident from the gross porosity that occurs in welding when the arc length is too long. Much of this porosity is attributable to changes in the solubility of nitrogen with temperature variation in the molten pool. (Fig. 2).

Nitrogen as a shielding and chamber gas in hyperbaric welding

The effect of the pressure of the head of water in shrinking the size of the SMAW protective gas shield is very marked and becomes evident even at a depth of only a few metres. At a depth of only 10 metres the equivalent gas shield flow rate is reduced to about half so that even at this shallow depth, lack of adequate protection may begin to show its effects. If an atmosphere of nitrogen can be regarded as inert from the point of view of fire risk, this gas must be recognised as extremely active during steel melting (11) (12) and this fact becomes very obvious as the normal SMAW shield volume contracts under the pressure of the head. Nitrogen begins to dilute the shield leading to gross porosity and impaired notch ductility. Pinford (13) et al remarked upon the severe reduction in charpy V energy caused by failure to maintain an extremely short arc under hyperbaric conditions which resulted in an escalation in nitrogen content to over 500ppm (Table 5). The very short arc length necessary to preserve shielding in an atmosphere of nitrogen, even under only slight hyperbaric pressure, can only be maintained in positional welding with extreme difficulty althoug the problem may not arise in welds made in the flat position. Hyperbaric welds made by the SMAW process in an atmosphere of nitrogen may be defective and this may not be detected because of limitations in inspection capability under water. If freedom from porosity and excellent notch ductility are required in the hyperbaric weld, shielding by means of the noble gases argon or helium should be considered. (Table 5).

The noble gases for shielding and in the chamber

Because of the logistics of providing supplies of argon, argonox, helium or heliox, and also owing to the increased cost, the temptation not to use these gases is great. The risks entailed in doing otherwise are ever presen and may, depending upon the quality of the weld required, be unacceptable. The use of these gases, sometimes called inert, provides a complete protection against atmospheric contamination. In some extreme cases, especially when the size of the welding chamber is small, the gradual build up of the proportion of gas exhaled by the diver may prove to be troublesome, requiring scrubbing of the atmosphere or other precautions but this is not usual except at high hyperbaric pressure. From the aspect of the achievement of weld deposits of excellent mechanical properties, the provision of a pure noble gas atmosphere is effective; however, the performance of some welding electrodes is impaired by the absence of oxygen. Welding electrode fluxes are designed to work in air and normally contain large additions of ferro alloys, the purpose of which is to combat oxidation and to supplement some of the alloy content, such as manganese, which may be lost thereby. In the absence of oxygen the recovery of some of these alloys may be excessive and, at the same time, the contribution which the oxidised alloy would otherwise

la conception de ce matériel est un autre problème ; quant au procédé de soudage lui-même, on a constaté que la simple élimination de l'eau n'était pas un facteur suffisant pour garantir la qualité de soudure requise. Les exigences simples formulées dans les codes pourraient être aisément satisfaites mais elles ne semblent pas capables de définir la qualité de soudure exigée pour des éléments de plates-formes ou de pipelines présentant une importance critique.

Soudage dans l'air, en conditions hyperbares (électrodes enrobées)

Le risque d'incendie en soudage dans l'air comprimé est omniprésent. Robinson (9) a signalé qu'à une profondeur de 46 m, les vêtements brûlaient dans l'air six fois plus vite qu'à la pression atmosphérique. Il a également indiqué que du fait des combustions catastrophiques, l'air comprimé n'offre pas de sécurité, même à des profondeurs de quelques pieds. Toutefois, dans une atmosphère de mélange héliox, suffisante pour la survie à cette même profondeur de 46 m, il a éprouvé des difficultés à enflammer des échantillons de tissu de coton, et il lui a été parfois impossible de maintenir la propagation de la flamme sur les quelques bandes de coton qu'il avait pu enflammer. Le mélange héliox assure donc une relative sécurité.

Outre le risque d'incendie auquel le plongeur est exposé, la contamination par l'air peut avoir des effets catastrophiques sur la compacité et les caractéristiques mécaniques du métal fondu. Pour ramener ces effets à des limites tolérables, le soudage manuel avec électrodes enrobées a été expérimenté. On a plus spécialement étudié les protections gazeuses susceptibles d'être asssurées par CO_2 et CO car ces gaz peuvent être aisément produits par calcination de la craie ou du carbonate de calcium contenus dans l'enrobage. En complément ou à la place du carbonate de calcium, il est possible d'ajouter des produits cellulosiques produisant des gaz similaires au moment de leur combustion à l'extrémité de l'électrode. La quantité de ces gaz semble réduite mais en réalité, elle demeure considérable. On a évalué (10) que le volume de gaz protecteur produit lors de la fusion et de la combustion de l'enrobage d'une électrode typique correspondait pratiquement à un débit de CO_2 de 12 litres par minute. A la pression ambiante, et sous réserve que l'arc soit maintenu raisonnablement court, cette quantité de gaz peut à la rigueur protéger le bain de fusion contre l'atmosphère ambiante. Cette protection est nécessaire pour éviter la formation de porosités et obtenir la résistance mécanique requise, mais elle est tout juste suffisante si l'on en juge par les soufflures qui se produisent lorsque l'arc est trop long. Une grande partie de cette porosité est imputable aux variations de solubilité de l'azote en fonction de la température du bain de fusion, Fig. 2.

Utilisation de l'azote comme gaz de protection et comme atmosphère de caisson en soudage hyperbare.

L'effet de la pression de l'eau, qui se traduit par une réduction du volume des gaz protecteurs en soudage avec électrodes enrobées, est très sensible ; il est perceptible même à une profondeur de quelques mètres. A seulement - 10 m, le débit de l'équivalent en gaz protecteur est réduit à environ 50 %, de sorte que même à cette faible profondeur l'absence de protection adéquate peut déjà commencer à se faire sentir. Bien que l'azote puisse être considéré comme inerte du point de vue des risques d'incendie, ce gaz est reconnu comme étant très actif lors de la fusion de l'acier (11, 12)

have made to the slag is lost. This may result in undesirable changes in
the properties of the slag, leading to changes in mobility, viscosity and
detachability. Sometimes these effects can result in poor weld coverage,
leading to bad weld shape and difficulties in handling. Often problems can
be overcome by adding a small dose of oxygen to the shield in the range
1 - 5%, depending upon the consumable and the pressure. (Table 3) (13).

Heliox and argonox provide ideal atmospheres for hyperbaric welding such
that the damaging effects of nitrogen are completely avoided. By avoidance
of the strain age damaging effects which bedevil the achievement of the
optimum crack tip opening displacement (CTOD), (24) properties often
specified for weld metal such as superb notch ductility may be displayed by
hyperbaric welds made in these atmospheres. However, the dangers of hydrogen
cold cracking are not eliminated thereby. Hydrogen is present in the hyperbaric chamber and in the electrode flux in the form of moisture and the
effect of this is important.

Hydrogen cold cracking in hyperbaric welding (SMAW)

Allum recently reported that an increase in ambient pressure has an important
effect upon the diffusible hydrogen content of welds deposited by the SMAW
process. (Fig. 3). An increase in pressure from 1 bar to 8 bar was accompanied by an approximate increase in hydrogen of two or three times. The
dampness of the electrode flux used in his tests was, of course, carefully
controlled and was not a variable. He demonstrated also that the humidity
of the welding chamber contributed very little, if anything, to this
increase. Choosing a low hydrogen basic covered electrode which consistently
displayed in standard tests, when welded at atmospheric pressure, a diffusible
hydrogen content of ca. 5ml/100grs, typical of a normal commercial product,
he measured in several hyperbaric experiments an increase in hydrogen
evolved to volumes of between 10 and 15 ml/100grs. (14). This important
observation was used to explain the marked increase in cold cracking in
Tekken tests performed under hyperbaric conditions at Cranfield for the
American Gas Association (Figs. 5, 6) (14). The main cause of this increase
in the efficiency of hydrogen recovery is presently attributed to the severe
effects that pressure is known to have upon the SMAW arc. A further important
observation is that this effect is not confined to very deep water conditions
but appears, in the tests so far performed to be almost fully developed at
the relatively shallow depth of only 80m. (Fig. 3). A suggested mechanism
for this will be discussed later. The significance of nearly doubling the
hydrogen content of SMAW deposits, even at a depth of say 60m, is of the
greatest importance to those considering the use of this process for the
welding of thick steels, especially when these are of comparatively high
carbon equivalent. (Fig. 5).

According to various Japanese workers (15) (16) the contribution that
hydrogen makes towards cold cracking in the welding of structural steels
can be expressed in rather strange terms of:= The Effective increase in
carbon equivalent = diffusible hydrogen per 100 grs deposited weld metal (J
                    ───
                                          60

Sometimes a divisor of 100 has been suggested or implied by other workers
and in more recent Japanese work rather more complicated factors have been
proposed to take into account the performance of steels of rather unsual
composition. At Cranfield the divisor of 60 was used to predict quite
accurately the behaviour of the Tekken tests which exhibited this unexpected
cold cracking propensity. Hydrogen was determined by the IIW not the JIS
method.

et ce phénomène devient très net lorsque le volume normal de gaz protecteur en soudage avec électrodes enrobées se contracte sous l'effet de la pression. L'azote commence à diluer la protection gazeuse avec, comme conséquences, la formation de soufflures et un abaissement de la ductilité sous entaille. Pinfold et al (13) ont signalé une forte réduction de l'énergie Charpy V imputable au fait qu'on avait négligé de maintenir un arc très court en conditions hyperbares, ce qui avait provoqué un accroissement de la teneur en azote à plus de 500 ppm (Tableau 5). L'arc très court, qui est nécessaire pour assurer la protection du bain dans une atmosphère d'azote même sous une légère pression hyperbare ne peut être maintenu que très difficilement en soudage en position, mais on peut dans certains cas souder à plat sans problème. Les soudures exécutées en conditions hyperbares avec électrodes enrobées et en atmosphère d'azote peuvent être défectueuses et il arrive que les défauts ne soient pas détectés à cause des limites des méthodes de contrôle sous l'eau. Si les soudures hyperbares doivent être exemptes de porosités et présenter une excellente ductilité sous entaille, il convient de recourir à une protection de gaz nobles, tels que l'argon ou l'hélium (Tableau 5).

Utilisation de gaz nobles comme gaz de protection et comme atmosphère de caisson

Compte tenu des importants moyens logistiques nécessaires pour l'approvisionnement en argon, mélange argonox, hélium ou mélange héliox, et également de l'augmentation du prix de revient dû au coût de ces produits, la tentation est grande de ne pas les utiliser. Avant de céder à cette tentation, il convient de ne pas oublier les risques inhérents à un tel choix, risques qui peuvent se révéler inacceptables eu égard à la qualité exigée des soudures. L'emploi de ces gaz, parfois appelés inertes, assure une protection totale contre la contamination atmosphérique. Dans certains cas extrêmes, notamment lorsque le caisson est de dimensions réduites, l'accumulation graduelle des gaz exhalés par le plongeur peut être gênante ; il devient alors nécessaire de purifier l'atmosphère ou de prendre d'autres précautions. Toutefois, ce problème n'est pas courant, sauf en conditions hyperbares sous pressions élevées.

Du point de vue de la réalisation de soudures ayant d'excellentes caractéristiques mécaniques, l'emploi d'une atmopshère constituée d'un gaz noble pur est efficace mais les caractéristiques de certaines électrodes sont altérées par l'absence d'oxygène. Les enrobages d'électrodes de soudage sont étudiés en vue d'une utilisation dans l'air et contiennent normalement de grandes quantités de ferro-alliages dont le rôle est de combattre l' oxydation et d'apporter le complément d'alliages, tels que le manganèse, qui peuvent être perdus pendant le soudage. En l'absence d'oxygène, le rendement en certains alliages peut être excessif et, en même temps, on perd le bénéfice de l'action que l'alliage oxydé aurait exercée sur le laitier. Cela peut se traduire par des modifications indésirables des propriétés du laitier, entraînant une altération de sa viscosité et de son aptitude à se détacher. Parfois, ces effets se traduisent par une mauvaise protection de la soudure avec, comme conséquences, des cordons de forme inadéquate et des difficultés opératoires. On peut souvent résoudre ces problèmes en ajoutant une petite quantité d'oxygène dans le gaz de protection, par exemple 1-5 % suivant l'électrode et la pression (Tableau 3) (Ref. 13).

From this work it can be inferred that the minimum preheat required to avoid cold cracking in a practical weld configuration is given by:

where:

$$T_c = 600p - 75Q - 180^{\circ} C$$

T_c = minimum required preheat $^{\circ}$C
H_c = diffusible hydrogen content ml/100grs (IIW)
Q = thermal input of deposit (kj/mm)
P = carbon equivalent $+\dfrac{H}{60}$

NB Plate thickness used in these tests = 25mm (combined thickness 50mm)

Space considerations make it impossible to attempt to justify these derivations here; however, reference may be made elsewhere. (14). The interrelationship of these factors is thought to be sufficiently accurate to allow for their use under practical welding conditions in the normal atmosphere. They can be used also for hyperbaric welding. However, when using helium or heliox in the chamber, especially at high hyperbaric pressure, the cooling rate below 100° C may be sufficiently enhanced to inhibit hydrogen diffusion. This may entail a requirement for additional preheating. Perhaps it is worth while to ponder over the significance of these conclusions. The equivalence of the various important factors affecting weldability can be presented in the following form:

C Eq. % 0.05 = a thermal input of 0.4 kj/mm = 3ml/100grs hydrogen = 30° C preheat.

Whether or not this conclusion is accepted as sufficiently accurate, the implication is clear. The effect of a small increase in the hydrogen content of the deposited weld metal can have a surprisingly severe effect upon the cold cracking propensity of a typical weld connection, especially in the root area. (Fig. 6). The difference in the reported hydrogen content of SMAW welds made at 1 bar and at 8 bar is equivalent in loose terms to an increase in the carbon content of the steel of about 0.1%. This difference is about equal to the range of carbon content between the least weldable and most weldable structural steels in common use today. This observation is equally important when welding at atmospheric pressure. Steel makers are often exhorted to produce steel by exotic methods involving the combination of micro-alloys and thermo-mechanical processes so as to achieve ever lower levels of C Eq. which, despite all their efforts, may be equivalent to only a tiny reduction in hydrogen content. On this basis a more rigid control of hydrogen in welds in structural steel might be far more rewarding than the imposition of such severe requirements for C Eq. that complications with HAZ notch ductility begin to be risked.

Complications in calculating thermal input for hyperbaric SMAW welds

Evidently, hydrogen cold cracking can be avoided, either by increasing the thermal input or raising the preheat temperature. (Fig. 4). Preheating to high temperature is not always favoured by welders confined in the small space of a hyperbaric chamber and surrounded by the heavy atmosphere of the compressed gas. An increase in the thermal input of the welding process is often desirable.   The normal methods in use for calculating thermal input for SMAW under normal atmospheric conditions appear to be equally appropriate to the hyperbaric case. (Fig. 7). The composition of the chamber gas had no discernable effect upon the cooling rate down to 100° C or upon the HAZ hardness. However, as already mentioned, a noticeable cooling effect below 100° C may have an adverse effect upon hydrogen diffusion rates.

L'héliox et l'argonox sont des atmosphères idéales pour le soudage hyperbare en ce sens que les effets nocifs de l'azote sont complètement éliminés. En évitant les effets défavorables du vieillissement par déformation qui gênent l'obtention de valeurs optimales de CTOD souvent spécifiées pour le métal fondu (22), on peut obtenir dans ces atmosphères des soudures hyperbares présentant une excellente ductilité sous entaille. Toutefois, les risques de fissuration à froid par l'hydrogène ne sont pas éliminés pour autant. En effet, l'hydrogène est présent dans le caisson hyperbare et dans l'enrobage de l'électrode sous forme d'humidité et ses effets sont importants.

Fissuration à froid par l'hydrogène en soudage hyperbare avec électrodes enrobées

Allum a récemment signalé qu'une augmentation de la pression ambiante exerçait un effet notale sur la teneur en hydrogène diffusible dans les soudures exécutées avec électrodes enrobées (Fig. 3). Un accroissement de la pression de 1 bar à 8 bar a multiplié la teneur en hydrogène par deux ou trois. Bien entendu, l'humidité de l'enrobage utilisé dans ces essais était soigneusement contrôlée et ne constituait pas une variable. Allum a également démontré que l'humidité du caisson ne jouait pratiquement aucun rôle dans cet accroissement de la teneur en hydrogène. En choisissant une électrode basique à bas hydrogène produisant régulièrement, au cours d'essais standards à la pression atmosphérique, une quantité d'hydrogène diffusible d'environ 5 ml/100 g, typique d'une électrode du commerce, il a pu mesurer lors de plusieurs essais en conditions hyperbares des quantités comprises entre 10 et 15 ml/100 g (14). Cette observation importante a été utilisée pour expliquer la forte augmentation de la fissuration à froid observée lors d'essais Tekken effectués en conditions hyperbares à Cranfield pour le compte de l'American Gas Association (Fig. 5, 6) d'après (14). La cause principale de cet accroissement de la teneur en hydrogène est actuellement attribuée aux effets connus exercés par la pression sur l'arc en soudage avec électrodes enrobées. Une autre observation importante est que cet effet ne s'exerce pas uniquement aux très grandes profondeurs ; en fait, dans les essais effectués jusqu'à présent, on a constaté qu'il s'exerçait pleinement à des profondeurs relativement faibles, de l'ordre de 80 m seulement (Fig. 3). Nous décrirons plus loin, un mécanisme qui a été suggéré pour expliquer ce phénomène. Le fait que la teneur en hydrogène dans les soudures exécutées avec électrodes enrobées soit pratiquement doublée à la profondeur de 60 m est d'une importance considérable pour ceux qui envisagent d'utiliser ce procédé pour souder des aciers de fortes épaisseurs, surtout lorsque le carbone équivalent est relativement élevé (Fig. 5).

Selon certains chercheurs japonais (15, 16), le rôle de l'hydrogène dans la fissuration à froid lors du soudage d'aciers de construction peut s'exprimer par cette relation un peu inhabituelle :

$$\text{Augmentation efficace du Ceq} = \frac{\text{hydrogène diffusible/100 g métal déposé (JIS)}}{60}$$

Parfois, un diviseur 100 a été suggéré ou utilisé implicitement par d'autres chercheurs, et, dans des travaux japonais plus récents, des facteurs un peu plus compliqués ont été proposés pour tenir compte des caractéristiques d'aciers ayant des compositions moins courantes. A Cranfield, le diviseur 60 a été choisi pour prédire avec une grande précision le

One other possible pit-fall was evident. It is common practice in some quarters, when detailing drawings and writing specifications, to express required thermal input in terms of weld size i.e. the throat thickness or leg length. The assumption that the relationships between run size and thermal input are independent of pressure may lead to serious error and to cracking. When using the SMAW process and reverse polarity Pinfold (13) noticed an increase in electrode burn off rate of up to 30% as the simulated depth of water was increased over the small depth increment of only 0 - 30m. Different behaviour was reported by Allum (14) when using straight polarity (electrode negative) when that same approximate increase in burn off rate was experienced over an even smaller depth increment of about 12m; then a fairly abrupt change in arc behaviour reversed the effect (Fig. 9). When the electrode is connected to the positive pole, which is normal for low hydrogen basic electrodes, weld penetration appears to be greatly affected by pressure. As the hyperbaric pressure is increased over the depth range 0-80m more heat is gradually generated at the electrode tip and less at the cathodic molten pool. The welding process, in terms of deposition rate, becomes in a sense more efficient. At the same nominal heat input the speed welding may increase by as much as 30% but such a weld will be significantly colder than its atmospheric counterpart. For given thermal input the weld size should be increased as the hyperbaric pressure is increased if it is required to maintain the thermal input at a fixed value.

The effects of this change in heat distribution between cathode and anode are several and one of these is thought to be responsible for the large increase in diffusible hydrogen noted. As the proportion of the heat developed at the anode increases, and perhaps as a consequence of it, the droplet size decreases and the droplet frequency increases. One explanation of the increase in hydrogen recovery noted in hyperbaric welds cites this same phenomenon. The increased surface area of the total sum of the droplets together with perhaps the longer time they spend in the arc atmosphere is thought to provide an explanation of this cause and perhaps also suggest a cure.

Because of this cold welding effect, together with the increased levels of diffusible hydrogen encountered, it is recommended that for equivalent freedom from cracking the weld size in the case of hyberbaric welding should be considerably larger than would be required for one atmosphere welding with the SMAW process.

Other effects of pressure upon the deposition characteristics of SMAW

Many SMAW electrodes which work well at atmospheric pressure perform so badly at even moderate hyperbaric pressure as to make them unusable. Not being designed for these conditions, this is not at all surprising; indeed it is a wonder that they work at all. As pressure is increased problems begin to be encountered, such as short circuits, arc outages, etc.; eventuall welding may become exceedingly difficult, resulting in deterioration in appearance and impossible slag detachability. Few electrodes are capable of satisfactory operation under hyperbaric conditions. (13). As depth exceeds about 150m, problems with weldability, especially for all position welding, may necessitate a progressive reduction in electrode size until at, say, a pressure of 20 bar the maximum usable size may be as small as 2.5mm or at most 3.25mm. This restriction, leading as it does to further reduced current and hence thermal input, compounds the hydrogen cracking problem just describ making preheat and control of interpass temperature even more essential. As a rough guide, the total effect of these variables is to necessitate an increase in preheat of about 100° C over and above that which would be requir

18

comportement des éprouvettes Tekken qui présentaient cette tendance
inattendue à la fissuration à froid. L'hydrogène a été déterminé par la
méthode de l'IIS et non pas celle du JIS.

De ces travaux, on peut déduire que la température minimale de préchauffage
pour éviter la fissuration à froid dans la pratique est donnée par
l'équation suivante :

$$Tc = 600p - 75Q - 180°C$$

dans laquelle

Tc = température minimale exigée pour le préchauffage (°C)
H = teneur en hydrogène diffusible en ml/100g (IIS)
Q = apport de chaleur (kJ/mm)
p = carbone équivalent + $\frac{H}{60}$

Nota : Epaisseur de tôle dans ces essais = 25 mm (épaisseur combinée
50 mm).

Faute de place, il est impossible d'essayer d'expliquer ici comment ces
conclusions ont été tirées, mais d'intéressants travaux ont été publiés
sur ce sujet (14). Les relations entre ces différents facteurs semblent
être suffisamment précises pour justifier leur emploi dans des conditions
pratiques de soudage en atmosphère normale. Elles peuvent être utilisées
également pour le soudage hyperbare. Toutefois, lorsque l'atmopshère du
caisson est constituée d'hélium ou de mélange héliox, surtout à des
pressions élevées, la vitesse de refroidissement au-dessous de 100°C
peut croître au point d'inhiber la diffusion de l'hydrogène et ceci
pourrait justifier un préchauffage supplémentaire. Il serait peut-être
intéressant de réfléchir sur la signification de ces conclusions. L'équi-
valence des divers facteurs importants influant sur la soudabilité peut
être présentée sous la forme suivante :
Ceq % 0,05 = apport de chaleur de 0,4 kJ/mm = 3 ml hydrogène/100 g =
30°C de préchauffage suivant qu'on estime ou non que cette conclusion est
suffisamment précise, les conséquences en sont claires. Une légère augmentation de la teneur en hydrogène dans le métal déposé peut avoir des effets
étonnamment défavorables sur la tendance à la fissuration à froid d'un
assemblage soudé typique, surtout à la racine du joint (Fig. 6). La
différence de teneur en hydrogène de soudures exécutées avec électrodes
enrobées à 1 bar et à 8 bar équivaut, très approximativement, à une
augmentation de 0,1 % de la teneur en carbone dans l'acier. Cette différence correspond, en gros, pour les aciers de construction courants, à la
fourchette des teneurs en C comprise entre celles des aciers les moins soudables et celles des aciers plus aisément soudables. Cette observation est
également importante en soudage à la pression atmosphérique. On incite
souvent les aciéristes à utiliser des méthodes compliquées, faisant appel
à des micro-alliages et des procédés thermomécaniques, dans le but d'obtenir des Ceq plus faibles alors que tous leurs efforts en ce sens ne
produisent qu'un résultat comparable à celui correspondant à une très
faible réduction de la teneur en hydrogène. De ce fait, il pourrait être
beaucoup plus rentable de contrôler plus sévèrement l'hydrogène dans les
soudures sur acier de construction, plutôt que de formuler à propos du
carbone équivalent des exigences si sévères que des problèmes de ductilité
en zone thermiquement affectée ne tarderaient pas à se poser.

for the same sized weld at normal atmospheric pressure.

In addition to all these effects, the unsuitability of commercially availabl[e] electrodes for the special conditions of the hyperbaric chamber results als[o] in an increase in the carbon content of the weld metal, accompanied by loss[es] of more desirable elements such as manganese. The various chemical reactio[ns] giving rise to these effects have been discussed at length elsewhere and the[re] is no need to re-examine them here. In any case, problems with the unsatis-factory composition of the weld metal are thought to be of a secondary natu[re] which could easily be solved by the development of special coatings, should this be thought to be worthwhile. (Fig. 8).

Manual metal arc welding using the SMAW process at hyperbaric pressure may seem to be simple but it is not. Many of its deficiencies could no doubt be solved if the incentive were strong enough. Fortunately other welding processes show less sensitivity to hyperbaric welding conditions and some o[f] these will now be discussed.

METAL INERT GAS WELDING (MIG)

The disparity between the relative maximum achievable rates of deposition between the SMAW process and the MIG process when the latter is operated in the dip-transfer mode becomes less as the maximum usable electrode size for positional welding diminishes as a consequence of pressure. The continuity of the MIG process leads to long arcing time and great efficiency, making the process very attractive when high speed welding is required at depth. Bare wire MIG is however somewhat sensitive to pressure. When connecting the feeder to the positive pole, as is usual with this process, the arc reacts quickly to pressure increase. The normal shape of the arc gradually contracts, penetration and spatter increase, accompanied by vast quantities of fume. Under normal welding arrangements at a depth as shallow as 30m frequent wire stubbing and excessive fuming make it impossible to produce an acceptable weld. In this range of pressure, negative polarity has even worse effects and the process may become unusable. When using negative polarity, however, at a pressure of about 7 bar and beyond, a remarkable change begins to take effect. The arc begins to stabilise and welding gradually becomes easier so that at 10 bar reasonable welding conditions ca[n] be established.

Power source design may do much to vary these effects and by concentrating on this aspect one reliable source reports excellent welding using the MIG process at a depth of 150m and perhaps deeper. Reticence born of commercial interests makes objective comparisons of competing welding processes for hyperbaric welding difficult. In our work at Cranfield the limitations in the usable parametric envelope for manual MIG welding at depth were found to be quite restrictive. In particular the process was found to be rather sensitive to 'stick out'. Perhaps these deficiencies could have been overcom[e] had we concentrated more actively upon the development of the power source as others are said to have done.

For fully automatic welding the regime of the tight limits on operating parameters is of little inconvenience and The Welding Institute has reported encouraging results in the application of bare wire MIG to fully automatic welding. This process is one candidate for fully automatic orbital welding for pipeline connections at great depth, should a requirement for such a system become established.

Difficultés inhérentes au calcul de l'apport de chaleur en soudage hyperbare avec électrodes enrobées

Bien évidemment, la fissuration à froid par l'hydrogène peut être évitée soit en augmentant l'apport de chaleur, soit en élevant la température de préchauffage (Fig. 4). Le préchauffage à des températures élevées n'est pas toujours apprécié des soudeurs confinés dans l'espace restreint d'un caisson hyperbare et entourés de l'atmosphère lourde du gaz comprimé. Il est souvent souhaitable d'augmenter l'apport d'énergie de soudage. Les méthodes couramment utilisées pour calculer l'apport de chaleur pour le soudage avec électrodes enrobées à la pression atmosphérique semblent également convenir pour le soudage hyperbare (Fig. 7). La composition de l'atmopshère du caisson n'exerce pas d'effet notable ni sur la vitesse de refroidissement jusqu'à 100°C ni sur la dureté de la zone thermiquement affectée. Toutefois, comme nous l'avons vu, un refroidissement sensible au-dessous de 100°C peut avoir une influence néfaste sur les vitesses de diffusion de l'hydrogène. Mais un autre piège existe : En effet, dans certains milieux, il est courant, lors de l'élaboration des plans de fabrication et des spécifications, de formuler les exigences d'apport de chaleur en termes de dimensions des soudures, c'est-à-dire de leur épaisseur ou de la longueur des côtés, dans le cas des soudures d'angle. Si on suppose que les relations entre la dimension des cordons et l'apport de chaleur sont indépendantes de la pression, on s'expose à des erreurs graves et à des problèmes de fissuration.

En soudage avec électrodes enrobées, polarité inverse, Pinfold (13), a constaté une augmentation de la vitesse de fusion allant jusqu'à 30 % à mesure que la profondeur simulée était portée progressivement jusqu'à la profondeur relativement faible de 30 m. Allum (14), effectuant des essais de soudage en polarité directe (électrode négative) a noté un comportement différent : tout d'abord pratiquement le même accroissement de la vitesse de fusion, mais jusqu'à une profondeur d'environ 12 m, puis un changement assez brutal du comportement de l'arc qui a inversé l'effet (Fig. 9). Quand l'électrode est reliée au pôle +, ce qui est normal pour les électrodes basiques à bas hydrogène, la pénétration de la soudure paraît fortement affectée par la pression. A mesure que la pression est augmentée de façon à correspondre à une profondeur s'échelonnant de 0 à 80 m, une quantité de chaleur accrue est produite à l'extrémité de l'électrode et une quantité moindre au droit du bain de fusion cathodique. Du point de vue de la vitesse de fusion, le procédé de soudage devient plus efficient, dans un certain sens. Avec le même apport de chaleur nominal, la vitesse de soudage peut croître de 30 % mais les soudures ainsi obtenues sont beaucoup plus froides que celles exécutées à la pression atmosphérique. Pour un apport de chaleur donné et devant rester fixe, la dimension de la soudure devrait être augmentée en même temps que la pression.

Les effets de la variation de la distribution de la chaleur entre cathode et anode sont multiples et on pense que l'un d'eux est responsable de la forte augmentation d'hydrogène diffusible qui a été constatée. A mesure que la chaleur produite à l'anode croît, et peut-être même comme conséquence de cette élévation de température, la dimension des gouttes décroît et leur nombre augmente. Ce phénomène a été cité parmi les explications retenues à propos de l'hydrogène dans les soudures exécutées en conditions hyperbares. On pense que l'accroissement de la surface globale de l'ensemble des gouttes et également leur plus long séjour dans l'atmosphère de l'arc pourraient fournir une explication de ce phénomène et peut-être suggérer un remède.

FLUX CORED ARC WELDING (FCAW)

The convenience of the continuous wire feed system offered by the bare wire MIG process can, by using FCAW, be combined with the relative insensitivity to process variables displayed by the SMAW process. FCAW operates naturally in the straight polarity mode over the full range of depth of present interest thus avoiding the complication of variable hydrogen recovery at depth suffered by the SMAW process. (Fig. 10). In out tests no measurable difference in diffusible hydrogen content was experienced over a depth range of about 0-20 bar. The inclusion in the flux core of readily ionisable substances stabilises the arc sufficiently to permit wide variations in operating technique. Manual welding at a pressure of 20 bar (200m) remains feasible although the consumable itself must be carefully chosen for such deep work.

The assistance of the slag to support and mould the molten metal, thereby improving the contour of the weld, is invaluable for positional welding and this is specially appreciated in deep locations. Weld deposition rates need not be curtailed on account of increased pressure, neither is there any need to reduce the welding speed or thermal input below that which can be used at atmospheric pressure, even when working at depths as great as 200m. These features confer several benefits. FCAW can allow, even when welding very thick steel, deep controllable penetration with little or no risk of the lack of sidewall fusion which so afflicts bare wire MIG welds. (13). Depending upon the choice of consumable, the process allows a maximum hydrogen content of ca. 10 ml/100 gr to be maintained over the whole range of depth of current commercial interest. (Fig. 10). Contrary to the dwindling thermal input of the SMAW process, the required thermal input can be maintained even at very deep sea depths. Depending upon the composition of the steel to be welded and its thickness, this allows preheat requirements for many applications to be discarded or to be usefully reduced with consequent improvement to the working conditions of the diver. Unfortunately, ideal conditions cannot always be obtained when using standard commercial consumables but the choice of usable alternative consumables for FCAW is even more limited than for SMAW.

The self shielded FCAW process has been operated in an argonox atmosphere for some time with notable success. Great skill and ingenuity were shown by the inventors of this process. The use of alkali metals to produce metallic vapour shields, together with alkali earth elements to stabilise the arc, represents one of the most notable innovations in consumable design over the last 30 years. The skilful use of lithium to reduce contamination by nitrogen is extraordinarily effective when working in normal atmospheric conditions. However, just as with SMAW, the compression of the protective shield as a result of the hyperbaric pressure limits its effectiveness. A supplementary shield consisting of a noble gas with or without oxygen addition has proved to be suitable. Indeed an additional shroud of helium, heliox, argon or argonox has been found to be essential to high quality hyperbaric welding with the FCAW process, even if the chamber atmosphere itself is of the same noble gas. (Table 4). This is because, with the FCAW process, the effect of humidity entrained in the gas shield affects the total diffusible hydrogen content adversely. This effect was not observed when using SMAW under similar conditions and points to differences in the density or permeability of the integral shields offered by the two processes.

Just as for the SMAW process, successful welding with the FCAW process at hyperbaric pressure requires specially designed consumables for optimum results, particularly when welding in deep waters. Once again, special consumables are difficult if not impossible to obtain and this difficulty

Compte tenu du fait que le régime de soudage hyperbare est "froid" et que les teneurs en hydrogène diffusible sont élevées, il est recommandé, pour obtenir des soudures ayant la même résistance à la fissuration, de leur donner des dimensions bien supérieures à celles des soudures exécutées avec électrodes enrobées à la pression atmosphérique.

Autres effets de la pression sur les caractéristiques de fusion en soudage avec électrodes enrobées

De nombreuses électrodes enrobées donnant entière satisfaction à la pression atmosphérique se comportent de façon si médiocre aux pressions hyperbares, même modérées, qu'elles deviennent inutilisables. N'étant pas conçues pour ces conditions spéciales, cela n'a rien d'anormal ; l'étonnant est qu'elles soient tant soit peu utilisables. A mesure que la pression croît, les problèmes apparaissent, tels que les courts-circuits, interruptions d'arc, etc. et finalement le soudage peut devenir extrêmement difficile, s'accompagnant de phénomènes tels que la détérioration de l'aspect des soudures et la formation d'un laitier impossible à éliminer. Peu d'électrodes conviennent au soudage hyperbare (13). Dès que les profondeurs dépassent 150 m, des problèmes de soudabilité se posent, notamment pour le soudage en position, et peuvent rendre nécessaire une réduction progressive du diamètre d'électrode ; à titre d'exemple, à la pression de 20 bar, ce diamètre utilisable peut être de 2,5 mm à peine, de 3,25 mm au maximum. Cette restriction, qui entraîne une réduction de l'intensité du courant, donc de l'apport de chaleur, règle en quelque sorte le problème de la fissuration par l'hydrogène décrit plus haut, en donnant encore plus d'importance au préchauffage et au contrôle des températures entre passes. A titre indicatif, l'effet global de ces paramètres se traduit par la nécessité d'augmenter la température de préchauffage d'environ 100°C par rapport à celle qui aurait été prescrite pour une soudure de même dimension exécutée à la pression atmosphérique.

En plus de ces inconvénients présentés par les électrodes du commerce, qui normalement ne sont pas conçues pour le soudage hyperbare, il convient de signaler un accroissement de la teneur en carbone du métal fondu, s'accompagnant de la perte en éléments plus utiles, tels que le manganèse. Les différentes réactions chimiques qui sont à l'origine de ces effets ont été décrites en détail par ailleurs et il est inutile de les répéter ici. Quoi qu'il en soit, les problèmes liés à la composition inadéquate du métal fondu sont considérés comme secondaires et susceptibles d'être résolus, au besoin, par la mise au point d'enrobages spéciaux (Fig. 8).

Le soudage avec électrodes enrobées en conditions hyperbares peut paraître simple, mais il ne l'est pas. Nombre de ses insuffisances pourraient sans aucun doute être surmontées si la motivation était assez forte. Heureusement, d'autres procédés sont moins sensibles aux conditions hyperbares ; nous allons en décrire quelques-uns.

SOUDAGE MIG

Les différences de vitesses maximales de dépôt entre le soudage avec électrodes enrobées et le soudage MIG à l'arc court s'atténuent puisque le diamètre maximum admissible pour le soudage en position diminue du fait de la pression. Le soudage MIG étant un procédé continu, les temps de fusion sont plus longs et le rendement plus élevé, ce qui rend le procédé

stimulates interest in those processes that do not depend upon flux. Another deficiency of FCAW is the difficulty experienced when making open backed root runs, especially in pipe butt welds. For this purpose TIG welding is of outstanding interest.

TUNGSTEN INERT GAS WELDING (TIG)

The nature and location of under-water welding tends to divide itself naturall into two rather distinct groups. The repair of offshore structures, because of the nature of the wave forces and the parts of the platform most exposed to collision and damage, tends to involve thick material at relatively shallow depth, whereas submarine pipelines and pipework are commonly relatively thin. The chemical composition of pipe used in these applications is usually extremely favourable to welding designed, as they often are, for down vertical welding with cellulosic electrodes, thus combining low thermal input with high hydrogen availability. From the metallurgical aspect, pipelines are generally assumed to be very much more weldable than are the structural steels used in making the platforms.

The relatively low thermal input offered by the conventional TIG process makes it less suitable for welding thick steel than are the competitive processes. Lap welding and fillet welding of thick steel is specially difficult with this process, leading to poor weld penetration and lack of fusion as well as to low welding speed. For butt welding, on the other hand, especially for root runs, the easy control of the heat source finds applications for TIG welding even on thick steel. For the butt welding of the thin steel typical of pipelines, the TIG process displays considerable advantage. (19). The accurate control of melting available to the operator is further enhanced by the use of a separate filler wire, the rate of use of which is not dictated by arc idiosyncracies but by deposition and penetration requirements only. In the hands of a competent welder the TIG process is ideal for pipe welding if time is not pressing. The slow rate of welding characteristic of the TIG process is inconvenient even in atmospheric welding but is often tolerated in the interests of excellence. Under hyperbaric conditions this slow welding speed may be of little inconvenience in those jobs where the welding time is only a small part of the total time taken in the whole operation. Getting to the site and preparing the work for welding often involve a very large part of the total time, but this is not always so. It would be useful, therefore, if the rate of welding achievable by the TIG process could be improved. TIG welding possibly with automatic hot wire feed, displays interesting possibilities, particularly for the fully automatic orbital welding of pipe at great depth should such a requirement become evident.

At Cranfield, techniques for the improvement of several features which presently detract from the usefulness of high pressure TIG welding are being actively studied. The serious lack of stiffness displayed by the arc at high pressure makes it vulnerable to wander under the influence of quite weak magnetic fields which are normally tolerable at atmospheric pressure. (22) (25 This gives rise to diver complaints of excessive arc blow. Additionally gas buoyancy effects in welding positions other than the overhead become a nuisance because of the weakened arc jetting force. (20). Divers working in saturation prefer to operate in helium atmospheres rather than argon even although equipped with separate life support systems. This is because the narcotic effects of argon are feared. However, catastrophic erosion of the tungsten electrode is experienced when TIG welding at high pressure in helium atmosphere. The release of tungsten vapour arising as a consequence reduces arc conductivity, giving rise to unwanted dynamic power source effects.

très intéressant pour le soudage sous l'eau à grande cadence. Toutefois, le soudage MIG avec fil nu est légèrement sensible à la pression. En reliant le fil au pôle positif, comme cela se fait couramment avec ce procédé, l'arc réagit rapidement à l'accroissement de la pression. Il se contracte graduellement, tandis que la pénétration et les projections augmentent et que d'importantes fumées se dégagent. A une profondeur d'à peine 30 m et dans les conditions courantes de soudage, de fréquentes plongées de fil et des fumées excessives empêchent la réalisation d'une soudure acceptable. A ces pressions, la polarité négative produit des effets encore plus défavorables et le procédé peut devenir inutilisable. Toutefois, en polarité négative à une pression d'environ 7 bar et au-delà, on commence à observer un changement remarquable. L'arc commence à se stabiliser et le soudage devient graduellement plus facile, de sorte qu'à 10 bar on peut obtenir des conditions de soudage raisonnables. La conception des sources d'énergie peut influer sur ces effets et, à ce propos, il existe au moins un document digne de foi qui décrit une excellente opération de soudage MIG à une profondeur de 150 m et peut-être même plus. La réticence née des intérêts commerciaux rend difficile toute comparaison objective des procédés de soudage applicables au soudage hyperbare. Dans nos laboratoires de Cranfield, nous avons constaté que le domaine des réglages limites utilisable pour le soudage MIG manuel était très restreint. En particulier, on a trouvé que ce procédé était plutôt sensible à la longueur de fil libre. Ces imperfections auraient peut-être pu être éliminées si nous avions consacré plus d'attention au développement de la source de courant, ce que d'autres ont fait, parait-il. Pour le soudage entièrement automatique, les limites étroites imposées aux paramètres ne sont pas gênantes et le Welding Institute a obtenu des résultats encourageants en automatisant entièrement le soudage MIG avec fil nu. Ce procédé est un des candidats possibles pour le soudage orbital automatique de pipelines à grandes profondeurs, pour le cas où le besoin s'en ferait sentir.

SOUDAGE AVEC FIL FOURRE

En utilisant le soudage avec fil fourré, on combine les avantages de l'alimentation continue en fil nu du procédé MIG avec ceux qui résultent de la relative insensibilité aux paramètres que présente le soudage avec électrodes enrobées. Le procédé avec fil fourré fonctionne naturellement en polarité directe sur toute la gamme des profondeurs présentant actuellement un intérêt pratique, ce qui élimine les complications liées aux variations de teneur en hydrogène, inhérentes au soudage avec électrodes enrobées (Fig. 10). Au cours de nos essais, nous n'avons décelé aucune différence mesurable de la teneur en hydrogène diffusible dans la gamme 0-20 bar. En incorporant dans le flux du fil fourré des substances facilement ionisables, on stabilise suffisamment l'arc, ce qui permet de nombreuses variantes de la technique opératoire. Le soudage manuel à la pression de 20 bar (200 m) reste possible, mais pour un travail à une telle profondeur, il faut choisir le produit d'apport avec le plus grand soin. Le laitier, qui aide à supporter et à modeler la zone de métal fondu, donc à améliorer la géométrie de la soudure, joue un rôle inestimable en soudage en position ; c'est un avantage certain pour le soudage à grande profondeur. Même à 200 m de profondeur, il n'est pas nécessaire de réduire les vitesses de fusion à cause de la pression, ou de réduire la vitesse de soudage ou l'apport de chaleur en-dessous des valeurs applicables au soudage à la pression atmosphérique. Ces caractéristiques sont particulièrement intéressantes. En effet, le soudage avec fil fourré permet, même pour l'assemblage de tôles très épaisses, d'obtenir une forte pénétration dans des conditions contrôlables, pratiquement sans les risques de collage inhérents au soudage MIG (13). Suivant le produit d'apport choisi,

Methods have been proposed whereby these and other difficulties might be overcome or at least reduced in importance; such methods are being studied as part of a programme of work sponsored by the Science and Engineering Research Council. (20) (26). Both the TIG variant 'Plasma Welding' and 'Plasma MIG' are being studied also, but both these processes show less promise for deep water welding than does TIG welding which seems to offer several opportunities for innovation.

OTHER WELDING PROCESSES AND TECHNIQUES

One of the difficulties in working in a strange environment is that traditional requirements appropriate to more normal conditions tend to be applied to novel situations with scant regard to their relevance. Established and entrenched non-destructive testing requirements, by their applications to new methods of welding, have delayed the widespread adoption of new welding techniques which could have been beneficial. Examples of processes thought to have been blighted by this are flash butt welding, friction welding and explosion welding, sometimes now called high energy bonding.

There are evidently difficulties with the behaviour of welding arcs at high pressure and one might wonder whether further development in this direction is realistic. References to the possibility of devising fully automatic orbital welding techniques for deep water pipe welding are frequently found. However, rarely if ever are proposals heard for tackling weld configurations of greater complexity. The simple geometry of a pipe lends itself to automatic fusion welding but is ideal for other processes also.

Not only are there problems with welding in very deep waters but there are difficulties also, both of a psychological and physical nature, with the diver himself. At very great depth even the diver in the one atmosphere suit may find great difficulty, on account of visual aberrations caused by the refraction of the compressed gas and the effect of this in the turbulent conditions around the hot arc and weld. Problems of this nature will need to be solved before successful fusion welding in deep water locations can be achieved. Naturally it is tempting to extrapolate one's experience and pre-conceived notions from the familiar to the novel situation. So far, down to depths of say 200 or 300m perhaps, it appears to have been possible to do so without too much of a penalty. It seems reasonable to suppose that at greater depths present problems will be exacerbated and it is timely to reconsider whether the present conception of how joints in pipelines in deep water locations ought to be made, and what properties they should display are really valid. Fusion welding is only one of the alternatives.

Friction welding could find wider application both in the normal atmosphere and in submarine conditions. Objections to the quench rate of wet joints have been countered by The Welding Institute by the introduction of collars and plugs of plastic material. The potential for friction welding in making attachments to platforms and pipelines is evident but the provision of a large power source on the sea bed sufficient to effect radial friction welds in large pipes is not attractive. There are easier methods of providing the required energy.

SOLID PHASE WELDING

Another alternative to fusion welding is explosive bonding and this process offers release from the vagaries of the hyperbaric arc behaviour and is said to be immune to metallurgical complications if properly done. The energy required for bonding is conveniently packaged in the explosive.

ce procédé permet de maintenir une teneur maximale en hydrogène d'environ 10 ml/100 g sur toute la gamme des profondeurs présentant un intérêt pratique (Fig. 10). Contrairement au soudage avec électrodes enrobées, le soudage avec fil fourré autorise le maintien constant de l'apport de chaleur même à de très grandes profondeurs. Suivant la composition et l'épaisseur de l'acier à souder, il est possible, pour de nombreuses applications, de s'affranchir du préchauffage ou du moins de le réduire, ce qui contribue à améliorer les conditions de travail du plongeur. Malheureusement, il n'est pas toujours possible de choisir les conditions idéales lorsque l'on utilise les produits d'apport du commerce, mais le choix de produits utilisables pour le soudage avec fil fourré est encore plus limité que pour le soudage avec électrodes enrobées. Pendant un certain temps, le soudage avec fil fourré a été mis en oeuvre avec succès en atmosphère de mélange argonox. Les inventeurs du procédé avaient fait preuve de beaucoup d'habileté et d'ingéniosité. L'emploi de métaux alcalins pour produire une protection de vapeurs métalliques et d'éléments alcalino-terreux pour stabiliser l'arc représente l'une des plus remarquables innovations dans le domaine des produits d'apport au cours des 30 dernières années. L'emploi judicieux du lithium pour réduire la contamination par l'hydrogène est extrêmement efficace en soudage à la pression atmosphérique. Cependant, comme en soudage avec électrodes enrobées, la compression de la gaine de gaz protecteur, due à la pression hyperbare, limite son efficacité. Une protection supplémentaire constituée d'un gaz noble, avec ou sans addition d'oxygène, a donné de bons résultats. En fait, une protection d'hélium, héliox, argon ou argonox s'est révélée particulièrement utile pour l'obtention de soudures de haute qualité en soudage hyperbare avec fil fourré, même si l'atmosphère du caisson est déjà constituée du même gaz noble (Tableau 4). Ceci s'explique par le fait qu'en soudage avec fil fourré l'humidité entraînée dans le gaz protecteur influe défavorablement sur la teneur totale en hydrogène diffusible. Cet effet n'a pas été observé lors du soudage avec électrodes enrobées dans des conditions similaires, et laisse supposer qu'il existe, dans le mode de protection de ces deux procédés, des différences de densité ou de perméabilité.

Comme en soudage avec électrodes enrobées, il est nécessaire, en soudage hyperbare avec fil fourré, d'utiliser des produits d'apport spécialement conçus, surtout lorsque les profondeurs sont importantes. Mais, répétons-le les produits d'apport spéciaux sont pratiquement introuvables et cette difficulté renforce l'intérêt présenté par les procédés qui ne font pas appel à un flux. Un autre inconvénient du soudage avec fil fourré réside dans la difficulté d'exécuter les cordons de pénétration sur des assemblages avec écartement des bords, notamment lors du raboutage de tubes. Pour cette application, le soudage TIG présente un intérêt certain.

SOUDAGE TIG

Compte tenu de la nature et du milieu dans lequel il est mis en oeuvre, le soudage sous l'eau se subdivise naturellement en deux groupes assez distincts. Du fait du type d'efforts dus aux vagues et de la nature des éléments de plates-formes les plus exposés aux chocs et aux endommagements, la réparation des structures offshore a tendance à s'appliquer à des matériaux épais et à des profondeurs relativement faibles. Par contre, les pipelines et autres tuyauteries immergées sont en général relativement minces. La composition chimique des tubes utilisés pour ces applications est en principe très favorable du point de vue du soudage puisqu'ils sont souvent conçus pour le soudage vertical en descendant avec électrodes cellulosiques, combinant ainsi un faible apport de chaleur avec de grandes quantités d'hydrogène.

For reliable welding it seems that the surfaces to be welded must be very clean and dry. The achievement of such conditions seems to have led to complication in what would otherwise seem to be a relatively simple process. It was mentioned earlier that the application of the principle of 'fitness-for-purpose' allows the use of wet welds which, if evaluated on a conventional basis, might fail to conform. Imperfect welds behave adequately in a host of applications. Indeed the cold expansion of tubes into tube sheets provides adequate strength and sealing for many purposes without any welding at all. An objective assessment of what is really required in order to provide reliably safe connections under submarine conditions in lieu of the automatic application of normal code requirements is thought to be lacking.

CONCLUSION

An attempt has been made to compare some of the features of those welding processes of most general interest for the purpose of welding under water. The subject is very large and this paper is not intended to cover the whole field of interest. Indeed discussion of some interesting processes such as water curtain, stud welding etc. (21) had to be curtailed altogether. During the last 15 years the number of papers published on this subject has been great however, many of these provide little information because of commercial interests which prevent free discussion. Because of this there are no doubt more deficiencies in this paper than there would have been otherwise. Less than 20 years ago it was said that more is known about welding in space than is known about under-water welding.

This remark may still be true because much secrecy attends that application also. We can claim to have made notable progress, especially during the last decade, but how far this knowledge has been generally disseminated and understood is less certain. The literature is still peppered with meaningless remarks such as 'a 20% increase in ductility and a 4% increase in impact strength'. This indicates that the real mechanisms of failure are not fully understood and as a result engineers may be striving to achieve results which are not strictly necessary whilst ignoring more important factors. Closer links between scientists working in this field and practising engineers are desirable, particularly in establishing meaningful requirements for very deep water working.

Discussion in this paper has been confined to hyperbaric welding. The alternative 'one atmosphere welding' has not even been mentioned. Clearly this technique offers relief from many of the woes described in this paper, albeit at some escalation in cost. No doubt the various one atmosphere systems are prone to difficulties peculiar to themselves and for very deep working it would be worthwhile to discuss these.

So far as the depth range 0 – 200m is concerned one atmosphere welding is not necessary as welding of superb quality can now be reliably performed within this region.

How far this satisfactory state of affairs will be maintained at greater depths remains to be seen.

During the last ten years enormous strides have been made in both the study and application of underwater welding. The casual use of the SMAW process in the wet environment is being gradually superseded by the application of advanced technology. Difficulties encountered with such matters as hydrogen cracking, porosity, brittleness and arc instability have gradually yielded to the application of scientific principles to a degree that is less evident

Du point de vue métallurgique, les aciers pour pipelines passent généralement pour être beaucoup plus soudables que les aciers utilisés dans la construction des plates-formes.

L'apport de chaleur relativement faible qui caractérise le soudage TIG classique le rend moins apte que ses concurrents au soudage des fortes épaisseurs. Les assemblages à recouvrement et en angle sur aciers de forte épaisseur sont particulièrement difficiles à exécuter avec ce procédé : la pénétration est insuffisante, des manques de liaison sont à craindre et la vitesse de soudage est faible. Par contre, pour le soudage bout à bout, notamment pour la réalisation des passes de fond, même sur tôles fortes, le soudage TIG peut trouver des applications grâce à la facilité de réglage de l'apport de chaleur. Pour le soudage bout à bout de l'acier de faible épaisseur typique des pipelines, le procédé TIG dispose d'un avantage considérable (17). La possibilité offerte à l'opérateur de contrôler la fusion de façon précise est encore renforcée par l'utilisation d'un fil d'apport dont la vitesse d'alimentation ne dépend pas des caractéristiques de l'arc mais uniquement des exigences concernant la fusion et la pénétration. Entre les mains d'un soudeur compétent, le procédé TIG est idéal pour le soudage de tubes, dans la mesure où le temps n'est pas un facteur critique. La lenteur du procédé TIG, même à la pression atmosphérique, est un inconvénient que l'on accepte volontiers car le procédé présente, par ailleurs, des qualités exceptionnelles. En conditions hyperbares, cette lenteur peut ne pas être rédhibitoire pour des travaux où le temps de soudage ne compte que pour une partie minime du temps total de fabrication. Le fait de se rendre sur le site et d'effectuer les préparations avant soudage représente souvent un très important pourcentage du temps total mais ce n'est pas toujours le cas. Il serait donc utile de pouvoir améliorer la vitesse du procédé TIG. Par exemple, le soudage TIG avec alimentation automatique en fil chaud, offrirait des possibilités intéressantes, notamment pour le soudage orbital entièrement automatique de tubes à grandes profondeurs.

A Cranfield, on étudie activement les méthodes permettant d'améliorer certaines caractéristiques du soudage TIG sous pression élevée pour le rendre plus facilement utilisable. Le sérieux manque de rigidité de l'arc aux pressions élevées le rend particulièrement erratique sous l'influence de champs magnétiques très faibles qui sont normalement tolérables à la pression atmosphérique (20, 23).Les plongeurs se plaignent alors d'un soufflage magnétique excessif. D'autre part, dans des positions de soudage autres qu'au plafond, la poussée des gaz devient une gêne du fait de la réduction de la pression exercée par l'arc (18). Les plongeurs en saturation préfèrent souder en atmosphère d'hélium plutôt que d'argon, dont ils redoutent les propriétés narcotiques,même étant équipés d'un masque respiratoire. Toutefois, on a constaté une érosion catastrophique de l'électrode de tungstène lors du soudage en atmosphère d'hélium à haute pression. Les vapeurs de tungstène ainsi produites réduisent la conductivité de l'arc, donnant lieu, au niveau de la source d'énergie, à des phénomènes dynamiques indésirables. Des méthodes ont été proposées pour résoudre ces problèmes, et même certains autres, au moins partiellement ; ces problèmes sont étudiés en Grande-Bretagne dans le cadre d'un programme de recherches subventionné par le Science and Engineering Research Council (18, 24). Les deux variantes du procédé TIG sont étudiées, à savoir le soudage plasma et le soudage plasma-MIG, mais pour le soudage à grandes profondeurs, toutes deux sont moins prometteuses que le procédé TIG qui semble offrir plusieurs possibilités d'innovation.

in more common applications. The alchemy which is still preponderent in the development of fluxes for welding and their secret formulations has limited progress in the use of fluxed processes for underwater welding. In fact the general absence of suitable consumables for deep water working has led to a diminished use of the SMAW process in favour of those processes which depend less upon the composition of the flux and are therefore more suitable for systematic study. FCAW, because it permits speedy welding, is widely used by those with access to a suitable consumable. TIG and MIG welding are available to all but considerable technical knowledge is required to achieve optimum performance in deep waters. Although the number of welds made under water are few in comparison with the number of those made in the normal atmosphere, their cost and value is very great. This has made the alloca-tion of funds for the study of fundamental technology justifiable. Escape from the despair of empirical experimentation has been achieved and the results have been rewarding.

Many problems remain to be solved. The four competing fusion welding proces-ses SMAW, FCAW, MIG and TIG are all at or approaching their limiting depth capability at the greatest depths presently encountered in the North Sea. All these processes can be cajoled to operate, perhaps marginally, at twice these depths but the need for significant improvement and perhaps radical modification is obvious. At these same depths the present capabilities of saturation divers appear to be strained also and the attractions of one atmosphere diving becomes alluring. As presently practised this will entail less diver dexterity than has been displayed by the free swimming saturation diver within the present diving regime. So far as fusion welding processes are presently visualised this invites automation. Automatic fusion welding by maintaining a fixed arc length for example gives rise to less arc insta-bilities than does manual welding. It has been shown that at 40 bar a change in arc length of only 1mm will entail a consequent change in arc voltage of 3 volts with significant transient effects. Successful automation can not only lead to improved arc stability but demands it also. Techniques such as magnetic pool stirring, gas shield swirling, secondary arcs and HF arc stiffening have all been proposed for the improvement of TIG welding and the achievement of much higher welding speeds. (22). The benefits of such innovations need not be confined to hyperbaric welding but might find application at atmospheric pressure. It is questionable whether one atmos-phere welding will be widely adopted. The expression 'automation' conjures visions of orbital welding much as is presently practised and subject to occasional manual intervention. This discounts the role that will be played by the introduction of 'play-back' and intelligent robots.

Japan's MITI is reported to be about to spend about US $ 70 million in the next seven years to speed the development of 'learning robots' for sea-bed and similar applications. (27). The Komatsu sea-bed cable controlled robot is only perhaps a hint of what might soon be in store. At the bottom of the sea political and social objections to the introduction of robots do not pertain and it is in this very environment where man is at such a disadvan-tage that the intervention of the robot would be so rewarding. The simplicity and high speed welding potential for TIG variants using high currents and incorporating hot wire addition, perhaps used in a narrow gap mode and controlled through the actions of a learning robot aided by some form of optical sensor, (28), might transform our capability for hyperbaric welding in the deep sea.

Present predictions of how we will be welding underwater in the future are thought to be highly speculative.

AUTRES PROCEDES ET TECHNIQUES DE SOUDAGE

L'une des difficultés de travailler dans un milieu inconnu réside dans le fait que l'on a tendance à appliquer aux conditions nouvelles les exigences traditionnelles sans vérifier suffisamment si elles sont adaptées. En appliquant directement à de nouvelles méthodes de soudage les exigences de contrôle non-destructif bien établies et bien ancrées dans les habitudes, on a retardé le développement de nouvelles techniques de soudage qui auraient pu se révéler intéressantes. Parmi les procédés qui semblent avoir pâti de cet état de choses, on peut citer le soudage par étincelage, par friction et par explosion.

De toute évidence, le comportement des arcs sous pression élevée pose des problèmes et on peut se demander s'il est réaliste de poursuivre les recherches dans cette direction. Les spécialistes citent souvent la possibilité de mettre au point des techniques de soudage orbital entièrement automatique pour le soudage de tubes à grande profondeur, mais les tentatives d'étudier des géométries de joints plus complexes ont été très rares. La géométrie simple d'un tube se prête bien au soudage automatique par fusion mais elle est idéale aussi pour d'autres procédés.

Le soudage en eaux très profondes pose non seulement des problèmes d'ordre technologique, mais aussi d'ordre psychologique et physique liés au plongeur lui-même. A très grande profondeur même le plongeur portant des vêtements conçus pour le travail à une atmosphère peut éprouver de grandes difficultés du fait d'aberrations optiques résultant de la réfraction dans le gaz comprimé et de son effet sur la turbulence autour de l'arc et de la soudure. Il sera nécessaire de résoudre les problèmes de cette nature avant de pouvoir réaliser des soudures par fusion satisfaisantes à grandes profondeurs. Naturellement, il est tentant d'extrapoler à une situation nouvelle l'expérience et les idées préconçues que nous avons d'une situation familière. Jusqu'à maintenant, il semble que nous ayons pu procéder ainsi impunément jusqu'à des profondeurs de 200 - 300 m. Il paraît raisonnable de supposer qu'à des profondeurs plus grandes, les problèmes actuels se poseront avec plus d'acuité, et il est temps de vérifier la validité de nos conceptions actuelles sur la façon d'exécuter les joints sur pipelines à grandes profondeurs et sur les propriétés qu'ils doivent présenter. Le soudage par fusion n'est qu'un des procédés utilisables.

Le soudage par friction pourrait trouver des applications plus étendues, aussi bien pour les travaux en atmosphère normale que dans les conditions sous-marines. Les objections liées à la vitesse de trempe des joints en pleine eau ont été éliminées par le Welding Institute qui utilise des colliers et des bouchons en plastique. Les possibilités offertes par le soudage par friction pour fixer des attaches sur des plates-formes et des pipelines sont évidentes mais il est encore difficile de disposer, au fond de la mer, d'une source de courant assez puissante pour permettre des soudures sur tubes de grandes dimensions. Il existe des méthodes plus simples pour fournir l'énergie nécessaire au soudage.

SOUDAGE EN PHASE SOLIDE

Parmi les procédés pouvant remplacer le soudage par fusion, on peut citer le soudage par explosion, qui permet de s'affranchir des caprices de l'arc hyperbare et semble ne jamais poser de problèmes métallurgiques lorsqu'il est

ACKNOWLEDGEMENT

Thanks are due to Chris Allum and John Nixon of the Cranfield Under-water Welding Team for their assistance in preparing this paper. Sub-Sea International Limited, by siting their hyperbaric chambers at Cranfield and allowing them to be used for the purpose of experiment, made the work of this team possible. (Fig. 11). Their enquiry into Hyperbaric SMAW was financed by the American Gas Association. Funds provided by the Science and Engineering Research Council allow the more fundamental aspects of hyperbaric welding to receive due attention. Such knowledge is essential for the expeditions solving of problems.

PROCEDE	HYDROGENE DANS LE METAL FONDU (+) ml/100 g de métal déposé	
	dans l'air	dans l'eau
MAG (CO_2, fil Ø 1 mm)	3,0	21,2
Manuel à l'arc (Electrode E 917 K, Ø4 mm séchée 2 h à 150°C) Sans vernis (++)	14,1	non déterminée
Avec vernis	17,9	62

(+) Moyenne de 3 valeurs

(++) L'enduit étanche a peu influé sur la teneur totale en hydrogène. De tels revêtements sont nécessaires pour éviter que l'eau ne pénètre dans les interstices de l'enrobage, provoquant ainsi sa désintégration et même son éclatement aux pressions hyperbares.

Tableau 1 - Influence du soudage en pleine eau sur la teneur en hydrogène diffusible (d'après Salter, Ref. 7)

exécuté correctement. L'énergie nécessaire au soudage est tout simplement emmaganisée dans l'explosif. Pour obtenir des assemblages satisfaisants, il semble que les surfaces doivent être très propres et sèches. Ces exigences semblent avoir légèrement compliqué un procédé qui, sans cela, paraissait relativement simple. Nous avons indiqué plus haut que l'application du principe de l'"aptitude à l'emploi" permet d'utiliser des constructions soudées en pleine eau qui, si elles étaient contrôlées de façon classique, pourraient être jugées non conformes. De telles soudures imparfaites se comportent convenablement dans de nombreuses applications. Ainsi, le dudgeonnage de tubes sur des plaques tubulaires assure dans de nombreux cas une liaison assez résistante et étanche sans qu'il y ait de soudure. Il semble que personne n'ait déterminé objectivement ce qu'il faudrait réellement exiger pour obtenir des joints fiables en conditions sous-marines, au lieu d'appliquer systématiquement les exigences normales des codes.

CONCLUSIONS

Nous avons tenté de comparer certaines caractéristiques des procédés de soudage présentant le plus d'intérêt pour le soudage sous l'eau. Le sujet est vaste et la présente communication n'a pas l'ambition de traiter de tous ses aspects. En fait, nous n'avons pas pu décrire en détail des procédés intéressants tels que le soudage avec rideau d'eau, le soudage des goujons (19), etc. Durant les 15 dernières années, les publications sur le sujet ont été nombreuses mais elles n'ont fourni que peu d'informations car les intérêts commerciaux ont empêché les auteurs de s'exprimer librement. De ce fait, on peut penser que la présente communication fait état de plus d'imperfections qu'il n'y en aurait en réalité. Il y a moins de 20 ans, on disait que le soudage dans l'espace était mieux connu que le soudage sous l'eau. Cette remarque est peut-être toujours valable, compte tenu du secret qui entoure cette dernière technique. Nous pouvons prétendre avoir fait des progrès notables - surtout depuis une dizaine d'années - mais dans quelle mesure ces connaissances ont-elles été diffusées et assimilées, nul ne le sait exactement. La littérature donne encore souvent des indications vides de sens telles que "20 % d'accroissement de la ductilité" et "4 % d'accroissement de la résilience". Cela signifie que les vrais mécanismes de rupture ne sont pas encore bien connus et, de ce fait, les ingénieurs s'évertuent parfois à obtenir des résultats qui ne sont pas strictement nécessaires, tout en négligeant des facteurs plus importants. Il est souhaitable qu'il y ait une coopération plus étroite entre les scientifiques et les ingénieurs travaillant dans ce domaine, notamment pour l'élaboration d'exigences réalistes pour les travaux en eaux très profondes.

Dans cette conférence, nous nous sommes limités au soudage hyperbare. La méthode dite de "soudage en chambre à la pression atmosphérique" n'a même pas été mentionnée. C'est un fait que cette technique permet de s'affranchir de nombreux inconvénients décrits plus haut, même si c'est au prix d'un accroissement des coûts de production. Toutefois, il est certain que cette méthode soulève des problèmes qui lui sont propres et, pour des travaux à très grandes profondeurs, il serait utile de l'étudier plus en détail.

Pour des profondeurs comprises entre 0 et 200 m, il n'est pas nécessaire de recourir à cette méthode de soudage car il est possible, maintenant, d'obtenir dans d'autres conditions des soudures de haute qualité.

Reste à savoir dans quelle mesure cet état de choses favorable sera préservé à des profondeurs plus grandes.

Energie de soudage	Epaisseur de tôle	TEMPS DE REFROIDISSEMENT			VITESSE DE REFROIDISSEMENT JUSQU'A 300°C	
		1000-300°C	800 - 500°C			
kJ/mm	mm	dans l'eau en s	dans l'eau en s	dans l'air (+) en s	dans l'eau °C/s	dans l'air °C/s
1	25	3,6	2,1	2,9	90	33
1	50	3,6	1,9	2,7	90	44
2	25	8,5	3,3	6,6	29	8,2
2	50	6,6	2,8	3,7	39	19

Tableau 2 - En comparaison avec le soudage en atmosphère sèche, l'influence de l'épaisseur est relativement insignifiante
en soudage en pleine eau.

(+) Valeurs calculées

cours de la dernière décennie, des efforts énormes ont été consacrés à
étude et à la mise en oeuvre du soudage sous l'eau. L'utilisation épisodi-
e du soudage en pleine eau avec électrodes enrobées est progressivement
mplacé par des techniques avancées. Progressivement, les problèmes liés
tamment à la fissuration par l'hydrogène, à la porosité, à la fragilisa-
on ou à l'instabilité de l'arc ont pu être progressivement résolus, grâce
l'application de principes scientifiques, à un degré qui est moins évident
ns les applications courantes. L'alchimie qui préside toujours à la mise
point de flux pour le soudage ainsi que les secrets qui entourent leurs
rmulations ont freiné le développement, pour le soudage sous l'eau, des
océdés utilisant un flux. En fait, l'absence de produits d'apport adéquats
ur le soudage à grande profondeur a réduit l'utilisation du soudage avec
ectrodes enrobées au profit de procédés dépendant moins de la composition
s flux et donc pouvant être étudiés de façon plus systématique. Le soudage
ec fil fourré, qui permet de grandes vitesses d'exécution, est largement
ilisé par ceux qui disposent du produit consommable adéquat. Les procédés
G et MIG sont à la disposition de tous, mais pour obtenir les résultats
timaux à grandes profondeurs, il est indispensable d'avoir des connaissan-
s techniques considérables. Bien que les soudures exécutées sous l'eau
ient peu nombreuses en comparaison avec celles qui sont effectuées en
mosphère normale, leurs coûts et leur importance sont considérables, ce qui
stifie les sommes importantes consacrées aux études fondamentales. Nous
ons pu, enfin, abandonner les expérimentations empiriques, et les études
ndamentales se sont révélées fructueuses.

nombreux problèmes restent à résoudre. Les quatre procédés courants, à
voir le soudage avec électrodes enrobées, avec fil fourré, TIG et MIG,
t pratiquement atteint leurs limites d'application aux plus grandes pro-
ndeurs actuellement atteintes en mer du Nord. Il serait éventuellement
ssible d'utiliser ces procédés, même marginalement, à des profondeurs
ux fois plus grandes mais il serait nécessaire, évidemment, de procéder
des perfectionnements importants et peut-être même à des modifications
dicales. A ces profondeurs, les possibilités actuelles des plongeurs en
turation semblent également avoir atteint les limites extrêmes et la plon-
e en chambre à la pression atmosphérique présente des aspects séduisants.
le qu'elle est pratiquée actuellement, elle exige moins de dextérité
la part du plongeur, en comparaison avec le plongeur en saturation sou-
nt en pleine eau, dans le cadre du régime actuel de plongée. Les procédés
soudage par fusion devraient être automatisés. Par exemple, en soudage
tomatique avec maintien d'une longueur d'arc constante, l'arc est moins
stable qu'en soudage manuel. Il a été démontré qu'à 40 bar une variation
la longueur d'arc d'un millimètre seulement entraîne une variation de la
nsion à l'arc de 3 volts, s'accompagnant d'importants phénomènes transi-
ires. L'automatisation peut non seulement conduire à une amélioration de
stabilité de l'arc, elle peut aussi l'exiger. Des techniques telles que
brassage magnétique du bain, l'emploi d'un gaz de protection tourbillon-
ire, d'arcs secondaires, ou le raidissement de l'arc par courant HF ont
é proposées pour améliorer le procédé TIG et obtenir des vitesses de sou-
ge beaucoup plus élevées (20). Les avantages de telles innovations n'ont
cune raison de se limiter au soudage hyperbare ; ils pourraient s'appli-
er au soudage à la pression atmosphérique. On peut se demander si le souda-
en chambre à la pression atmosphérique sera largement adopté. Le mot
utomatisation", évoque l'image du soudage orbital tel qu'il
t pratiqué actuellement, avec des interventions occasionnelles de l'opéra-
ur. C'est faire peu de cas du rôle que joueront les robots programmables
intelligents.

35

Gaz	Profondeur en pieds	Ceq	Epaisseur mm	Détails sur la couverture de laitier	Détails sur la stabilité de l'arc	Analyse des enregistrements de la tension, de l'intensité et du type de transfert	Tension (V) à 150 A (voltmètre)	Autres observations
Air	0	0,45	25,4	Couverture totale	Arc très stable	Pas d'enregistrement	22 - 23	
Azote	120	0,45	25,4	Couverture totale de la racine et des cordons en général, mais parfois couverture partielle. Couleur grise	Arc court, sinon extinction	Globulaire/courts-circuits , 3-4/s 27 V max, 20 V min	23 - 25	
Argon	500	0,45	12,7	Couverture partielle. Chapelets de laitier des 2 côtés du cordon. Couleur noire	Petites longueurs d'arc malgré tension élevée	Globulaire/courts-circuits , 5-6/s 29 V max, 23 V min	26 - 32	Fortes projections
Argon	60	0,45	6,35	Couverture partielle. Chapelets de laitier des 2 côtés du cordon.	Très stable	Globulaire/courts-circuits , 4-5/s 19 V min, 26 V max	22 - 26	
Héliox	500	0,45	25,4	Couverture totale. Laitier auto-détachant	Stable	Globulaire/courts-circuits , 4-5/s 29 V, principalement globulaire	26 - 32	
Azote	60	0,45	6,35	Couverture partielle	Arc court, sinon extinction	Globulaire/courts-circuits (principalement globulaire) 4/s enregistrement tension cyclique 32 V max, 21 V min	24 - 26	Dégagement gazeux en fin de cordon
Azote	60	0,36	6,35	"	"	Globulaire/courts-circuits (principalement courts-circuits) 3/s, 23 V en moyenne	24 - 26	Certaine porosité à la surface des cordons
Azote	120	0,36	6,35	"	"	Globulaire/courts-circuits (principalement courts-circuits) 3/s , enregistrement tension cyclique 28 V max, 23 V min	23 - 25	"
Argon	120	0,45	6,35	Couverture totale. Laitier noir. Cordons couleur argent	Bonnes conditions de soudage	Pas d'enregistrement	24 - 28	
Héliox	250	0,45	6,35	Couverture totale. Laitier noir et dépôt brun sur les bords du joint.	L'arc a tendance à s'éteindre ; difficultés d'amorçage	Pas d'enregistrement	30	Quelques projections
Héliox	250	0,36	6,35	"	"	Pas d'enregistrement	30	Quelques projections
Argon	500	0,36	6,35	Couverture partielle mais très épaisse	Stable	Pas d'enregistrement	24 - 30	Moins de projections qu'avec l'hélium
Héliox	500	0,45	6,35	Couverture totale. Couleur noire	Stable	Globulaire/courts-circuits (principalement courts-circuits) 8/s, 28 V en moyenne	26 - 35	

Profondeur mm	Procédé de soudage	Gaz protecteur (+)	Gaz du caisson	Humidité relative	Hydrogène diffusible (ml/100g)(valeur moyenne)
0	Arc avec fil fourré auto-protecteur	Argon	Air	Sec	10,2
0	"	"	"	Humide	13,8
76	"	"	Héliox 10%	Sec	10,2
76	"	"	"	Humide	11,5
150	"	"	Héliox 5%	Sec	11,8
150	"	"	"	Humide	9,7
76	"	Néant	Héliox 10%	Humide	37,0

(+) Protection supplémentaire

Tableau 4 — La teneur en hydrogène dans le métal fondu ne semble pas influencée par la composition du gaz du caisson ou par l'humidité relative si une bonne protection gazeuse est assurée (d'après Réf. 13)

ATMOSPHERE		Ceq (%)	Epaisseur tôle (mm)	Joint N°	Métal déposé Traction longitudinale				RESILIENCE CHARPY V (Joules)									Traction transversale Re (N/mm²)	Essais pliage	Texture	Observation
Gaz	Profondeur (pieds)				Re (N/mm²)	R (N/mm²)	A %	Striction %	-90 °C	-60 °C	-45 °C	-30 °C	-15 °C	0 °C	+30 °C	+60 °C	+90 °C				
Air	0	0,36	19,0	A1	491	559	24	73	15	18	-	123	-	151	207	207	-	618	Acceptables		
Air	0	0,41	19,0	A5	460	525	24	65	12	15	-	125	-	184	198	199	-	614	Acceptables		
Air	0	0,45	6,35	A2	-	-	-	-	-	-	-	-	-	-	-	-	-	690	Acceptables	pas de défauts	
Air	0	0,45	12,7	A3	-	-	-	-	-	-	-	-	-	-	-	-	-	651	Acceptables	Acceptable	
Air	0	0,45	19,0	A4	510	544	22	72	10	16	-	142	-	170	194	201	-	638	Acceptables		
Air	0	0,45	25,4	A6	469	522	30	75	8	11	-	115	-	156	175	193	-	607	Acceptables		
Azote	60	0,36	19,0	A12	493	559	10	15	-	8	-	35	-	48	122	151	152	574	Acceptables		(c)
Azote	60	0,36	19,0	A11	513	571	10	14	-	16	-	34	-	60	100	117	120	618	Acceptables		(c)
Azote	120	0,36	19,0	A13	499	503	6	6	-	14	-	26	-	52	80	96	119	588	Acceptables		(c)
Azote	120	0,45	25,4	A7	566	647	14	43	-	12	-	41	-	56	99	122	137	(a)	Acceptables		(c)
Argon	60	0,45	19,0	A9	440	500	24	70	11	20	122	152	-	169	181	-	-	599	Acceptables		
Argon	60	0,36	19,0	A14	449	513	24	68	12(b)	134	136	187	-	182	-	-	-	590	Acceptables		
Argon	500	0,36	19,0	A17	476	525	24	73	8	39	-	90	-	104	130	122	-	587	Acceptables		
Argon	500	0,36	12,7	A8	-	-	-	-	-	-	-	-	-	-	-	-	-	652	Acceptables	Pas de défauts	
Héliox(10%)	250	0,36	19,0	A16	431	460	22	70	-	15	54	109	-	122	138	138	-	562	Acceptables		
Héliox(10%)	250	0,45	19,0	A15	428	460	24	53	15	54	-	106	-	138	128	119	-	588	Acceptables		
Héliox (5%)	500	0,45	19,0	A18	426	469	10	23	12	30	-	57	-	98	113	104	-	582	Acceptables		(d)
Héliox (5%)	500	0,45	25,4	A10	454	493	12	18	-	12	-	39	92	87	99	95	-	565	Inacceptables		(e)

Notes = (a) Métal fondu en qualité insuffisante ; non déterminée

(b) 54 J à - 75°C

(c) Valeurs d'allongement et de striction: réduites, à cause d'une forte porosité

(d) Valeurs d'allongement et de striction réduites, probablement à cause d'un incident matériel (fuite d'huile hydraulique sur le joint). Essai à refaire

(e) Importante fissure amorcée à partir d'un collage à la racine du joint

'après une récente publication (25), le MITI, au Japon, serait sur le point
e dépenser environ 70 millions de dollars US au cours des 7 prochaines
nnées pour accélérer le développement de robots à apprentissage pour des
pplications au fond de la mer ou des travaux similaires. Le robot Komatsu
commande par câble n'est peut-être que le précurseur annonçart la venue
rochaine d'une autre génération de robots. Au fond de la mer, les objections
olitiques et sociales à l'introduction de robots n'ont pas cours, et c'est
ans ce milieu-là, où l'homme est si désavantagé, que l'intervention des
obots serait si payante. Nos possibilités dans le domaine du soudage hyper-
are à grandes profondeurs pourraient être élargies grâce à la simplicité
t aux grandes vitesses de soudage présentées par les variantes du procédé
IG à fortes intensités et faisant appel au fil chaud, éventuellement
tilisées avec faible écartement des bords et commandées par un robot à
pprentissage aidé d'un capteur optique (26).

es prévisions actuelles relatives au développement du soudage sous l'eau
elèvent de la spéculation.

REMERCIEMENTS

'auteur tient à remercier Chris Allum et John Nixon de l'équipe "Soudage
ous l'eau" de Cranfield pour leur assistance dans la préparation de la pré-
ente communication. La société Sub-Sea International Ltd, en prêtant ses
aissons installés à Cranfield, a permis à cette équipe d'effectuer ses
xpériences (Fig. 11). Leur étude du soudage hyperbare avec électrodes enro-
ées a été financée par l'American Gas Association. Les fonds alloués par
e Science and Engineering Research Council ont permis d'étudier plus en
étail certains aspects fondamentaux du soudage hyperbare. De telles recher-
hes sont la condition nécessaire du progrès dans ce domaine.

REFERENCES

1. MASABUCHI K., MELONEY M.B. "Underwater welding of low carbon and high strength (HY80) steel / Soudage sous l'eau de l'acier HY80 à haute résistance et à bas carbone". 6th Offshore Tech. Conf. OTC 1951 / 6è Conf. Technologie Offshore, OTC, 1951

2. HART P., HARASAWA H. "TWI report no. 40/1977/M June 1977 / Rapport The Welding Institute, no. 40/1977/M, juin 1977

3. OZAKI H., NAIMAN J., MASABUCHI K. "A study of hydrogen cracking in under water steel welds / Etude de la fissuration par l'hydrogène dans des soudures sur acier réalisées sous l'eau". Welding Journal, August 1977, pp. 231s-237s / Welding Journal, août 1977, pp. 231s-237s.

4. GRUBBS C., SETH O. "Multi-pass all position wet welding / Soudage multi-passe en pleine eau, en toutes positions". Offshore Tech. Conf. 1972. OTC 1620 / Conf. Technologie Offshore, 1972, OTC 1620

5. GOOCH T., DEFOURRY J. "Overall appraisal of wet welding investigations carried out at CRM and TWI / Evaluation des recherches sur le soudage en pleine eau effectuées au CRM et au Welding Institute. Doc. 3539/7/78 TWI

6. HELBURN S. "Underwater welders repair drilling rigs / Réparation par soudage sous l'eau de plates-formes de forage". Welding Design and Fabrication. July/Juillet, 1979, pp. 53-59

7. COTTON H. "Application of underwater welding to offshore structures / Application du soudage sous l'eau aux structures offshore". Underwater Conf. Techn. Conf. April 1975. University College, Cardiff

8. SILVA E.A. "Welding processes in deep oceans / Procédés de soudage en eaux profondes". Naval Eng. Jnl. Vol. 80, n° 4, 1968

9. ROBINSON G. "Underwater welding in a dry environment / Soudage sous l'eau en atmosphère sèche".

10. CHEW "Prediction of weld metal hydrogen levels obtained under test conditions / Prediction des teneurs en hydrogène dans le métal fondu obtenues dans des conditions expérimentales. Welding Journal, 1973, pp. 386s-391s.

11. BLAKE P.D. "Nitrogen in steel weldments / L'azote dans les métaux fondus en acier". Metal Construction, April/Avril 1979, pp. 196-197.

12. JENKINS N., STEVENS S. "Nitrogen in ferrous welding / L'azote dans les soudures sur métaux ferreux". Research Bull., The Welding Institute, Janv. 1977.

13. PINFOLD B.E., NIXON J., DORLING D., MOTTA. "Hyperbaric shielded metal arc welding / Soudage à l'arc avec électrodes enrobées en conditions hyperbares". 1981 Final report to the Welding Supervisory Committee of the American Gas Association / Rapport final 1981 au Comité de Surveillance"Soudage" de l'American Gas Association.

4. ALLUM C.J. "Hyperbaric shielded metal arc welding - "Cold cracking' 1982 / Soudage à l'arc avec électrodes enrobées en conditions hyperbares. Fissuration à froid. 1982". Report PR-147-135. Welding Supervisory Committee. Pipeline Research Committee. American Gas Association / Rapport PR-147-135. Comité de Surveillance "Soudage". Comité de Recherche "Pipelines". American Gas Association.

5. SUZUKI H., YURIOKA N., OKUMURA M. "A new cracking parameter for welded steels considering local accumulation of hydrogen / Nouveau paramètre de fissuration des aciers soudés, prenant en considération l'accumulation locale d'hydrogène". IIW Doc. IX-1195-81/ Doc. IX-1195-81 de l'IIS

6. YURIOKA N., SUZUKI H., OHSHITA S., SAITO S. "Determination of necessary preheating temperature in steel welding / Détermination de la température de préchauffage lors du soudage des aciers". Nippon Steel Corporation, April/ Avril 1982

7. KNAGENHJELM H. "Hyperbaric welding at 320 MSW / Soudage hyperbare à une profondeur de 320 mètres". Underwater Tech. Conf. Inst. Pet.

8. ALLUM C.J. "Gas flow in the column of the TIG arc / Ecoulement de gaz dans la colonne de l'arc TIG". J. Phys.D. Appl. Phys. 14, 81. IIW Doc. 212-495-80/ Doc. 212-495-80 de l'IIS.

9. HAMASAKI M., TATEIWA F., SHIGENO H. "Bubbles ensure successful underwater stud welding / Les bulles assurent le succès du soudage de goujons sous l'eau". Welding and Metal Fabr., may/mai, 1980, pp. 241-245.

10. ALLUM C.J. "Underwater welding with TIG / Soudage TIG sous l'eau". IIW doc. 212-522-82 / Doc. 212-522-82 de l'IIS.

11. WELLS A.A. "The meaning of fitness for purpose and the concept of defect tolerance / La signification de l'aptitude à l'emploi et la notion de tolérance de défauts". Fitness for purposes. International Conf. / Conf. Internationale sur l'Aptitude à l'Emploi. The Welding Institute, nov. 1981

12. HARRISON J.D. "The state of the art in Crack Tip Opening Displacement testing / L'état de la question des essais CTOD". TWI Res. Report 108/1980 / Rapport de recherche The Welding Institute, 108/1980

13. GAUDIANO A.V. "Offshore Tech. Conf. Texas 1975 / Conf. Technologie Offshore, Texas, 1975

14. ALLUM C.J., PINFOLD B.E., NIXON J.H. "Some effects of shielding gas flow on argon tungsten arcs in high pressure environments / Influence du débit d'argon sur l'arc TIG dans un milieu à pression élevée". Welding Journal, july/juillet, 1980, pp. 199s-207s.

15. CUSUMANO M. "Robots step out of the factory / Les robots sortent de l'usine". New Scientist, Janv. 6, 1983

16. JONES S., POPE J., WESTON J. "A review of optical methods for adaptive control in arc welding / Revue des méthodes optiques utilisées pour la commande adaptative du soudage à l'arc". TWI Report 200/1982 / Rapport The Welding Institute 200/1982, nov. 1982

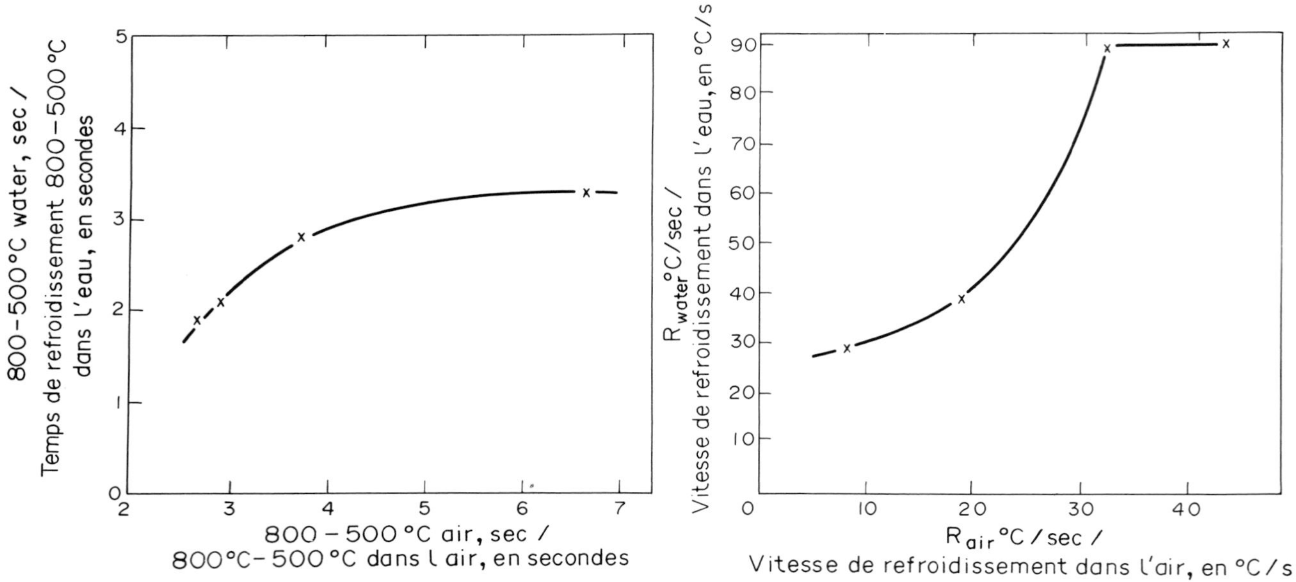

Fig. 1. Comparative quench rates in dry and wet welding / Comparaison des vitesses de trempe en soudage en atmosphère sèche et en pleine eau (d'après Salter – The Welding Institute)

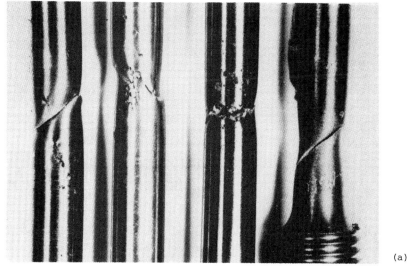

(a)

12   11   13   7

(b)

Fig. 2. Porosity in all weld metal tensiles made at hyperbaric pressure with a slightly long arc (SMAW) in nitrogen chamber gas / Porosité d'éprouvettes de traction en métal fondu non dilué obtenu en soudage hyper-bare avec électrode enrobée, avec un arc assez long, en atmosphère d'azote (d'après Pinfold et al. Réf. 13)

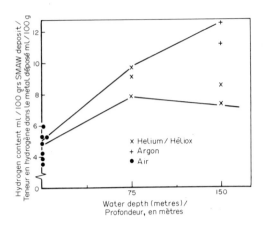

Fig. 3. The effect of water depth upon dissolved hydrogen content (SMAW) /
Influence de la profondeur de l'eau sur les teneurs en hydrogène dissous
(soudage avec électrodes enrobées)

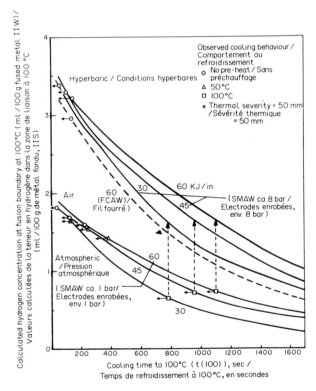

Fig. 4. The effect of welding process, thermal input and ambient pressure
upon the hydrogen content at the fusion boundary / Influence du procédé de
soudage, de l'apport de chaleur et de la pression sur la teneur en hydrogène
dans la zone de liaison (d'après Allum, Réf. 14)

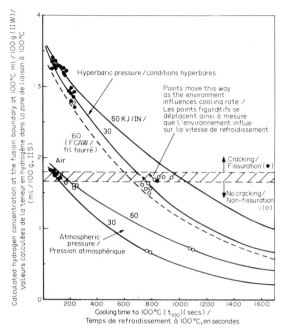

Fig. 5. The effects of pressure and cooling time upon the calculated hydrogen concentration at the critical location of fusion boundary at 100°C / Influence de la pression et du temps de refroidissement sur la teneur calculée en hydrogène dans la zone critique constituée par la zone de liaison à 100°C 25mm Oblique Tekken Tests (thermal severity 50mm) / Essais Tekken 25mm (séverité thermique 50mm) (d'après Allum, Réf. 14)

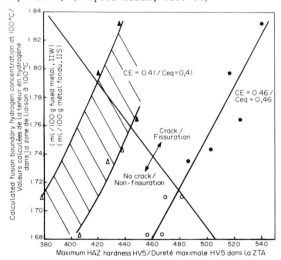

Fig. 6. Relationship between HAZ hardness, calculated hydrogen concentration at the fusion boundary at 100°C (t$_{100}$) and cracking observed in hyperbaric Tekken Tests / Relations entre la dureté en ZTA, la teneur (calculée) en hydrogène dans la zone de liaison à 100°C et al fissuration observée lors d'essais Tekken en conditions hyperbares (d'après Allum, Réf. 14)

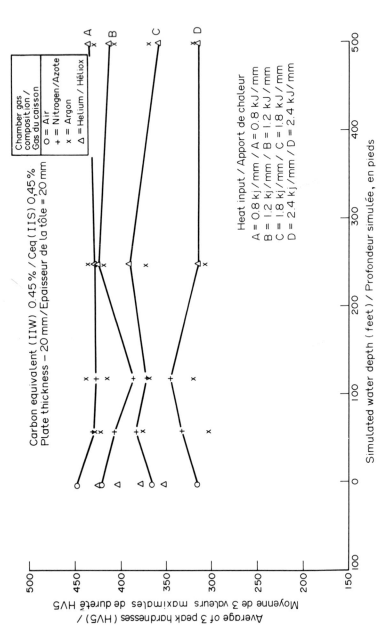

Fig. 7. Peak hardness is a function of carbon equivalent, thermal input and mass and is not affected significantly by the composition of the chamber gas or the pressure / La dureté maximale est fonction du carbone équivalent, de l'apport de chaleur et de la masse de l'assemblage soudé; elle n'est pas sensiblement influencée par la nature du gaz du caisson (d'après Pinfold et

Carbon equivalent (IIW) 0.45% / Ceq (IIS) 0,45%
Plate thickness – 20 mm / Epaisseur de la tôle = 20 mm

Chamber gas composition / Gas du caisson	
O = Air	
+ = Nitrogen / Azote	
x = Argon	
Δ = Helium / Héliox	

Heat input / Apport de chaleur
A = 0.8 kj/mm / A = 0.8 kJ/mm
B = 1.2 kj/mm / B = 1.2 kJ/mm
C = 1.8 kj/mm / C = 1.8 kJ/mm
D = 2.4 kj/mm / D = 2.4 kJ/mm

Simulated water depth (feet) / Profondeur simulée, en pieds

Average of 3 peak hardnesses (HV5) / Moyenne de 3 valeurs maximales de dureté HV5

46

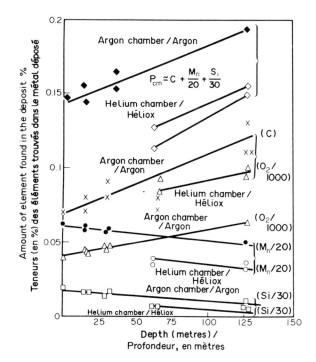

Fig. 8. The influence of chamber composition and pressure upon the composition of weld metal deposited by a basic SMAW electrode / Influence de la nature du gaz du caisson et de la pression sur la composition chimique du metal déposé par une electrode basique

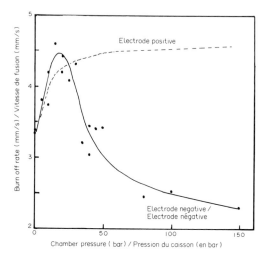

Fig. 9. The influence of pressure upon the burn off rate of SMAW electrodes and the significance of polarity / Influence de la pression sur la vitesse e fusion d'électrodes enrobées; effet de la polarité (d'après Réf. 13 et 14)

47

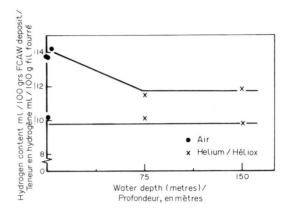

Fig. 10. The dissolved hydrogen content of a FCAW deposit V pressure / Tene en hydrogène dans le métal déposé par un fil fourré, en fonction de la pression (d'après Réf. 13 et 14)
Basic hydrogen content depends upon flux composition and moisture content / La teneur en hydrogène dépend surtout de la composition du flux et de la teneur en humidité

Fig. 11. FCAW welding by remote handling methods in one of the hyperbaric welding chambers at Cranfield Institute / Commande à distance du soudage au fil fourré dans l'un des caissons hyperbares du Cranfield Institute

GENERAL SURVEY   REVUE D'ENSEMBLE

Technologies and Practices of Underwater Welding

K. Masubuchi*, A. V. Gaudiano** and T. J. Reynolds***

*Ocean Engineering and Materials Science, Massachusetts Institute of Technology, Cambridge, MA 02139, USA
**Taylor Diving and Salvage Co., Belle Chasse, LA 70037, USA
***Sea-Con Service, Inc., New Iberia, LA 70560, USA

FOREWORD

It is our great pleasure to present this paper on technologies and practices of underwater welding at this International Conference on Underwater Welding. This paper has been prepared through the combined efforts of three authors, one in academia and two in industry. This paper is intended to cover: (1) an overview of the current state-of-the-art technologies of underwater welding and (2) some specific information on a few widely used techniques. The first part of the overview has been written by Masubuchi. The second part, which is about the practices of hyperbaric dry welding, has also been written by Masubuchi and based on the information supplied by Gaudiano. The third part, which is on wet shielded metal arc welding, has been prepared by Reynolds, with minor editorial changes to make a cohesive document.

ABSTRACT

This paper describes the present state-of-the-art of technologies and practices of underwater welding. The first part of this paper presents a review of various underwater welding technologies. The techniques currently in use may be classified depending upon the environment in which the welding takes place -- dry chamber technique, portable dry spot technique, and wet technique. Shielded metal arc (SMA), gas tungsten arc (GTA), and gas metal arc (GMA) processes are the most widely used in underwater application. Stud welding has been used to a limited degree. Studies have been made of underwater applications of various processes including submerged arc, plasma arc, etc. Although all practical applications and most of the studies so far have been on manual welding, limited efforts have been made to automate welding operations by use of remotely controlled manipulators.

The later parts of this paper discuss the practical aspects of two techniques which are widely used -- hyperbaric dry chamber habitat welding and wet shielded metal arc welding. The second part of this paper, on hyperbaric dry welding, is based on actual experiences obtained at the Taylor Diving and Salvage Co. The third part, on wet welding process, is based on actual experiences obtained at Sea-Con Services, Inc.

IIW-C

KEYWORDS

Underwater welding, stud welding, hyperbaric dry welding, wet shielded
metal arc welding.

PART I AN OVERVIEW OF TECHNOLOGIES OF UNDERWATER WELDING

1.1. Introduction and General Discussion

Underwater welding can be defined as "welding produced underwater". Difficulties associated with underwater welding, especially those with "wet"
shielded metal arc welding, have been experienced since the 1900's.
Compared to welds made in air, underwater welds are plagued by increased
hardness due to rapid quenching caused by the surrounding water, increased
occurrence of porosity and other defects, and increased susceptibility to
hydrogen-induced cracking. As a result, underwater welds tend to have infe-
rior mechanical properties, especially in ductility, compared to welds made
in air. Therefore, until recently, uses of underwater welding had been
limited to temporary repairs and salvage operations.

Bare-wire electrodes coated with varnish for waterproofing were used during
World War I.[1] In 1946, Van der Willigen developed covered electrodes with
waterproofing. This technique significantly improved arc stability and the
soundness of the weld.[2] During the next 15 years, some investigators made
further efforts to improve the quality of underwater welds. However, the
study and application of underwater welding were done by a relatively small
number of people. Therefore, the subject of underwater welding was not the
interest of a sizable segment of the welding industry and technical commu-
nity.

A significant increase in interest in underwater welding started around 1960,
primarily to meet the needs of the offshore oil-drilling industry. Increased
interest in underwater welding can be seen in the sudden surge in the number
of publications at that time. Figure 1 shows the number of articles on
underwater welding published each year since 1930.[3] A sudden increase in
the number of publications occurred around 1960. Although no special effort
was made in preparing this paper to survey the number of articles on underwater welding published during the last few years, it is believed that a
large number have been published in recent years.

To show how the industry and professional societies have become increasingly
interested in underwater welding, a few examples are cited here. In 1974,
the American Welding Society established the D3b Subcommittee on Underwater
Welding within the D3 Committee on Welding in Marine Construction. As a
major task of this subcommittee, the first AWS Specification for Underwater
Welding[4] has been prepared. In 1977, the International Institute of Welding
formed the Select Committee on Underwater Welding. It is indeed very significant that I.I.W. has recognized the importance of underwater welding and
decided to hold a two-day International Conference on Underwater Welding
during its 1983 Annual Assembly in Trondheim, Norway.

Figure 2 shows how the size of the world fleet of offshore oil-drilling rigs
has expanded since 1960.[3,5] In 1962 there were only 62 offshore oil-drilling
rigs in the world. That number had increased to 470 in 1978. The offshore
oil-drilling rigs constructed so far can be classified into several types
including jack-up type, semi-submersible type, ship/barge type, and the

submersible type, as shown in Fig. 2.

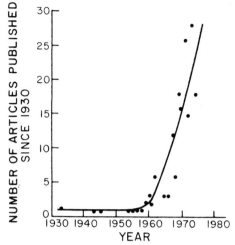

Fig. 1. Articles on underwater welding published each year since 1930.[3]

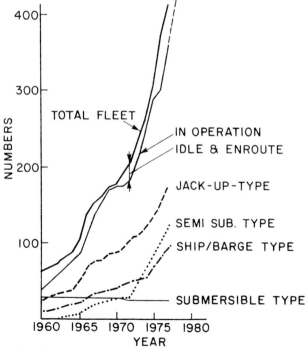

Note: This figure has been prepared based on the survey at M.I.T. Although some articles on underwater welding may not have been included, the figure clearly shows the trend.

Fig. 2. Changes of the size of the fleet of offshore oil-drilling rigs.[3,4,5]

51

The geographical distribution of offshore oil-drilling rigs is quite interesting. The United States, primarily in the Gulf of Mexico and Alaska has the largest fleet. Following in order are the North Sea, South America, Southeast Asia and Australia, Persian Gulf, and Mediterranean Sea. Approximately 50% of all offshore oil-drilling rigs currently in operation are located in the United States and the North Sea.

Although most of the actual offshore oil production comes from waters less than 400 feet (120 m) deep, exploratory drilling has been conducted in water depths greater than 3,000 feet (900 m).[3,6,7] As the development of offshore oil fields expands, drilling and production will go into deeper and deeper waters. Although offshore oil-drilling rigs represent the major ocean engineering structures in operation today, many other structures for various other applications are being built or considered. They include structures for ocean bottom mining, ocean exploration, ocean thermal energy conversion (OTEC), salvaging, etc. As the number of ocean structures increases, there will be an increased demand for underwater construction, inspection, and repair of these structures. Consequently, some very large structures may need to be joined underwater.

The above points to an obvious growing demand for improved technologies on underwater welding and inspection. The needs can be grouped as follows:

1. Increasing numbers of offshore structures are getting old enough to require repairs of varying degrees.

2. Increasing numbers of ocean structures are going to be constructed in the near future, and we must have adequate technologies for constructing and repairing them.

3. There is an increasing demand for extending the underwater welding and inspection capabilities to deeper waters.

There is an interrelationship between the demand for new and improved technologies and their potential uses.[3] New or increased demand results in the improvement of existing technology, even the development of new technology. At the same time these new or improved technologies increase or create different markets for themselves. For example, underwater inspection and repair technologies (which include welding and cutting technologies) have been improved during the last decade, primarily to meet the demand for inspection and repair of offshore oil drilling rigs. These new technologies may be used for other applications. With some modifications, they might be used for inspecting and repairing ships while they are afloat instead of putting them in dry docks. Similar technologies could be used for underwater inspection and repair of steel structures in bridges or harbors.

Another unique application is the use of underwater welding and cutting techniques for repairing nuclear reactor components that are immersed in water. As this area has certain problems, such as the protection of diver/welders from radiation, this paper will not discuss this use.

1.2. Classifications of Underwater Welding Techniques

Many types of underwater welding techniques have been proposed and used. These may be classified depending upon the environment in which the welding operation takes place: wet or dry. They may be classified on the basis of the welding process used: arc welding, stud welding, etc. The different properties required of the welds lead to various welding techniques. Also, as the depth of water increases, fewer techniques become practicable.

Although recent studies cover many joining processes, arc welding processes (especially shielded metal arc (SMA), gas metal arc (GMA), and gas tungsten arc (GTA) processes) are most widely used in underwater welding applications. Underwater welding techniques currently in use may be classified into five groups, as shown in Table 1, depending upon the environment in which the welding operation takes place.[3,4] Distinctions among these techniques are not strict, and a variety of techniques exist which can be regarded to be between each of the techniques shown in Table 1.

TABLE 1   Classification of Underwater Welding Techniques
on the Basis of Environment in which Welding Takes Place

A. **Dry Chamber Techniques**   Welding takes place in a dry environment.

1. One-Atmosphere Welding.   Welding is performed in a pressure vessel in which the pressure is reduced to approximately one atmosphere independent of depth.

2. Hyperbaric Dry Habitat Welding.   Welding is performed at ambient pressure in a large chamber from which water has been displaced. The welder/diver does not work in diving equipment.

3. Hyperbaric Dry Mini-Habitat Welding.   Welding is performed in a simple open-bottom dry chamber which accommodates the head and shoulders of the welder/diver in full diving equipment.

B. **Portable Dry Spot Technique**   Only a small area is evacuated and welding takes place in the dry spot.

C. **Wet Technique**   Welding is performed in water with no special device creating a dry spot for welding. In manual wet welding the welder/diver is normally in water.

Dry chamber techniques are capable of producing high quality welds, but they are costly, especially when used in very deep sea. Mechanisms of welding in the one-atmosphere welding are essentially the same as those used in ordinary welding in air on land. The one-atmospheric welding can be performed at any depth as long as a diving system is developed. However, this technique is most expensive, primarily due to the cost of construction and deployment of the pressure vessels involved. In these pressure vessels, the pressure is maintained at approximately one atmospheric pressure independent of depth. In the hyperbaric dry habitat welding technique, the pressure inside the habitat is at ambient, same as the pressure of water outside. It is widely used for making underwater welds of high quality.

Compared to dry techniques, wet welding techniques, especially those using shielded metal arc process, are less expensive and more versatile; however, the quality of welds are rather poor and their uses are generally limited to repairs.

Qualities of welds may be classified as follows:

1. High quality welds for underwater construction,

2. Permanent repairs of underwater structures,

3. Emergency repairs of underwater structures.

The AWS specification for underwater welding classifies the weld types as follows:

Type A welds intended for structural applications and are made in accordance with a qualified welding procedure.

Type B welds intended for limited structural applications and are made in accordance with a qualified welding procedure.

Type C welds, which are crack-free welds, made in accordance with a qualified welding procedure for service applications where structural quality is not critical.

Type O welds which are of quality equivalent to that produced on the surface in compliance with a code or standard applicable to the particular construction involved. Type O welds have metallurgical and mechanical properties which are essentially the same as those of welds made in air.

1.3. Depth Effects on Underwater Welding Technologies

As the depth of water increases, fewer joining techniques become practical. There are basically two factors which limit the use of certain joining processes for deep-sea applications as follows:

1. Diving system limitations and costs,

2. Depth related technical problems.

Discussions given below come primarily from a report by Masubuchi.[3]

Diving System Limitations. Underwater welding technologies are highly dependent on the diving systems with which they are used. Unfortunately, the arc welding processes now being widely used require a high degree of manipulative ability which can only be provided by a skilled welder/diver in direct contact with the workpiece. Even in shallow water, poor visibility and a lack of body stability and extreme cold may hamper welder/diver performing wet welds or welds in a mini-habitat. As the depth increases, even harsher constraints are imposed. Unless saturation diving techniques are employed, allowable diver working time decreases drastically as depth increases. Even saturation diving techniques cannot be used beyond a diver's physiological limits and submersibles or remotely controlled work vehicles must be used.

Diving systems of interest may be divided into two groups as follows:

a. Systems with direct man-work interface

    1. Conventional diving

    2. Saturation diving

    3. Ambient pressure chambers

4. Constant pressure chambers

b. Systems with remote man-work interface

5. Manned submersibles

6. Remotely operated work vehicles.

The systems in the first group have a direct man-work interface in which the diver/operator gets hands on the work. The system in the second group have a remotely controlled interface in the form of manipulators, television cameras and other devices.

Surface diving is suited for short-time missions in shallow waters. Decompression is required after only a few minutes when working below depths of 100 feet (30m). When air is used to breathe, the safe depth limit is about 200 feet; with a helium-oxygen mixture it is less than 400 feet. Using saturation diving techniques, divers can be kept in a living chamber beyond 1000 feet and transported to a job site for work shift periods of approximately 8 hours.

Several commercial diving companies that serve the offshore oil industry use underwater welding chambers to obtain a dry environment at an ambient pressure for construction and repair of undersea pipelines and other structures. Some details of practices employed by Taylor Diving and Salvage Company are presented in Part II of this paper. A welding chamber is attached to the structure being reported. Next, the water in the chamber is displaced by pressurized gas. Then divers enter from the bottom and fold down gratings for a work platform. A helium-oxygen mixture with an oxygen partial pressure of 6-8 psi (0.42-0.56 kg/cm^2 or 41-55 KPa) has been found suitable. The worker may or may not wear breathing gear while working inside the chamber. At the onset of welding, however, the welder/diver breathes through a mask using a separate system of gases. The atmosphere in the chamber tends to be humid, therefore, special equipment is used to reduce the humidity to create an environment suitable for obtaining good-quality welds.

A submerged chamber maintaining a constant pressure of one atmosphere is another solution for underwater welding. A personnel transfer capsule moves the welders or other personnel to the chamber. The benefits of this system are: the workers are not exposed to pressure or other diving hazards; diving skill is not required of the welder; it is the only system with a direct man-work interface which is not severely limited in depth capability. The primary drawbacks are its limited application and very high cost. Work can only be done in very small areas, enclosed by a specially designed work chamber which can only be mated with a custom designed transfer capsule. Special plugs will have to be developed that are able to resist great differential forces and also must be able to be released externally. Further, they will have to be flexible enough to negotiate the 5 d radius of a riser elbow and have override systems should the externally activated release systems fail.

Within the scope of this paper a manned submersible is considered to be any undersea vehicle capable of transporting a man or men at a constant pressure and capable of performing some manipulative work underwater.

Submersibles of a variety of designs and capabilities (as deep as 20,000 feet) have been built and used for various purposes.[3,8] The limitations, as they affect joining techniques, are not depth-related but determined by

the manipulative devices of the submersibles. The vehicle itself may be
remotely operated. Several remotely operated maintenance systems intended
to perform predetermined work on undersea structures are being designed, for
example, the Seafloor Pipeline Repair System (SPARS) by Shell Oil Company
and the Submerged Production System (SPS) by Exxon Co.[3,9,10] However, under
water welding using manipulators operated by either manned submersibles or
remotely operated work vehicles is still in the development stage. Currentl
basic research on underwater welding and cutting by remote manipulation tech
niques is being conducted at M.I.T[11]

Depth-Related Technical Problems. As the operating depth increases, the
effects of pressure on joining processes become of greater importance.
Pressure effects on the arc are common to all arc welding processes. Other
processes also have problems caused by increased operating depths.

A welding arc is a sustained electrical discharge through a high temperature
highly electrically conductive column of plasma and is produced by relativel
large current and low voltage. Welding arcs in air are constricted to some
extent by electromagnetic forces. An underwater welding arc is additionally
compressed by external forces and cooling effects. In wet welds cooling is
caused by the surrounding water and hydrogen is dissociated from the steam i
the bubble near the arc. Similar arc restriction is observed in dry environ
ment welding due to the gas pressure. In order to maintain the rate of
current transfer, core temperatures must increase. The very high arc-core
temperatures found at greater depths increase penetration.

As depth and hydrostatic pressure increase, the current density of the arc
increases and therefore a higher voltage is required to maintain a constant
arc length.[12] Increased pressure also affects the behavior of the shielding
gas. The density of the gas is increased and higher flow rates are required
It has been observed that porosity of the weld metal is reduced in welds mad
under high pressure. This is perhaps due to suppression of bubble formation
rather than a decrease in the amount of gas in the weld.

Only limited information has been obtained regarding the effects of pressure
on processes other than arc welding. Kataoka[13] investigated the effects of
pressure on arc stud welding in air and water. He found that increased
pressure caused the arc to bend and become unstable. He was able to produce
"good quality" welds with studs 1/2 inch (12.5 mm) diameter in air and underwater, up to a pressure of 50 psig (3.5 kg/mm^2 or 340 KPa). Welds made at
100 psig were not satisfactory however.

Some underwater joining processes require a source of electrical power and
some of these processes require shielding gas as well. In relatively shallo
waters, these items can be provided by cables and hoses from the surface.
These problems become more complicated as the depth increases. Requirements
for power supply and shielding gas become less restrictive if submersibles
and remotely operated work vehicles are used.

1.4 Various Types of Underwater Welding Techniques and Processes

During this two-day International Conference, a number of papers will cover
various subjects related to underwater welding. Obviously, various types
of welding techniques and processes will be discussed. The following pages
describe some of the underwater welding techniques and processes. These
examples are selected to show a variety of techniques and processes, and
they do not necessarily represent those which are widely used or promising.

Today, many joining processes are commercially available.[14] In an extremely
broad sense, most of these processes are potentially useful for underwater
applications, especially "dry" techniques. In fact, many investigators have
studied potential uses of a variety of processes under "wet" conditions. As
far as actual uses of these processes are concerned, however, only three
processes including shielded metal arc (SMA), gas tungsten arc (CTA) and gas
metal arc (GMA) processes, have been widely used. Arc stud welding has only
been used to a limited degree.

Arc Welding Processes. Arc welding processes include such processes as SMA,
GTA, GMA, submerged arc, plasma arc, etc.

Engineers at Mitsubishi Heavy Industries of Japan have developed a method of
creating a local dry atmosphere near the welding arc by a high-speed jet of
water, as shown in Fig. 3.[15] The ends of the double torch form trumpet-
shaped nozzles. When shielding gas is supplied from the central part of the
torch and a high speed water jet is injected from the outer nozzle, a stable
local dry zone (cavity) is formed around the welding arc. Thereby, gas
metal arc welding processes can be performed in the gaseous region.
Mitsubishi engineers conducted extensive studies on this method.

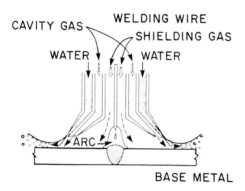

Fig. 3. Principle of local drying underwater welding method. [15]

Underwater wet welding using plasma arc and submerged arc processes have
been studied by several researchers.[16-19] M.I.T. investigators[20] found that
submerged arc welding provides an effective way for reducing the hardness of
the weld metal and the heat-affected base metal since the weld region is
covered by flux.

Stud Welding. Underwater uses of stud welding have been studied by investigators in various countries, including the U.S.A., Japan, France, and the
Netherlands.[3,20-25] In fact, stud welding may be the only process, other
than arc welding processes, which has been used to some extent for actual
construction underwater. COMEX Services in France and several other companies have used stud welding for some underwater construction and repair
works. Figure 4 shows schematically how a steel stud may be attached to an
underwater structure. Since this process is being used for actual construction, it may be worth considering the development of a standard and/or
recommended practice of underwater stud welding.

Stud welding on land has been used since the 1930's.[26] The first recorded

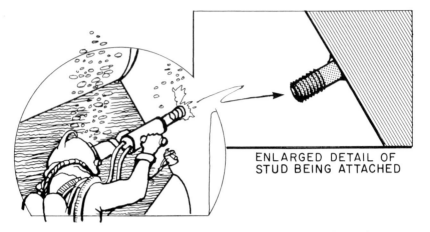

ENLARGED DETAIL OF
STUD BEING ATTACHED

Note: This figure has been drawn based on the information
given in an article prepared by COMEX Services[24]

Fig. 4.   Schematic figure showing how a steel stud may be attached to an
underwater structure.

underwater stud welding field tests occurred in 1944.   Conventional stud
welding equipment consists of a stud welding gun, an electric controller,
and a D.C. power supply.   To make a weld, the operator simply loads a stud
into the gun, presses it against the workpiece, and pulls the trigger.   The
electric controller then controls the sequence of the welding process
according to pre-determined settings.   Two types of power supplies are being
used.   In the case of capacity discharge stud welding, a bank of charged
capacitors are used as the power source.   This process is limited to studs
smaller than 1/4 inch (6.4 mm) in diameter.   The second type of stud welding,
known as arc stud welding, can weld studs of more than one inch in diameter.

Presented in the following is a short summary of the work performed at M.I.T.
where researchers have studied underwater stud welding since 1974.[20,22]
Initially, simple tests using a capacitor discharge type gun were conducted.
Further tests were made with arc stud welding guns using steel studs up to
3/4 inch (19 mm) in diameter.   Welds with satisfactory metallurgical and
mechanical properties were obtained underwater.   Hardness tests on cross
sections revealed interesting results.   Although the maximum hardness occurs
in the heat-affected base metal (location 4 in Fig. 5), it is little affected
by the presence of water.   Efforts were made to develop a stud welding system
which would be used underwater.   Figure 6 shows the design of a prototype
model which can be operated by a diver with no welding skill.[22]   Efforts are
currently in progress to develop a fully automated and integrated stud welding system which can be remotely operated from an underwater vehicle.[11]

M.I.T. researchers believe that stud welding is fundamentally suited for
underwater wet welding as compared to other welding processes such as
shielded metal arc welding.   In wet shielded metal arc welding, molten metal
is transferred from the electrode in the form of small particles.   These
particles travel in an arc atmosphere surrounded by water which has a
tremendous quenching effect on the weld.   In the case of stud welding, on
the other hand, metals in the central portions of the stud are not severely
affected by the presence of water.   This can be effected provided that

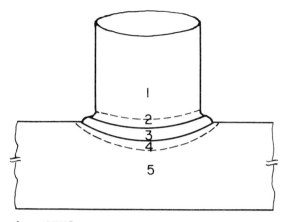

1. STUD
2. HEAT AFFECTED ZONE (STUD SIDE)
3. WELD METAL
4. HEAT AFFECTED ZONE (BASE PLATE SIDE)
5. BASE PLATE

Fig. 5. Cross section of a stud weld.

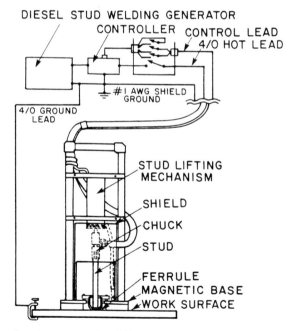

Fig. 6. An arc stud welding system which can be operated by a diver.[22]

welding conditions employed are such that the small amount of water that existed between the stud and the base plate is squeezed out when the stud is pushed against the plate at the end of the welding operation.

Other Welding Processes. Investigators also studied possible underwater applications of several other processes including electron beam welding, explosive welding, and thermite welding.[28,30]

Automation and Robotization. At present almost all underwater welding jobs in both dry and wet conditions are performed manually. In very recent years, however, there has been tremendous interest in the welding industry in automation and robotization. Many types of robots have been developed. There has been a surge of interest in automatic control of welding operations. Until recently, all automatic welding machines which were developed perform welding operations using predetermined welding conditions with no adaptive control. Efforts are being made to develop "smart" welding machines which have sensing and control capabilities in real time.[11] These new advances in welding technology may soon be adapted in underwater welding.

PART II TECHNOLOGIES AND PRACTICES OF HYPERBARIC DRY HABITAT WELDING

2.1 Introduction

This section discusses practices of hyperbaric dry habitat welding as used by Taylor Diving and Salvage Company. The concept of underwater dry habitat welding on pipelines essentially began with the 1954 patent by Osborn. He developed a design of an underwater enclosure that was to become the predecessor of welding habitats used today.[31-39] The patent was ahead of its time and received little attention until shortly before it expired. Starting around 1965 several companies, including Taylor Diving, that were involved in diving and related activities built various types of underwater welding habitats and made repairs on offshore platforms and pipelines.

Taylor Diving recognized the problems arising from the increasing depths at which marine pipelines were being constructed.[32] While pipe could be laid successfully in considerably deep waters, the flexibility of their long risers made the latter very difficult to set. Existing pipelines were often damaged when attempts were made to bring ends to the surface for connection or repair. Large diameter pipe, or those under other pipelines, could not be raised to the surface.

Taylor's solution to these problems was to develop techniques for welding pipe in place in a dry environment so that joint properties obtained could be comparable to those of welds made on the surface. The first successful underwater hyperbaric dry habitat welding was performed by Taylor in 1968 at a depth of 62 feet (19 m). The techniques originally developed for pipeline tie-in and repair have since been adapted to riser tie-ins on concrete platforms, repair of damaged offshore structures, and many other applications. The techniques and equipment have been refined to the point where crews are doing hyperbaric welding in more than 500 feet (150 m) of water on a routine basis. Welding at a depth of 1,030 feet has been performed in the North Sea. Most of the work thus far has been done from Brown and Root barges in the Gulf of Mexico and North Sea.

2.2 Equipment and Operational Procedures

The basic technique involves the use of an open-bottom enclosure or Underwater Welding Habitat (UWH), as shown in Fig. 7. It can be placed over
opposing ends of the members to be welded. Newest welding habitats measure
approximately 10 feet (3 m) square and are designed to handle pipe up to 54
inches (1,370 mm) in diameter.

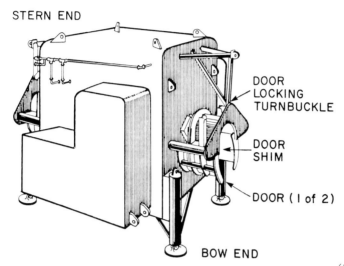

STERN END

DOOR
LOCKING
TURNBUCKLE

DOOR
SHIM

DOOR (I of 2)

BOW END

Fig. 7. Underwater welding habitat (UWH) as used with SPAR.[32]

Most jobs included the use of a Submarine Pipe Aligning Rig (SPAR). Figure
8 shows the SPAR Model 2, which is 65 feet (20 m) long. It weighs 175 tons in
air and 58 tons in water. As an example of explaining how the SPAR and the
UWH may be used, Fig. 9 shows a typical riser/pipeline connection at the base
of an undersea steel structure.

The habitat is lowered into place over the pipe ends through clamping ends
located on each end of the habitat. The SPAR is remotely guided into place
over the pipe ends. Next, divers manually operate the hydraulic drive units
to coarsely align the pipe ends. Once the habitat is lowered into place
divers enter it and install seals at the intersection of the pipe with the
habitat walls. With the seals installed, divers remove the remaining water
from the chamber by displacing it with helium gas, and complete the final
alignment of the pipe.

Figure 10 shows schematically the welding operation inside the habitat. The
habitat contains an environmental control system. Since the habitat is also
filled with high percentage of helium gas, and an environmental control
system is used, the divers can spend significant amounts of time to performing useful work. Humidity control of the environment gas is an important
consideration during welding. Equipment is provided to dry the environment
and to maintain it in the dry condition.

In order to perform welding jobs which cannot be done by using the SPAR-UWH,
a Modular Underwater Welding Habitat (MUWH) has been developed. As an

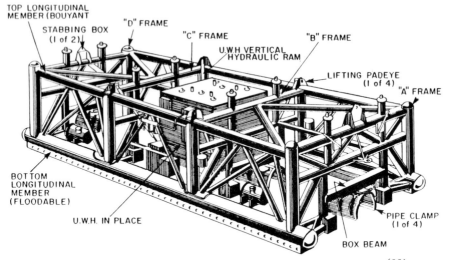

Fig. 8. Submarine pipeline alignment rig (SPAR, Model 2).[32]

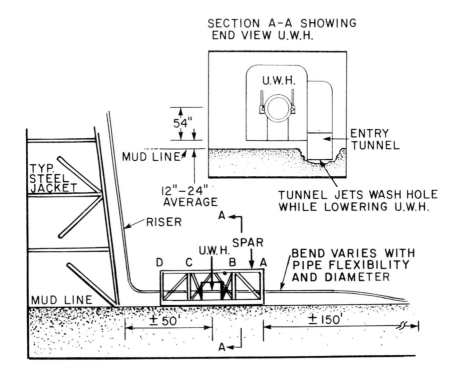

Fig. 9. Typical riser/pipeline connection at base of steel structure.

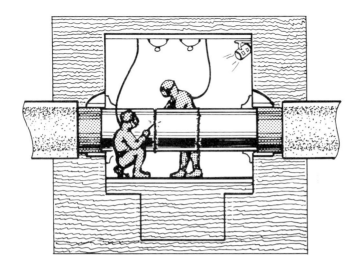

Fig. 10. Sketch of welding operation in a hyperbaric dry habitat.

example of explaining how the MUWH may be used, Fig. 11 shows a set-up for repairing a diagonal member of a tubular structure. Details of operational procedures vary to some extent from one job to another depending upon the pipe size, water depth, sea bottom condition, etc. Reference (31) presents a summary of 26 early underwater dry habitat welding jobs. Taylor has completed more than 250 dry habitat welding jobs to date.

PURPOSE BUILT HABITAT 8' x 8' x 12'

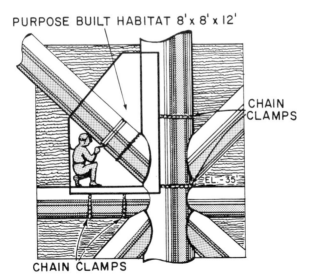

CHAIN CLAMPS

EL. - 35'

CHAIN CLAMPS

Fig. 11. Set-up for repairing a diagonal member of a tubular structure.(32)

2.3 Discussions on Some Subjects Related to Welding

Depth Simulator. To ensure that welds made in a dry habitat have satisfactory metallurgical and mechanical properties, various welding processes
and consumables need to be tested under conditions which simulate those
encountered in the Underwater Welding Habitat (UWH). At its headquarters in
Belle Chasse, Louisiana, Taylor Diving operates a hyperbaric research and
training facility. The facility is used to test equipment, train personnel
to perform a variety of underwater tasks, and to study man's ability to work
and live at great depths. The most important equipment in the facility is
the depth simulator. With this, the present facility has been able to
conduct tests at pressures equivalent to a depth of 335 meters (1100 feet)
of sea water. The main chamber of the depth simulator has an inside diameter
of 12 feet (3.6 m).

Evolution in Welding Processes.[33] In the early years of evaluating welding
processes in shallow water depths, less than 60 feet (18 m), few differences
from surface welding were experienced.

Gas tungsten arc (GTA) welding was initially used by Taylor for the entire
weld as it met the requirements for producing a root bead of inside a convex
contour with satisfactory metallurgical and mechanical properties. However,
GTA is a very slow process, and simulated hyperbaric welds were made with
the gas metal arc (GMA) process. Although the welding speed of the GMA
process is high, it has tended to produce a concave-shaped root pass; also,
lack of fusion possibilities and wire feed problems tended to offset the
gain and deposit rate.

Shielded metal arc (SMA) process using low hydrogen type electrodes was then
investigated. It was found that good penetration with a reasonable welding
speed could be achieved under hyperbaric conditions. As a result, the SMA
process is most widely used at present. Since the SMA process with low
hydrogen electrodes tends to produce marginally acceptable roots in open butt
joints, the GTA process is still exclusively used for the root bead and
increasingly for the hot pass and first fill pass also.

Semi-Automatic and Automatic Welding Processes. Taylor Diving is actively
pursuing use of manual semi-automatic flux cored arc welding (FCAW) process
as a means of shortening the time it takes to repair offshore structures.
As regards fully automatic weld processes such as the use of orbital
machines for hyperbaric welding of pipelines, the following conditions must
be met:

1. Develop efficient means of aligning and fitting pipe joints;

2. Develop an effective means of obtaining positive inside root welds with
    no linear indications.

Changes in Pipeline Material Since 1968. Since Taylor Diving's first dry
habitat welding in 1968, the yield strengths of line pipe has, in general
increased from 42,000 to 65,000 psi (30 to 46 kg/mm^2 or 290 to 450 MPa).
It has been necessary for them to weld, under hyperbaric conditions, steels
such as DIN 17100, St 52-3N material, in addition to quenching and tempering
high yield strength steels. Initially, the choice of consumables was made
to match the properties of the pipe and to exceed slightly its tensile
properties, but with North Sea installations, toughness and hardness specifications of classification societies also came to be required. The Charpy
V-notch toughness requirements for 1-inch (25 mm) thick line pipe hyperbaric

welds have been 35 ft-lb (4.8 kg-m or 47 J) at -10°C (14°F). These requirements have had to be made with the Charpy V-notch located in the weld cap,
weld root, at the fusion line and at 2 and 5 mm (0.08 and 0.2 in.) from the
fusion line.

PART III TECHNOLOGIES AND PRACTICES OF UNDERWATER WET WELDING

3.1. Introduction

Underwater wet welding was first developed many years ago as a method to
effect a fast, temporary and inexpensive repair until the repaired weldment
could be raised from the water and have a more permanent dry welded repair
made. In those days, the temporary wet welded repair was of suspect quality.

The offshore industry created a need for quality wet welded repairs. Wet
welding is a fast method of repair providing sound, structural quality welds.
It requires less support equipment than a similar underwater dry welded
repair or the alternative mechanical connections. For these two primary
reasons wet welding is a more inexpensive repair method. The most significant change for wet welding, and the primary reason behind using this
joing procedure now more than ever before, is the fact that more than 100
structures have been repaired successfully without failure. As a result of
this, wet welding is now considered to be a means of effecting permanent
repairs to offshore structures.

3.2. Applications in Wet Welding

The use of wet welding is not limited to repair work. Wet welding has been
used for several years as a method of attaching new sacrificial anodes to
marine structures. Strain gage supports have also been installed on critical
areas of marine structures. Lifting lugs or pad eyes can be wet welded to a
submerged structure for salvage; however, the primary function and application of wet welding is repair work.

Wet welding is primarily used to repair damaged offshore structures. Wet
welding is also used to repair damaged submarine pipelines, dock steetpile
corrosion or damage, and cooling water inlet and dispersal systems for power
generating facilities.

Developing a Wet Welding Procedure. Quality repair work by wet welding
requires the same engineering "know how" and in some instances, more than
that, for repairing worn out equipment parts, overloaded bridge supports and
many other above the surface repair situations. The Underwater Welding
Engineer must ascertain why the failure occurred. He must design a repair
that will eliminate or minimize stress concentrations or any metallurgical
notches. He must take into consideration the fact that a wet weld has less
ductility than a dry weld. The material to be joined underwater by wet
welding must be identified by either certified test reports from the as-
built records, or by obtaining a metal sample and having its chemistry
analyzed.

A Welding Procedure Qualification Record (PQR) must be established to prove
that a sound weld can be made under conditions similar to the actual repair
situation. Typical physical tests on the PQR weldment would be visual
inspection (VT), radiography (RT), macroexamination including a hardness

transverse, reduced section tensiles (RST's) bend tests, and as supplementary
tests, Charpy V-notch impact testing if required.

Anatomy of a Wet Welded Repair. Once a wet welding procedure is qualified,
a Welding Procedure Specification (WPS) is written, experienced welder/divers
are qualified in accordance with the WPS, plans then can be made to carry
out the desired repair in accordance with the WPS. The following is an
account of several wet welded repairs to a platform in the Gulf or Mexico
during the summer of 1982.

Scope of Work. Sea-Con Services, Inc. was contracted to replace two heavily
damaged horizontal diagonals at a depth of 28 Feet Sea Water (FSW). This
damage had been discovered the previous year during a platform inspection.
(See photographs 1 through 4 at end of paper). The contract included the
removal of the damaged diagonals and the field fabrication and installation
of the replacement members. The final installation was accomplished using
wet welding. The entire contract was completed in a total of 88 working
hours using an 8-man crew.

Structure Description. This 8 pile platform had five (5) underwater levels
and stood in 195 FSW. The damaged braces were at the 28 FSW level in the
A-B Bay, as shown in Fig. 12.

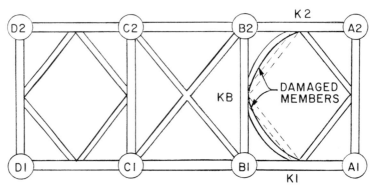

Fig. 12.   Schematic showing location of damaged members.

Extent of Damages. The braces were badly bent and cracked. The one extending from K2 to KB was split for most of its length. The structural braces
were made from welded pipe which has a tendency to corrode excessively in
the HAZ when not adequately protected by sacrificial anodes. Some of the
welds were also broken.

Design of New Members. First, all individual parts of the replacement structure (see Fig. 13) are fitted together and tack welded underwater. The
importance of proper fit-up prior to production welding cannot be overempha-
sized. Thus, less welding is required because good fit-up means less welder/
diver bottom time, resulting in finishing the contract on or ahead of schedule. The structure is then completely welded topside into two portions
separated by the split sleeve. Thus, all wet welds (i.e., doublers to horiz-
ontals and split sleeves to diagonals) are fillet welds. (See photographs
5 through 12 at end of paper).

66

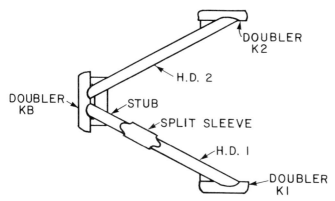

Fig. 13. Replacement horizontal diagonal structure.

3.3 Summary

Wet welding technology has developed from a temporary repair method to an acceptable, more economical, and permanent repair method. It has been shown by employing the correct technology for electrode development and having the proper welding engineering background for electrode selection, that high quality underwater wet welds can be made. In addition, comprehensive welder/diver training is essential to fully achieve high quality, structurally sound, underwater wet welds.

REFERENCES

1. Tsai, C. L. and K. Masubuchi (1977). WRC Bulletin No. 224, Welding Research Council.
2. Silva, E. A. (1971). The Welding Journal, 50(6), 406-415.
3. Masubuchi, K. (1981). Development of joining and cutting techniques for deep-sea applications, MIT Sea Grant Report No. MITSG 81-2.
4. AWS D3.6. Specification for underwater welding. American Welding Society, to be published in 1983.
5. Kuriyama, T. (1979). Bulletin of the Society of Naval Architects of Japan, No. 603, 2-8.
6. Iimura, O. (1978). Underwater fabrication and construction systems. M.S. thesis, MIT, Department of Ocean Engineering.
7. Snyder, L. J. (1977). Offshore, 315-322.
8. Penzias, W., and M. W. Goodman (1973). Man Beneath the Sea. (John Wiley & Sons, Inc.) New York.
9. Tubb, M. (1974). Ocean Industry, 9 (11), 91.
10. Alberstadt, M. (1974). Preparing for the Deep. Profile - Exxon Company, U.S.A., pp. 17-24.
11. Masubuchi, K., and V. J. Papazoglou (1982). Automation in Underwater Welding. Proceedings of the Third International Conference on Behavior of Offshore Structures, Hemisphere Publishing Corp., New York, 756-766.
12. Perlman, M., A. W. Pense, and R. D. Stout (1969). Welding Journal, 48 (6), 231s-238s.
13. Kataoka, T. (1978). Applications of arc stud welding for deep-sea salvage operations. M.S. Thesis, MIT, Department of Ocean Engineering.

14. Welding Handbook (1976). American Welding Society, 1, 7th edition.
15. Shinada, K., M. Tamura, Y. Nagata, T. Omae, H. Wada, Y. Manabe, and T. Nobushige (1979). Mitsubishi Heavy Industries, Technical Review, 16 (2); also Welding Research Abroad (1980), 26 (1), 13-22 Welding Research Council.
16. Hasui, A. (1971). Journal of the Japan Welding Society, 40 (7), 622-631.
17. Hasui, A., Y. Suga, and N. Sekimizu (1980). Transactions of the Japan Welding Society, 11 (1); also Welding Research Abroad (1981), 26 (3), 2-8.
18. Hasui, A., Y. Suga, S. Kishi, and M. Teranishi (1980). Transactions of the Japan Welding Society, 11 (1); also Welding Research Abroad (1981), 26 (3), 9-13.
19. Masubuchi, K., and C. L. Tsai (1978). Patent #4,069,408.
20. Tsai, C. L., H. Ozaki, A. P. Moore, L. M. Zanca, S. Prasad, and K. Masubuchi (1977). Development of new, improved techniques for underwater welding, MIT Sea Grant Report No. MITSG 77-9.
21. Masubuchi, K., H. Ozaki, and J. Chiba (1978). Oceans '78. The Marine Technology Society and the Institute of Electrical and Electronics Engineers, 125-219.
22. Masubuchi, K., D. W. Schloerb, and H. L. Gustin (1983). OTC 4601, Offshore Technology Conference.
23. Hamasaki, M., and F. Tateiwa (1979). Transactions of the Japan Welding Society, 10 (1); also Welding Research Abroad (1979), 25 (10), 2-6.
24. COMEX Services, personal communication.
25. Personal communication.
26. Shoup, T. E. (1976). Welding Research Council Bulletin No. 214.
27. Personal communication.
28. Electron Beam Pipe Welding Ready to Go Offshore (1982). Offshore Engineering, 23-24.
29. Redshaw, P., A. W. Stalker, and K. Allen (1978). OTC 3349, Offshore Technology Conference.
30. Masubuchi, K., and A. H. Anderssen (1973). OTC 1910, Offshore Technology Conference.
31. Gaudiano, A. V. (1975). OTC 2302, Offshore Technology Conference.
32. Taylor Diving Hyperbaric Welding, a brochure prepared by Taylor Diving and Salvage Co.
33. Gaudiano, A. V., and D. Groves. Underwater Dry Environment Habitat Welding.
34. Gaudiano, A. V., personal communication.
35. Delaune, T. P. (1979). The Welding Journal, 59 (8), Part I, 17-25, and Part II (9) 28-35.
36. Gaudiano, A. V. (1976). Oil in Deeper Waters (Birmingham Metropole).
37. Delaune, T. P. (1978). OTC 3348, Offshore Technology Conference.
38. Delaune, T. P., Welding operations at sea in 165 meter (540 ft) water depth, a draft of a speech.
39. Franckpane, J. A case history of an offshore underwater structure repair in the North Sea, a document supplied by Taylor Diving and Salvage Co.

1. View of damaged braces from below.

2. Close-up view of KB-K4 brace.

3. Horizontal split due to corrosion.

4. Broken weld at KB.

5. Horizontal diagonal ready to be lowered and set in place.

6. Scalloped split sleeve pipe splice welded out.

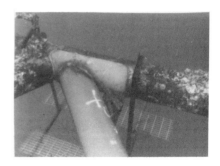

7. Fitting doubler K1 and H.D.1.

8. Scaffolding at K1.

9. Topside welding of stubs and gussets to doubler KB.

10. Doubler KB welded completely.

11. K2 doubler welded completely.

12. Doubler KB and scalloped split sleeve splice for H.D.1 welded completely.

The Metallurgy of Underwater Welding

N. Christensen

SINTEF, N-7034 Trondheim-NTH, Norway

ABSTRACT

Wet and dry welding processes are surveyed in terms of compositional changes, cooling programme and weld joint performance. Introduction of hydrogen and oxygen in wet welding may lead to porosity and slag inclusions. High cooling rates, mainly controlled by convection, may cause cracking. Significant improvements can be achieved by local protection of the site of welding.

Effects of pressure predominates in dry welding, leading to increased absorption of hydrogen, oxygen and carbon and to reduced contents of silicon and manganese in MMA welding. The increased impurity level limits the application of hyperbaric MMA welding to about 300m depth. Inert-gas protected processes are shown to be much less affected by pressure. Examples of performance and methods of predicting the chemistry and properties are given.

KEYWORDS

Wet welding chemistry; cooling rate; hyperbaric welding; pressure-dependent chemistry; depth limits in hyperbaric welding.

INTRODUCTION

It is convenient to consider separately the two main procedures of underwater welding:

Wet Welding, characterized by the presence of water.

Dry Welding, characterized by an increased pressure.

Wet welding is restricted to shallow water, of a depth down to 40 or 50m, because the diver must be saturated to do an efficient job at greater depth. Once the expensive system for saturation diving has been established for a given task, the advantages of welding in the absence of water will clearly be used.

Intermediate cases are represented by developments where water is excluded from the site of welding by means of a local protection. Such methods are in the main used at shallow depth, and hence a high pressure is not a predominating factor. Although in principle "dry", they will therefore be considered separately.

WET WELDING

The presence of water will affect the quality of the welded joint in several ways. Disregarding for the present discussion problems caused by limited visibility and other operational difficulties, the presence of water may be considered from two main points of view:

- effects on weld metal chemistry,
- effects of rapid cooling of the weld and its surroundings.

The soundness and the mechanical properties of the weld metal will in the main be controlled by absorption of hydrogen and oxygen. The incidence of cracking will depend both on absorbed hydrogen and the rate of cooling. The grade of parent metal is of course as important in wet welding as in other applications.

Effects of Water on Weld Metal Chemistry

Nearly all arc welding methods have been tested in wet underwater welding, as may be seen from a literature survey by Dadian (1975). Covered elec-trodes (MMA) have been employed in the majority of practical applications. Gas shielded processes using inert gas or CO_2 shielding (MIG or MAG) have the advantage of continuous operation; however, the flux-cored wire process (FCAW) applied without gas shielding has further advantages of easier handling and better visibility. Examples of weld metal composition obtained with these methods are given in Table 1. The elements listed are those expected to be affected by the presence of water: hydrogen, oxygen, carbon, silicon and manganese. The composition obtained on welding in open air has been included for comparison when available.

Hydrogen is seen to be present in quantities corresponding to the "high" range as classified by Coe (1973) on the basis of H_{DM}. Exceptions are found, for the dead-soft MMA deposit reported by Mathisen and Gjermundsen (1979) and for the self-shielded FCAW deposit described by Savitch (1974), both belonging in the "medium" range. A rather unexpected effect of the arc voltage was reported by Savitch, lower hydrogen contents being observed at higher arc voltages. It is stated that the average life of drop existence is reduced at increased arc voltage and current, thereby reducing the time available for hydrogen absorption.

TABLE 1   Examples of Weld Metal Composition in Air and Shallow Water

Method and type of consumable	Diffusible hydrogen a) H_{DM} ml/100g	H_{FM} ppm	Remark	Oxygen ppm air/water	Carbon % air/water	Silicon % air/water	Manganese % air/water	Ref.
Covered electrodes (MMA) — Ferritic								
Dead-soft	8-21	3-8	-	-/900	0.05/0.05	0.00/0.00	0.04/0.00	1
E 6020	24	-	Resid. 14ml/100g	-	0.01/0.01	0.00/0.00	0.04/0.03	2
Exper.(oxide)	40	22	-	-/840	0.11/0.08	0.11/0.04	0.70/0.48	1
E 6013	54	-	Resid. 3ml/100g	-	0.07/0.03	0.20/0.06	0.45/0.19	2
E 6024	43	-	3ml/100g	-				2
E 7014	58	30	-	410/625	0.07/0.07	-	-	1
Austenitic EXX18	31	-	b)	-	-	-	-	3
E 310-16	38	-	b)	-	-	-	-	3
E 312-15	40	-	Resid. 40ml/100g	-	-	-	-	2
Solid wire — MIG Unprotected	27-38	-	Arc volt. 40-25	-	-	-	-	4
Ar shielded	24-34	-	40-25	-	-	-	-	4
CO_2 shielded	23-31	-	40-25	-	-	-	-	4
MAG CO_2 shielded	-	-	-	-	0.13/0.13	0.00/0.00	0.51/0.13	5
CO_2 shielded	-	-	-	-	0.12/0.13	0.23/0.00	0.65/0.30	5
CO_2 shielded	-	-	-	-	0.11/0.10	0.30/0.00	0.95/0.51	5
Flux-cored wire (FCAW) Self-shielded	14-23	-	Arc volt. 40-25	-	-	-	-	4
Self-shielded	41	17	30V	-/350	0.12/0.11	0.20/-	1.20/-	1
Self-shielded	45	20	30V	-/550	-/0.09	-	-	1
Rutile core; CO_2 shielded	55	23	30V	-/570	0.05/0.07	-	-	1

References: /1/-Mathisen and Gjermundsen (1979); /2/-Stalker, Hart and Salter (1975);
/3/-Ozaki, Naiman and Masubuchi (1977); /4/-Savitch (1974);
/5/-Shlyamin and Dubova (1961).

a) H_{DM} reported on deposited metal; H_{FM} on mass of fused metal

b) Measured acc. to JIS Z 3113-1975 and converted to ISO 3690

In wet MMA welding the surrounding water may not be the only source of hydrogen. Ozaki, Naiman and Masubuchi (1977) found essentially unchanged hydrogen contents in rutile electrode deposits after various periods of storing up to one hour in water, regardless of the presence of water-proofing layers applied to the electrode. Possibly, the amount of water remaining in the coating from the fabrication process may be high enough to account for the observed hydrogen content (41 to 53 ml/100g deposit), additional water absorbed by the coating prior to welding being expelled by resistance heating and therefore less important. It is also possible, however, that the controlling factor is water vapour introduced directly into the arc atmosphere, which would not depend on the moisture content of the coating. The former interpretation would seem to be supported by blistering of the water-proofing varnish observed by Mathisen and Gjermundsen; moreover, the lower content shown for the basic EXX18 types could indicate that the initial content of strongly bound water is an important factor.

Neither one of these interpretations can account for the extremely low hydrogen content of the dead-soft deposit from electrodes that have not been baked to a low initial water content. In this case the absence of deoxidants would allow evolution of carbon monoxide from the liquid metal during cooling, and hence an efficient purging of dissolved hydrogen would be expected both in normal and wet welding.

It appears thus that hydrogen absorption in wet welding may be controlled by several factors, and that more information is needed for a proper understanding of these factors. From the user's point of view it must be accepted that the majority of methods and consumables tested will give hydrogen levels of the order of 30-40 ml/100g deposited metal, roughly corresponding to 20 ppm on a fused-metal basis. The presence of such amounts could cause porosity, and would also reduce the static ductility of the weld and its nearest surroundings.

Examples of porosity as affected by operational variables are shown in Fig. 1 for two types of covered electrodes and for CO_2 welding, from Masumoto and colleagues (1971). It is seen that the ilmenite type has a fairly wide range of current for non-porous welds regardless of water-proofing, while the safe range for the basic type is extremely narrow and also dependent on the correct type of protective layer. The minimum of porosity observed at 26V arc voltage in CO_2 welding would seem to confirm the trend observed by Savitch.

The effects of hydrogen absorption on HAZ cracking will be discussed in connection with cooling rates. The influence on static and dynamic ductility will be seen from the mechanical properties to be considered below.

Oxygen is absorbed to a greater extent in wet than in open-air welding. Analytical data, as shown by one example in Table 1, do not reveal the amount of oxygen absorbed and subsequently rejected as silicate slag. In addition, the oxygen contents shown have been measured in single beads deposited at a water depth of 0.2m under carefully controlled conditions. Multi-run joints made by a diver-welder under realistic conditions are likely to contain more oxygen, both due to lack of dilution and because it is very difficult to avoid inclusions of entrapped welding slag.

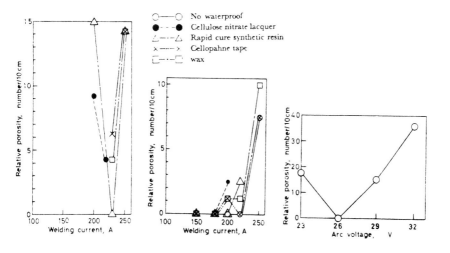

Fig. 1.   Examples of Weld Porosity, (from Masumoto and
colleagues).

Deoxidants remaining in the weld give a better indication of the amount of
oxygen that has passed through the liquid metal. The burn-off of silicon,
manganese and in some cases carbon shown in Table 1 corresponds to 0.2-0.5%
equivalent oxygen. Similar trends are seen from Fig. 2, showing the contents
of silicon, manganese and carbon in MMA welds made at various depths: the
silicon and manganese contents decrease, and the carbon content decreases
or remains constant with increasing depth. The electrodes tested are of
acid type (Mathisen and Gjermundsen) or of types SV08 and SV08G wire with
EPS-5 or EPS-52 coatings (Shlyamin and Dubova).

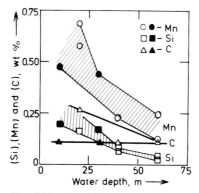

Fig. 2.   Alloying Elements in Wet Welds.
Open symbols - Shlyamin and Dubova
Filled symbols - Mathisen and Gjermundsen

Mechanical properties are shown in Table 2 for weld metal produced at various depths by acid, rutile and oxidising type electrodes (Mathisen and Gjermundsen). As would be expected from the trends of chemical composition the strength is normal, the ductility is low to medium (better in the case of the low-hydrogen dead-soft deposit), and the notch impact toughness is medium to low (poorer in the case of the high-oxygen dead-soft deposit). A high hydrogen content does not affect the notch impact toughness, as is well known from the testing of welds deposited in open air.

TABLE 2   Examples of Wet Weld Metal Properties
(Mathisen and Gjermundsen)

Electrode		R_{el}, MPa	R_m, MPa	A, %	$CV/0^{o}C$, J
Acid	4m	440	460	10	40
	10m	400	450	8	40
	30m	390	440	5	35
Rutile	4m	430	460	8	30
	10m	410	440	5	30
	30m	370	410	9	30
Oxidising	4m	360	450	22	20
	10m	370	450	17	20
	30m	350	430	17	25

The Programme of Cooling in Wet Welding

The presence of water is not necessarily equivalent to "water quenching", which requires a rather vigorous movement of water relative to the quenched object. It is probable that excessive hardness values frequently observed in the vicinity of wet welds are due to convection currents set up by the welding process itself. This is illustrated in Fig. 3 after Mathisen and Gjermundsen, showing measured cooling times from 800 to $500^{o}C$. For welding in open air the cooling time in seconds is roughly equal to 4.5 times the gross arc energy in MJ/m, as would be expected with carbon-manganese steel. For wet welds in the flat position the factor is reduced to about one-half this value (between 1.5 and 2.5). Cooling times for wet welding in the vertical position are virtually independent of the arc energy, higher inputs of heat leading to stronger convection currents.

It is possible to the reduce the free convection by simple mechanical devices. Fig. 4 shows an example of such effects in wet welding with a self-shielding flux-cored wire at an arc energy of 2.4 MJ/m, giving a cooling time of 4s in free convection and of about 14s under conditions of reduced convection. In this case the increased cooling time will lead to an improved safety against hydrogen-induced cracking both on account of a non-martensitic heat-affected zone and becuase of a reduced content of diffusible hydrogen (Mathisen and Gjermundsen).

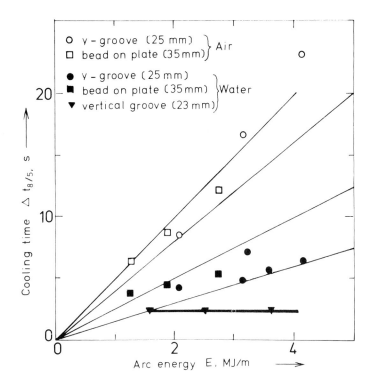

Fig. 3.   Free Cooling in Air and Water.

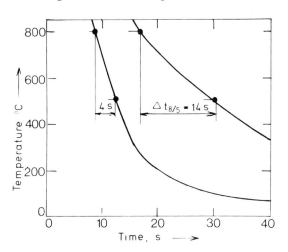

Fig. 4.   Cooling of Wet Welds under Free and Reduced Convection.

The risk of hydrogen-induced cracking can in principle be analysed on the basis of hydrogen content, welding conditions and steel grade. Since, by Table 1, non-basic electrodes produce roughly the same level of hydrogen in air and underwater welding, the risk of cracking with a given steel would presumably be equal at equal programmes of cooling. According to Fig. 3, the cooling time $\Delta t_{8/5}$ at 1 MJ/m in air would require an arc energy about 2.5 MJ/m in wet welding in the flat position. Wet welding in the vertical rising position would be roughly equivalent to air welding at 0.5 MJ/m regardless of the wet welding current.

Crack testing by means of 12.7mm Tekken specimens has been reported by Ozaki, Naiman and Masubuchi, using an arc energy of 1.5 MJ/m both in air and underwater welding. Table 3 from their work shows the cracking ratio (projected crack length to weld bead height) and the maximum hardness observed in the HAZ and the weld.

TABLE 3   Cracking Ratio and Hardness in Air and Wet Welding
After Ozaki, Naiman and Masubuchi (1977)

Steel	Electrode	Cracking ratio, %		Max. hardness HAZ		Weld hardness	
		Air	Water	Air	Water	Air	Water
Mild[1]	E 7014	0	0	235	420	215	240
	E 7018	0	0	220	400	220	210
	E 6024	0	0	220	410	175	200
St.52[2]	E 8018	2	30	200	380	160	210
A537A[3]	E 8018	21	26	245	430	200	230
HY 80	E 11018	95	100	470	470	400	400
	E 7018	18	100	440	460	290	350
	E 310-16	1	80	400	425	200	190

1)%C=0.20;C.E.=0.29 2)%C=0.19;C.E.=0.33 3)%C=0.16;C.E.=0.45

It is seen that the mild steel welds are free from cracks both in air and wet welding, in spite of a HAZ hardness above 400 in the latter case. The cracking ratio of the medium strength steels welded in air increases with increasing carbon equivalent, as would be expected. In wet welding there is little difference of cracking ratio. The hardness is not a good indicator, 30% cracking being observed at a hardness number of 380 in St.52 compared to zero cracking at a hardness of 420 in mild steel. The high strength material HY80 is sufficiently deep-hardening to give identical hardness numbers in air and underwater welding. The advantage of a filler metal that is under-matched with respect to strength is shown in the air weld made with E7018 electrodes; in wet welding this effect is not sufficient however to prevent complete cracking. The use of an austenitic filler metal (25% Cr - 20% Ni) is beneficial in air welding but not in water, where hot cracking is also observed.

cracking close to the fusion line has also been observed with austenitic filler metals due to a narrow zone in the weld itself. This layer, which has not been mixed with the bulk of the weld, will easily be diluted into the range of austenitic-martensitic or even purely martensitic micro-structures. It appears advisable to select over-alloyed filler metals, where the zone of critical dilution would be very narrow. Electrodes giving a deposit of more than 60% Ni have been used succesfully in laboratory underwater welding of St.52 material.

Effects of Local Protection of the Welding Site

A water curtain type of local protection used with Ar shielding has been shown by Arata, Hamasaki and Sakakibara (1981) to produce deposits contain-ing hydrogen of the order of 2 ml H_2/100g fused metal, determined by the glycerine method according to JIS Z313-1975. Welding was also done under shielding gas mixtures of Ar and H_2 in the absence of water, from which the following relationship was established:

$$H_{JIS} = 25.0\sqrt{pH_2} + 0.5 \text{ (ml/100g fused metal)} \tag{1}$$

An observed H_{JIS} content of 2 ml/100g fused metal has thus been obtained at a partial pressure of about 0.004. It has been shown on thermodynamic grounds (Christensen, 1949) and confirmed experimentally by Salter (1963) that gaseous hydrogen and water vapour are equally effective sources of dissolved hydrogen, and the partial pressure pH_2 can therefore be replaced by a combined partial pressure $p_w = pH_2 + pH_2O$. The amount of water vapour resulting from a liquid film in the protected area is thus about 0.4 per cent by volume.

The preferred procedure for hydrogen determination in welds is the "IIW mercury method" now standardised as ISO 3690-1977. For conversion of H_{JIS} data to the corresponding ISO figures on the basis of deposited metal H_{DM} the relationship recommended by IIW (1974) is:

$$H_{JIS} = 0.79H_{DM}-1.73 \text{ (ml/100g deposited metal)}^{1)}.$$

Assuming the ratio of deposited to fused metal equal to 0.5, an H_{JIS} level of 2 ml H_2/100g fused metal would correspond to a fused-metal IIW content $H_{FM} \approx 3ppm$. It may be interesting to compare the measured relationship from eq. (1) with values expected from Sieverts' law, taking into account an unavoidable loss of hydrogen in the period from solidification to quenching in the ISO procedure. The retained fraction has been estimated to 85% (Christensen, 1961). Introducing the maximum solubility $S_{max} \approx 38$ ppm from levitation melting experiments by Lakomski (1962), the expected ISO analytical content would be:

$$H_{FM} \approx 0.85S_{max}\sqrt{p_w} \approx 32 \text{ ppm}\sqrt{p_w} \tag{2}$$

The combined partial pressure expected to give an H_{FM} content of 3ppm is thus $p_w = 0.009$, as compared to 0.004 from eq. (1).

The cooling conditions obtained are comparable to those in air during the first stages until the bead is hit by the water curtain jet. From then on accelerated cooling is observed, as shown in Fig. 5(a) from Satoh and colleagues (1981).

$^{1)}$ A slightly different equation has been proposed by Suzuki (1978).

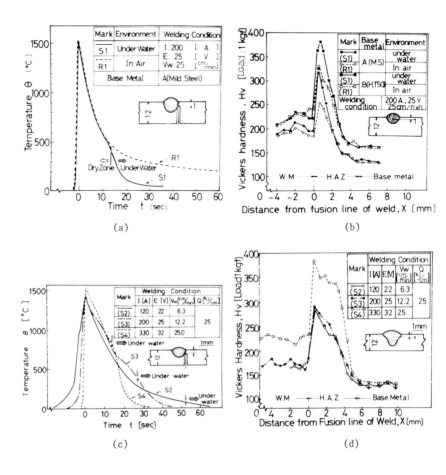

Fig. 5. Thermal Cycles and Hardness at Constant Welding Speed
(a. and b.) and at Constant Arc Energy (25MJ/m) (c. and
d.). After Satoh and colleagues (1981).

This difference of cooling rates is reflected in the hardness values, as
shown in Fig. 5(b). It is possible, however, to select operational parameters that will allow the bead and its nearest surroundings to remain for
a longer time under the gas shield, with corresponding beneficial effects
on the peak hardness (Fig. 5c and d). The authors claim a better corre-
lation of observed peak hardness with the cooling time from 800 to 300°C
than with the conventional parameter $\Delta t_{8/5}$.

Hoffmeister, Küster and Fusshöller (1982), using a similar kind of protection in double-V CO_2 butt welding of 15mm medium strength steels, observed
a more fine-grained microstructure close to the surface in underwater than
in air welds. They interpreted this as a result of a different cooling
programme during the stage of austenite transformation. Higher hydrogen

contents were observed than in the work by Satoh and colleagues, in the range from 5 to 12 ml H_2/100g deposit. Hydrogen fissuring was encountered close to the upper limit, and also substantial amounts of weld porosity. The Charpy V energy absorption was nevertheless above a level of 30J at -20°C, the presence of porosity being in the main reflected in a reduced energy for fracture propagation.

The water curtain local protection has been applied in butt welding of platform supporting legs of 700 to 800mm dia by 12mm thickness at a depth of 5 to 6m, using a shielding gas of 75%Ar-25%CO_2 (Shinada and Tamura, 1982).

The examples quoted indicate that substantial improvements of weld joint quality can be achieved by means of a local protection. It seems probable that excessive hydrogen absorption, resulting in fissures and porosity, is mainly a problem of improving the technology of protection.

DRY WELDING

So far, dry welding has been carried out down to sea depths of 320m, as reported by Knagenhjelm (1980). Simulated dives including manual welding have been performed under pressures corresponding to 500m depth (Knagenhjelm and colleagues, 1982). As would be expected, the weld metal chemistry is strongly influenced under such conditions. The cooling programme is also affected by the presence of high-pressure helium, but to much less extent than by water in wet welding. From a metallurgical point of view dry underwater welding could properly be described as hyperbaric welding, so much more as the habitat atmosphere is far from "dry".

Effects of Pressure on Weld Metal Chemistry

Liquid weld metal is known to react with its surroundings unless these are totally inert. The extent of such reactions are not always easily measured in normal welding. It is readily observed, however, that the weld metal composition is displaced when welding is performed under pressure. Reactions consuming gaseous species will be favoured, and those producing gaseous products will be suppressed. Important representatives of such reactions are the increased absorption of hydrogen from the arc atmosphere, and the reduced evolution of carbon monoxide. In both cases the impurity level will increase with increasing pressure, carbon being considered as an undesirable element in hyperbaric welds. Increased contents of oxygen in the metal will result in heavier losses of silicon and manganese during cooling.

Hydrogen absorption in hyperbaric welds has been reported in exploratory work by Christensen and Gjermundsen (1976, 1977a), by Delaune (1978), and by Coriatt and Bellamy (1978). Subsequently, more detailed information has been released from restricted reports summarized by Knagenhjelm (1980, 1982). Pinfold and colleagues (1981) and Allum (1982) have published extensive data. It has been shown that much lower impurity levels can be achieved with the inert-gas shielded processes than in MMA welding.

Some of the results from the early work are illustrated in Fig. 6 showing the impurity level versus pressure in single MMA beads. The hydrogen content (H_{FM}) is much lower than expected from Sieverts' law which,

assuming the composition of the arc atmosphere independent of pressure, may
be written:

$$[H]_{P \text{ bar}} = [H]_{1 \text{ bar}} \sqrt{P} \tag{3}$$

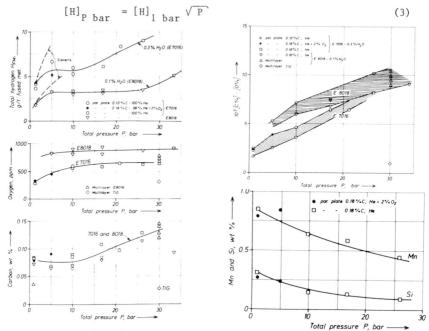

Fig. 6.   Chemical Composition vs. Pressure for MMA Single Beads.
(a) Hydrogen   (b) Oxygen   (c) Carbon
(d) Product %C %O   (e) Silicon and manganese.
After Christensen and Gjermundsen.

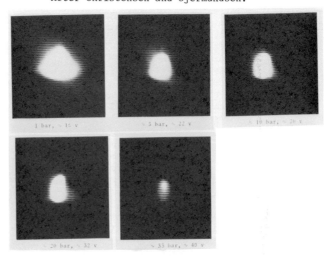

Fig. 7.   TIG Welding Arc in 85% Ar - 15% He at a Flow Rate of
10Nl/min. (Gjermundsen, 1979).

A possible explanation of the observed deviation has been sought in an increasing constriction of the arc column with increasing pressure, as illustrated in Fig. 7 for a TIG arc in 85%Ar-15%He (Gjermundsen, 1979). The surface of the weld pool does not change much, and hence an annular area will be exposed to the surrounding helium atmosphere rather than to the active gases from the electrode crater. This situation is equivalent to dilution of the arc atmosphere, and could be formally described by a dilution factor $1/f = (D_{pool}/D_{arc})^2$.

The observed contents of oxygen and carbon support this interpretation, measured products m_p = [%C][%O] at any pressure P being considerably smaller than the expected value Pm_1. Taking dilution factors $1/f = Pm_1/m$ from the scatter bands of Fig. 6(d), corrected values of the expected hydrogen content may be written:

$$[H]_{calc.} = [H]_1 \sqrt{Pf} = [H]_1 \sqrt{m_p/m_1} \tag{4}$$

The shaded bands in Fig. 8 superimposed on the measured values from Fig. 6(a) indicate that the effective partial pressures of carbon monoxide and hydrogenous gases follow the same general pattern. It appears thus that arc constriction is an important feature of hyperbaric welding.

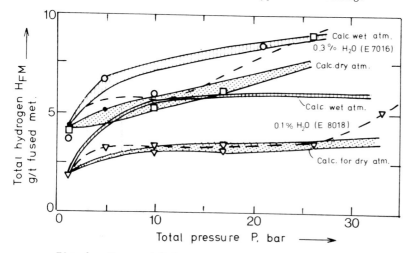

Fig. 8. Measured Hydrogen Contents from Fig. 6(a) for Comparison with Calculated Values in Dry and Wet Atmospheres.

The data of Fig. 6 have been obtained by deposition under dry helium or (He + 2%O$_2$) in a small simulator. Higher hydrogen contents may be expected when welding in a humid habitat, due to moisture absorption in the electrode coating and to direct absorption into the exposed annular surface of the pool. The latter mechanism is easily analysed on paper by adding to eq. (4) a term representing a constant vapour pressure, corrected by the solubility ratio $S_{1900°C}/S_{max} \approx (0.75)_2$ for a lower reaction temperature at the surface than in the arc. The effective partial pressure is thus $P_{eff} = f \, P_{arc} + 0.75 \, (1-f) \, P_{hab}$, and

$$[H]_{calc} = [H]_1 \sqrt{m_p/m_1 + (0.75)_2 (P - m_p/m_1)(p_{hab}/p_{arc})} \tag{5}$$

83

This equation has been plotted in Fig. 8 for comparison with the calculated and measured values in a dry environment. It is seen that direct absorption is not a major contribution unless the electrode has been baked to a very low initial water content. This conclusion is in agreement with results reported by Pinfold and colleagues (1981) for welding in wet and dry argon at about 16 bar (500 ft depth).

The risk of increased hydrogen absorption under realistic conditions of a humid habitat is, however, related to the time of electrode exposure. The rate of moisture pick-up depends strongly on the relative humidity and on the initial water content of the coating. Corriat and Bellamy (1978) reported an initial rate of 0.04% per minute for electrodes containing 0.05% water prior to exposure. Knagenhjelm and colleagues (1982), using electrodes pre-baked to various levels, have observed slightly lower rates in a manned simulator at 80% relative humidity and pressures up to 33 bar. Subsequent examination of the kinetics of moistening indicates that water absorption is in the main controlled by diffusion of water into the coating (Kvaale, Gjermundsen and Christensen, 1983). Introducing the moisture content from this model, they were able to compare hydrogen contents obtained with as-baked and stored electrodes at various contents of moisture m_w, as shown in Fig. 9. In spite of considerable scattering it is seen that both sets of data follow the same main trends.

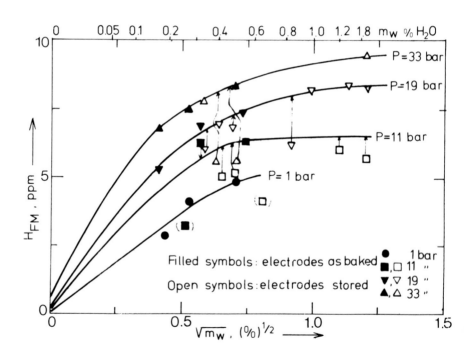

Fig. 9. Hydrogen in Welds Deposited in Manned Simulator.
(Gjermundsen and Kvaale, 1983).

The inert-gas shielded processes are able to produce welds of a much lower impurity content, as indicated for oxygen and carbon in TIG deposits in Fig. 6b, c and d. Hydrogen absorption is also much smaller. Table 4 shows the result of deposition of a flux-cored wire in helium (Gjermundsen, 1981). Higher values have been reported by Allum at 500 ft depth (16 bar), an average of about 5 ppm being virtually independent of the habitat humidity.

TABLE 4   Hydrogen Contents in FCAW Single Beads

Pressure, bar	C, %	Si, %	Mn, %	O, ppm	H_{FM}, ppm
1 (CO_2)	0.05	0.30	1.40	150	2.5
33 ($He+O_2$)	0.06	0.53	1.60	135	2.5

Oxygen absorption has been shown in Fig. 6(b) for MMA welding in helium. Similar values have been presented by Pinfold and colleagues for welding in Ar up to 16 bar, while deposition in helium with addition of 5 or 10% O_2 resulted in a discontinuous rise of about 340 ppm. The product m = [%C][%O] is also higher in these gas mixtures than shown in Fig. 6(d). The carbon contents are roughly as shown in Fig. 6 (c).

Since both oxygen and carbon are undesirable beyond certain limits, it is important to examine the parameter m = [%C][%O]. In particular, it would be interesting to see whether a systematically improved deoxidation will lead to increased carbon contents. Fig. 10 from Grong (1979) shows a series of such experiments, starting with the commercial E 8018-C1 electrode referred to earlier (marked S) and replacing stepwise iron powder by ferrosilicon (A, B and C). The general trend of increasing impurity content with increasing pressure is confirmed at all levels of deoxidation, low oxygen contents being mirrored by high contents of carbon. The product m = [%C][%O] shown in Fig. 10(c) may be represented by a single straight line within the precision of measurement. If this result could be confirmed for electrodes of different type, it would serve as a basis for a deliberate choice between two evils. It is possible to have a weld that is low in carbon and high in oxygen and vice versa, but not one that is low in both.

Again, the inert-gas shielded processes are superior, as was indicated in Fig. 6(d) for TIG welding at 30 bar at a value of m only one-tenth of that obtained in MMA welding.

Silicon and manganese are both lost to an increasing extent with increasing pressure, because more oxygen is left for reaction during cooling. The trend shown in Fig. 10(d) and (e) is representative for the E 8018 type examined, and also for E 7016 electrodes illustrated in Fig. 6(d).

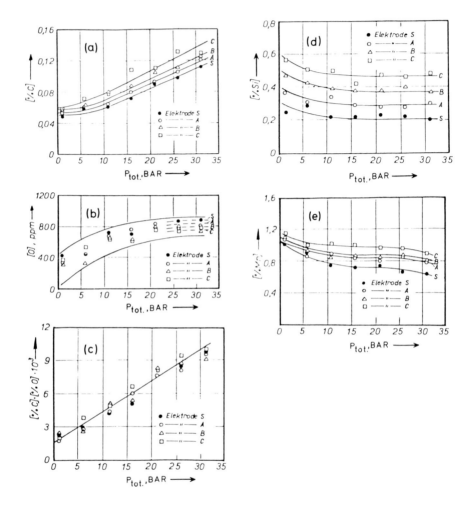

Fig. 10.   Effect of Deoxidation (Grong, 1979).
(a) Carbon (b) Oxygen (c) Product %C %O
(d) Silicon (e) Manganese.

The Programme of Cooling in Dry Welding

Cooling times in the range from 800 to 500°C have been reported for MMA welding in helium up to 16 bar pressure by Berthet and Gaudin (1976). Their curves indicate slightly shorter cooling times in high-pressure helium than in air.

Mathisen and Gjermundsen (1979) recorded temperature vs. time for TIG welding in helium up to 33 bar. From such records, and also from a comparison of peak hardness values of bead-on-plate weldments in open air and under pressure, they concluded that the cooling programme was virtually independent of the atmosphere down to 500°C. From there on there was a trend of faster cooling under high-pressure helium.

Allum (1982) found no influence of He pressure up to 16 bar on the cooling behaviour between 800 to 500°C. This was confirmed by hardness and micro-structural studies. On the other hand, the time from 1500 to 100°C was significantly shorter in He/16 bar than in air: 20% at an arc energy of 1.4 MJ/m and 50% at 1.8 MJ/m.

It may be concluded that increased cooling rates in the range below 500°C encountered in high-pressure helium will reduce the extent of hydrogen diffusion normally occurring at this stage of cooling. More hydrogen will therefore remain in the weld metal.

Effects of Increased Impurity Level

Hydrogen cracking has been reported both in the weld metal and in the HAZ of hyperbaric welds. Although these problems are very important in practical applications, they are not specific for hyperbaric welding. The comprehensive literature on hydrogen cracking in general may clearly be applied to hyperbaric conditions if the hydrogen content can be predicted with a reasonable precision.

Recent work on hydrogen-induced cracking has in the main been based on one of the three methods of crack testing:

- the Implant test
- the Tekken test
- the RRC or IRC test.

As is well known from surface welding, the test result will depend on the microstructure of the HAZ (controlled by the steel grade and the cooling programme), the supply of hydrogen, and the level of stress. A discussion of these factors is outside the scope of the present survey. Suffice it to illustrate the application of these principles by means of a few examples.

Fig. 11, from the early work by Berthet and Gaudin (1976), shows the extent of cracking vs. applied stress for a medium strength steel E36Nb after welding in air and under 15 bar pressure (He+3%O_2) at a relative humidity of 25 and 100%. It can be seen that a critical stress slightly more than one-half the yield stress is attained in surface welding, while the stress limit in hyperbaric welding is about 15% of the yield stress in a fairly dry atmosphere, and zero at saturation. No details are available on the hydrogen content.

Fig. 12, based on the Scandinavian concept of implant rupture stress R_{IR}, shows examples of the linear relationship between R_{IR} and log H_{FM} for two normalised grades of type St.52 (HT1 and HT2), one controlled cooled steel (QT1) and one low-alloy quenched and tempered type (QT2). Provided that the line is known for the steel to be welded, the weldability index R_{IR} can be interpolated for the expected absorption of hydrogen.

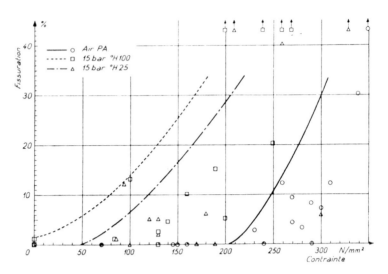

Fig. 11.   Implant Testing of a Medium Strength Steel E36Nb.
(Berthet and Gaudin).

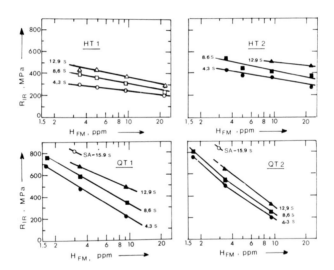

Fig. 12.   Implant Rupture Strength R_{IR} vs. Hydrogen Content.
(Christensen and Simonsen, 1981).

The cooling programme in the low-temperature range was shown to depend on
the presence of high-pressure helium; in addition, more significant effects
will of course be expected from preheating. Such effects can be taken into
account by replacing the actual cooling curve by an equivalent time of
isothermal holding at some chosen temperature prior to quenching, as

illustrated in Fig. 13 for three quenched and tempered steels. The equivalent holding time at 60°C (333K) is given by

$$\Delta t_{333} = \exp (2000/333) \int_{T_1}^{T_2} \exp (-2000/T) dt,$$

corresponding values of temperature T and time being taken from the cooling curve. It is seen that steels QT1 and QT2 welded under conditions

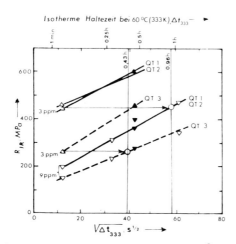

Fig. 13. Comparison of Preheating to 100°C (filled symbols)
and Isothermal Holding at 60°C (333K).
(Christensen, 1980).

giving a hydrogen content of 9ppm, which may be realistic under hyperbaric MMA welding, can obtain the same critical implant stress as with 3ppm without preheating if the weldment is maintained at 60°C for about 1h. Preheating to 100°C is equivalent to isothermal holding at 60°C for about ½ h.

Critical conditions assessed from Tekken testing have been reported by Allum for a pressure of 16 bar in comparison with welding in air. He established a cracking parameter P = CE + H/60 + R/200 000, where CE is the IIW carbon equivalent, H is the hydrogen content in ml H_2/100g fused metal, and R is the intensity of restraint in MPa. Using a preheating temperature T (°C), the critical arc energy Q (kJ/in) may be read from a graph shown in Fig. 14.

Methods for predicting hydrogen contents and for assessing their effect on safety against cracking are thus being developed as a necessary background for hyperbaric MMA welding. At depths greater than 300m it will probably require very rigorous precautions to maintain the hydrogen content at a safe level. At such depths the inert-gas shielded methods have an obvious advantage, the only source of hydrogen being direct introduction of humid habitat atmosphere into the arc.

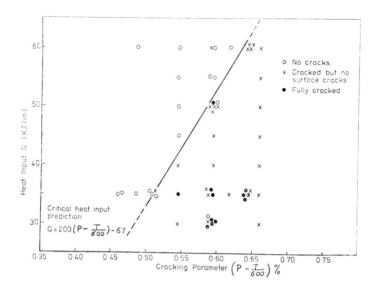

Fig. 14.   Chart of Safe Welding Conditions based on Tekken
Testing (After Allum, 1982).

Oxygen and carbon have been shown to increase with increasing depth in
hyperbaric MMA welding. The combined effects of these elements is a genuine
hyperbaric problem, because high concentrations of both elements are not
encountered in surface welding. An example of the Charpy V energy absorption is given in Fig. 15 from Berthet and Gaudin, showing a steady deterioration of toughness with increasing depth. The joint tested is a butt weld
in 20mm pipe of grade X65 welded with basic electrodes giving a nickel
content of 2.5%. At the highest pressure employed the contents of carbon

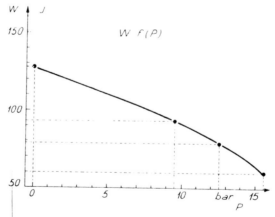

Fig. 15.   Charpy V Energy Absorption at -10°C vs. Pressure.
(Berthet and Gaudin).

and oxygen were about 0.12%C and 600ppm O, as compared to 0.05%C and 300ppm O in surface welds. The product m = [%C][%O] is about $7 \times 10^{-3}(\%)^2$, in a reasonable agreement with Fig. 10(c) giving about 6×10^{-3}. At 300m depth the expected product is about $0.01(\%)^2$. Limiting the carbon content to 0.10% would thus give 1000ppm O, and limiting the oxygen content to 400ppm would give 0.25%C.

Effects of oxygen in the absence of excessive amounts of carbon have been examined by Tuliani, Boniszewski and Eaton, by Kirkwood and by Abson, Dolby and Hart (1979). There is a general agreement that the upper shelf Charpy energy will be reduced at oxygen contents significantly higher than about 400ppm. Similar data have recently been reported by Terashima and Tsuboi (1982) for high-strength submerged-arc welds, as shown in Fig. 16.

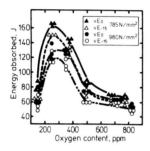

Fig. 16.   Charpy V Notch Toughness of Submerged-Arc Weld Metal. (Terashima and Tsuboi).

Since both carbon and oxygen contribute to a reduced notch toughness, as indicated by Fig. 15, it appears that these elements would restrict the range of depth for MMA welding even more than hydrogen. It is possible to meet current quality specifications by means of MMA welding down to medium depths of 150 to 200m. At 300m depth difficulties may be expected both with regard to hydrogen fissuring and notch toughness, and at even greater depth the inert-gas shielded processes are the only ones that can be expected to produce joints of a high quality.

SUMMARY AND CONCLUSIONS

The presence of water in wet underwater welding introduces metallurgical problems due to:

(1)   Absorption of hydrogen in amounts corresponding to the "medium range" as defined by the IIW. It is possible to obtain "low" contents in strongly oxidised weld metal, which has, however, a rather low notch toughness.

(2)   High cooling rates, requiring roughly twice the arc energy of air welding to give the same $\Delta t_{8/5}$ in the flat position. Vertical welding gives a value of $2 - 3$s regardless of arc energy due to convection currents.

(3)   Combinations of increased hydrogen absorption and rapid cooling, resulting in HAZ cracking of medium strength steel.

(4)   Porosity, caused by absorption of hydrogen.

The risk of cracking can be reduced by the use of austenitic filler metals of a substantial excess of alloying elements.

Local protection of the welding site may improve the quality substantially.

Dry welding is characterised by the effects of pressure. In MMA welding under pressures up to about 30bar metallurgical problems arise due to:

(1)   Hydrogen contents considerably below the values predicted by Sieverts' law, but nevertheless of the order of 2.5 times the content in surface welds. The risk of hydrogen-induced cracking can be evaluated by methods developed for this purpose in surface welding.

(2)   Oxygen and carbon contents controlled by the product $m = [\%C][\%O]$, increasing in proportion to the pressure and attaining a value about $0.01(\%)^2$ at 30bar. The notch impact toughness will be reduced by the presence of carbon and oxygen, which cannot both be maintained at a low level.

These problems appear to set a limit for MMA hyperbaric welding at depths of about 300m.

Inert-gas shielded processes will in general not introduce additional carbon into the weld. Hydrogen may be picked up from a humid habitat atmosphere, but to much less extent than in MMA welding.

ACKNOWLEDGEMENT

The author would like to thank his colleagues on the metallurgical staff of SINTEF for valuable discussions in preparing this survey. He is indebted to Messrs. Norsk Hydro a.s. for their permission to use data from restricted reports as illustrations.

REFERENCES

Abson, D.J., R.F. Dolby and P.H.M. Hart, (1979). The role of non-metallic inclusions in ferrite nucleation in carbon steel weld metals. In Trends in Steels and Consumables, London 1979, 75-101.
Allum, C.J. (1982). Hyperbaric shielded metal arc welding and cold cracking. Amer. Gas Assoc. No. L 51416.
Arata, Y., M. Hamasaki and J. Sakakibara (1981). Hydrogen content in arc atmosphere of water curtain type underwater argon arc welding. Trans. JWRI, 10, 19-24.
Berthet, P. and J.P. Gaudin (1976). Raccordement de canalisations par soudage sous-marin en conditions hyperbare, Soud. Tech. Conn.,Vol. 30, 246-259.
Christensen, N. (1949). Metallurgical aspects of welding mild steel. Weld. J., 28, 373s-380s.
Christensen, N. (1961). Hydrogenets rolle ved buesveising med dekkede elektroder. Svetsen, 20, 112-122.
Christensen, N. and K. Gjermundsen (1976, 1977a). Effects of pressure on weld metal chemistry. IIW Doc. II-212-384-76; supplement II-212-395-77.
Christensen, N. (1977b). Metallurgisk forskning ved NTH. Jernkont. Ann. Nr. 5.

Christensen, N. (1980). Schweissbarkeitsprüfung mit dem Implantverfahren. In DVS, Ber. 64, Düsseldorf, 51-55.

Christensen, N. and T. Simonsen (1981). Assessment of weldability by the implant method. Scand. J. Met. 10, 120-126.

Coe, F.R. (1973). Welding Steels Without Hydrogen Cracking, The Weld. Inst., Abington.

Coriatt, G.M. and G.P. Bellamy (1978). Hyperbaric welding for connection and repair of deep sea pipelines. In Tenth Offshore Technology Con ference, OTC 3351, 2601-2610.

Dadian, M (1975). Synthèse bibliographique sur la soudabilité des aciers sous l'eau. IIW Doc. IX-931-75.

Delaune, P.T., Jr. (1978). On-site weldes repairs to offshore structures using dry underwater habitats. In Tenth Offshore Technology Conference, OTC 3348, 2571-2579.

Gjermundsen, K. (1979). Personal communication (1979b) SINTEF Rep. STF34 F81004.

Gjermundsen, K. (1981). Inertgassbeskyttet rørtrådsveising under hyperbariske forhold. SINTEF Rep. STF34 F81004.

Grong, Ø. (1979). Metallurgical Engineer's Thesis, The Norwegian Institute of Technology,

Hoffmeister, H., K. Küster and J. Fusshöller (1982). Zum Einfluss der Porosität und des Gefüges auf die Zähigkeit unterwasserschutzgasgesch weisstes Stumpfnähte on StE 36 und StE 39. Schweissen Schneiden 34, 315-319.

Knagenhjelm, H.O. (1980). Hyperbaric welding at 320 msw, development of adequate welding procedures. In Underwater Technology Conference, Bergen 1980.

Knagenhjelm, H.O., K. Gjermundsen, P. Kvaale and D. Gibson (1982). Hyperbaric TIG welding to 500m simulated depth. In Underwater Technology Conference, Bergen 1982.

Kvaale, P.E., K. Gjermundsen and N. Christensen (1983). Unpublished.

Lakomski, V. (1962). Dokladi Akademii NAUK SSSR 147, 623-629.

Masumoto, I., Y. Nakashima, A. Kondo and K. Matsuda (1971). Study on the underwater welding (Report 1). J. Japan Weld. Soc., 40, 683-693.

Mathisen, U. and K. Gjermundsen (1979). Undervannssveising. SINTEF Rep. STF16 F79049.

Ozaki, H., J. Naiman and K. Masubuchi (1977). A study of hydrogen cracking in underwater steel welds. Weld. J.,56, 231s-237s.

Pinfold, B.E., J.H. Nixon, J.M.F. Mota and D.V. Dorling (1981). Hyperbaric shielded metal arc welding. Amer. Gas Assoc. No. L 51401.

Salter, G.R. (1963). Hydrogen absorption in arc welding. Brit. Weld. J., 10, 89-101.

Satoh, K., M. Tamura, K. Kawakami, K. Shinada, T. Ohmae and Y. Manabe (1981). Study of the cooling characteristics and weld hardness in the locally drying underwater welding of mild and 50 legf/mm^2 H.T. steels. IIW Doc. XII-B-17-81.

Savitch, I.M. (1974). Underwater welding of metals. In Welding in Offshore Construction, The Weld. Inst., pp. 217-220.

Shinada, K. and M. Tamura (1982). Trockenschweissen unter Wasser mit örtlichen Hohlraum-Entwicklung einer vollmechanisierten Schweissein victung und Versuchsergebnisse. Schweissen Schneiden.

Shinada, K. and M. Tamura (1982). Trockenschweissen unter Wasser mit örtlichen Hohlraum-Entwicklung einer vollmechanisierten Schweissein victung und Versuchsergebnisse. Schweissen Schneiden 34, 193-195.

Shlyamin, A.I. and T.N. Dubova (1961). Semi-automatic underwater welding. Weld. Prod., 8, 25-28.

Stalker, A.W., P.H.M. Hart and G.R. Salter (1975). A preliminary study of underwater manual metal arc welding. The Weld. Inst. Rep. RR-SMT-7466, Febr. (1975).

Suzuki, H. (1978) Cold cracking and its prevention in steel welding. Trans. JWS 9, 82-91.

Terashima, H. and J. Tsuboi (1982). Submerged arc flux for low oxygen and hydrogen weld metal. Met. Constr. 14, 648-654.

Contrôle et comportement en service des soudures exécutées sous l'eau

P. Amiot

Société Nationale Elf Aquitaine, Paris La Défense, France

RESUME

Dans une première partie, on décrit les méthodes de contrôle non destructif applicables aux soudures sous l'eau, en mettant en évidence les différences qui existent avec le contrôle des soudures exécutées à terre et en tentant de prévoir quels développements sont probables dans un proche avenir. Dans une seconde partie, on s'attache à rechercher quel est le comportement en service de soudures réelles exécutées sous l'eau en partant d'informations publiées et d'une enquête auprès des sociétés de service et des compagnies pétrolières exploitant des installations en mer.

MOTS—CLES

Soudage sous l'eau, dans l'eau, hyperbare, à la pression atmosphérique ; contrôles non destructifs sous l'eau, visuels, par rayons X, gammagraphie, ultra-sons, magnétoscopie ; comportement en service de soudures exécutées sous l'eau ; exploitation pétrolière en mer.

INTRODUCTION

Cette contribution est destinée à être présentée, à la séance publique de l'Assemblée Annuelle 1983 de l'Institut International de Soudure, à Trondheim. Elle fera suite aux conférences de K. Masubuchi, A.V. Gaudiano et T.J. Reynolds sur la technique et la pratique de soudage sous l'eau et de N. Christensen sur la métallurgie du soudage sous l'eau. Pour qu'une revue de cette technique nouvelle soit complète, il était nécessaire que le problème du contrôle des soudures sous l'eau soit traité. Ce sera l'objet de la première partie de cet exposé, qui tiendra compte des deux conférences précédentes et ne reviendra pas sur les sujets qu'elles abordent. Une soudure étant faite et contrôlée, la question qui vient à l'esprit est de savoir quel comportement elle aura dans les conditions de service, pendant la durée de vie de la construction dont elle fait partie. La deuxième partie de cet exposé tentera de répondre à cette question.

Des publications de plus en plus nombreuses traitant du soudage sous l'eau - près d'une centaine par an (Masubuchi, 1981) - il n'a pas paru nécessaire de donner une bibliographie extensive dans le cadre de cet exposé, considérant que les lecteurs spécialistes du sujet traité étaient parfaitement au courant de ce qui est publié. Pour les autres, ils trouveront dans les ouvrages cités en tête de la bibliographie l'occasion de se familiariser avec le sujet : ce sont des comptes rendus de conférences récentes où le soudage sous l'eau est évoqué, dans le cadre plus vaste des problèmes liés aux structures pétrolières marines.

PREMIERE PARTIE : CONTROLE DES SOUDURES EXECUTEES SOUS L'EAU

Le contrôle non destructif des soudures exécutées sous l'eau est l'étape normale après l'exécution proprement dite des soudures.

On remarquera que ce qui peut être dit sur les pratiques actuelles et sur les développements en cours chez plusieurs sociétés de service est en partie valable pour le contrôle sous l'eau des soudures exécutées à terre. On remarquera aussi que les travaux sous l'eau, en général, et à la mer, en particulier, nécessitent des moyens infiniment plus lourds que ceux utilisés à terre. Il en résulte que les opérations à la mer doivent être très soigneusement planifiées longtemps avant l'intervention proprement dite.

On doit rappeler aussi que le soudage sous l'eau et, par conséquent, les contrôles subséquents, ont évolué très rapidement dans les dix ou quinze dernières années pour répondre aux besoins nouveaux de l'exploitation pétrolière marine, c'est-à-dire qu'ils ont dû s'adapter à des exigences de plus en plus sévères, en proportion des besoins de sécurité et de sûreté de cette industrie.

Ceci s'est traduit par des spécifications émises par les compagnies pétrolières exploitantes des ouvrages, puis par des codes et règlements sanctionnant l'état de la technique, comme celui du Det norske Veritas (1976).

Conditions d'examen

Comme pour le soudage sous l'eau, plusieurs situations peuvent se présenter. Elles sont schématisées sur la Fig. 1.

Ces conditions influent sur la qualité des résultats obtenus. En effet, le le matériel et les inspecteurs sont confrontés aux problèmes de plongée : pression hydrostatique, solubilité des gaz dans l'organisme, froid, courants, tenue du matériel, etc... . La taille minimale des défauts que l'on peut espérer détecter à coup sûr, en dépend notablement.

Dimension des défauts à détecter

Lorsque l'on met en oeuvre des essais non destructifs pour le contrôle d'une soudure exécutée sous l'eau, on applique généralement des spécifications, codes ou règlements qui précisent non seulement la méthode à employer, mais aussi les types et dimensions acceptables des anomalies décelées. Bien entendu, les exigences sont très variables puisque les ouvrages concernés sont eux aussi très divers : il y a peu de points communs entre le soudage en pleine eau d'une anode sacrificielle de protection cathodique, pour lequel un

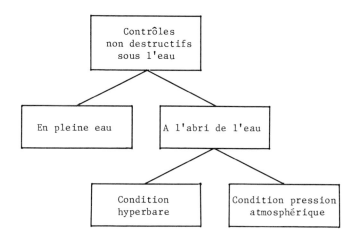

Fig. 1. Schéma des différentes conditions de contrôle
non destructif sous la surface de l'eau.

simple examen visuel est exigé, et les soudures hyperbares de réparation de
plates formes marines ou de raboutage de pipelines sous-marins dont la qualité n'a pas de raison d'être différente de celle des soudures exécutées à
l'air libre.

Il pourrait être tentant de demander une plus grande qualité aux soudures
faites sous l'eau qu'aux soudures faites à terre, en prescrivant un niveau
de détection plus élevé que celui exigé de ces dernières. Cette tendance
semble devoir être combattue, car, en l'état actuel des choses, il y a peu
d'espoir que les méthodes de contrôle sous l'eau – même si ce sont les mêmes
que celles utilisées à terre – aient la même efficacité, c'est-à-dire puissent découvrir à coup sûr des anomalies de mêmes dimensions ou plus petites.
Ceci est dû, non pas à la qualité intrinsèque des moyens, mais à tous les
problèmes liés soit à l'environnement, soit à la transmission des informations entre les opérateurs et la surface.

Qualification des opérateurs et interprétation des résultats

Dans les circonstances actuelles où un opérateur sous la mer est indispensable, celui-ci doit être qualifié pour les opérations de contrôle dont il est
chargé. Cette qualification doit évidemment être double : qualification de
plongeur et qualification d'opérateur de contrôle non destructif, conformes
aux codes et règlements applicables pour chaque cas d'espèce. De plus,
l'environnement spécifique impose que les opérateurs aient reçu un entraîne-
ment spécial à la communication puisqu'ils ne peuvent être livrés à eux-mêmes
et doivent obligatoirement communiquer avec la surface, où se trouvent les
décideurs, auxquels il faut transmettre à chaque instant les contatations.
D'autre part, le coût de telles opérations, la nécessité d'employer des opérateurs appartenant éventuellement à la société de service qui a effectué
la soudure examinée et qui sera amenée à faire les éventuelles réparations
de défauts, les difficultés de transmission des résultats, sont autant de
facteurs qui obligent à définir les responsabilités de chaque partie concer-
née par l'opération. Sans entrer dans les détails de cette définition, il est

évident que la responsabilité de la décision finale doit incomber à une auto-rité mandatée par toutes les parties concernées. Celle-ci, dûment informée par l'opérateur et au moyen des techniques audio-visuelles actuellement dis-ponibles pour retransmettre en surface ce qui se passe sous l'eau, doit pren-dre les décisions qui s'imposent, en temps réel et en connaissance de cause.

Méthodes de contrôle non destructif utilisées sous l'eau

En principe, les méthodes de contrôle non destructif utilisées à l'air libre peuvent aussi être utilisées sous l'eau moyennant les adaptations nécessitées par l'environnement. La réalité est moins simple et l'on verra plus loin qu'il existe d'importantes limitations d'emploi, en particulier, pour la gammagraphie.

Examens visuels. Ce sont les examens les plus simples et ils peuvent être appliqués à tous les types de soudures sous l'eau. Ce sont même les seuls à être pratiqués dans le cas des soudures dans l'eau, d'une part, parce que celles-ci sont souvent utilisées pour des joints très difficiles à examiner par d'autres méthodes, d'autre part, parce qu'on ne saurait généralement demander à de telles soudures d'être exemptes de défauts importants. Toujours dans ce cas, la qualité de l'examen dépend beaucoup des conditions locales, telles que profondeur, visibilité due à la turbidité de l'eau, courants, etc... A noter cependant qu'une instrumentation lourde peut éventuellement être mise en oeuvre, incluant télévision et relevés photographiques.

Lorsque les examens visuels sont appliqués aux autres types de soudures, c'est-à-dire celles exécutées à l'abri de l'eau, ils peuvent avoir la même efficacité que ceux utilisés à terre et ils peuvent être instrumentés de diverses manières. En particulier, l'information peut être renvoyée en sur-face, grâce à un circuit de télévision, et diverses mesures à l'aide de calibres peuvent être faites, la prise de photographies étant assez générale-ment la règle.

Examens magnétoscopiques. Ils sont couramment utilisés aussi bien dans l'eau qu'en atmosphère sèche, à pression normale ou en condition hyperbare. Bien entendu, comme à terre, ils ne peuvent révéler que des défauts de surface ou situés à faible profondeur, pourvu qu'ils aient des dimensions relativement importantes.

Les champs magnétiques peuvent être produits soit par des aimants permanents, soit par des enroulements parcourus par des courants : le choix est fonction des performances désirées et, bien entendu, des conditions locales d'environ-nement. En particulier, les enceintes de soudage hyperbares les plus impor-tantes disposent de moyens de génération de courant qui permettent l'emploi d'enroulements producteurs de champs magnétiques.

Il est de pratique courante de contrôler le résultat des examens par photo-graphie et par renvoi des images en surface par télévision. Dans le cas des examens en pleine eau, une meilleure définition est obtenue en utilisant des particules magnétiques excitables en lumière ultraviolette.

Examens radiographiques et gammagraphiques. La gammagraphie est celui des deux procédés qui est actuellement employé. Cependant, les sociétés de service, qui opèrent généralement dans le monde entier, ont éprouvé de grandes difficultés pour l'approvisionnement de leurs chantiers en sources radioactives, du fait des règlements de plus en plus sévères sur le transport des matières radioactives. Cette contingence, assez imprévue, associée au développement de générateurs de rayons X à refroidissement à air et d'un assez faible encombrement pour passer par les sas des chambres hyperbares, fait que l'on assistera dans un futur probablement proche au développement de la radiographie X au détriment de la gammagraphie pour le contrôle des soudures exécutées à l'abri de l'eau.

Aucune de ces techniques n'est actuellement applicable au contrôle en pleine eau, du fait de l'absorption du rayonnement par le milieu. Cependant, la société Comex Services, par exemple, étudie des appareils à haute intensité de rayonnement pour détecter, en pleine eau, certains défauts d'épaisseur provoqués par la corrosion.

Ces méthodes sont appliquées usuellement au contrôle des raboutages de tubes de pipelines ou de tubes ascenseurs (risers) qui relèvent essentiellement du soudage hyperbare. Ce procédé de soudage implique que les tubes soient vides d'eau. Cette circonstance permet la prise de gammagraphies dans les mêmes conditions que celles utilisées à terre, sauf, comme cela nous a été signalé, lorsque les obturateurs n'assurent pas une étanchéité parfaite et que le pipeline se remplit progressivement d'eau, rendant ainsi impossible la gammagraphie des points bas de la soudure.

Contrôles par ultrasons. On connait les avantages et les inconvénients de cette méthode de contrôle non destructif dans les applications à terre : du côté des avantages, on reconnait que c'est la meilleure méthode pour détecter des défauts minces, d'orientation quelconque et non débouchants que les examens par rayonnement électromagnétique risquent de ne pas révéler. De plus, pourvu que les circonstances soient favorables, elle est théoriquement capable de fournir des indications assez précises sur les dimensions et sur la position des défauts, données indispensables à l'évaluation de la nocivité des anomalies détectées, en utilisant les concepts de la mécanique de la rupture.

Du côté des inconvénients, il faut reconnaître que les industriels, d'abord séduits, éprouvent quelque répugnance à utiliser cette technique de contrôle, probablement parce que, dans le passé, ils leur est arrivé d'être victimes de résultats erronés aux coûteuses conséquences. A l'origine de ces malencontreuses erreurs, on trouve souvent, soit des difficultés opératoires, comme un couplage défectueux entre palpeurs et pièce à examiner, soit un manque de qualification des opérateurs. Mais, le reproche fait le plus souvent à cette méthode, est de ne laisser aucune trace objective probante que l'on puisse examiner contradictoirement et classer dans un dossier pour référence ultérieure.

Pour remédier à cette lacune, plusieurs dispositifs de mémorisation de l'information ont été développés, couplés ou non à une automatisation de l'examen. Parmi d'autres, on doit citer l'appareil P-scan développé il y a quelques années par Svejsecentralen, l'Institut de Soudure danois.

L'utilisation des ultrasons en contrôle non destructif sous l'eau est évidemment possible, moyennant l'adaptation à la mer des appareillages et la prise en compte des lourdes contraintes de l'environnement.

Dans le cas d'opérations en pleine eau, il y a même un certain avantage par rapport à l'utilisation en atmosphère gazeuse : le couplage palpeur-pièce est toujours correctement assuré, pour peu que les surfaces aient été préalablement débarrassées des salissures marines, ce qui est, bien sûr, le cas du contrôle d'une soudure fraîche. Il va de soi que cet avantage est perdu dans le cas d'examens de soudures effectuées en caisson hyperbare et en condition atmosphérique.

Un autre inconvénient des contrôles par ultrasons sous l'eau résulte du fait que les opérateurs devraient être non seulement parfaitement qualifiés pour interpréter ce qu'ils voient sur l'écran de leur oscilloscope et agir en conséquence, mais aussi, avoir toutes les qualifications de plongeurs. Comme ce n'est pas forcément toujours le cas et, ainsi qu'on l'a déjà signalé, comme les opérateurs appartiennent souvent à une société de service qui ne saurait prétendre être juge et partie, la transmission des informations de l'opérateur vers l'autorité mandatée pour décider et vice-versa, prend une importance décisive, ce qui n'est pas d'une réalisation facile.

Ceci a deux conséquences :
. d'une part, le contrôle par ultrasons des soudures effectuées sous l'eau, quel que soit leur type, n'est utilisé que lorsqu'il n'y a pas d'autre contrôle possible et que la sécurité de l'ouvrage l'exige,
. d'autre part, le besoin existant pourtant bien, les sociétés de service s'emploient actuellement à développer des systèmes aussi automatisés que possible et capables de mémoriser les résultats du contrôle.

Comme on le verra dans la deuxième partie, nous avons procédé à une enquête d'ampleur limitée sur le comportement en service des soudures exécutées sous l'eau. Le questionnaire établi à cet effet comportait une question relative aux moyens de contrôle mis en oeuvre après exécution des soudures. Les résultats sont donnés dans le tableau 1.

On ne doit pas conclure trop hâtivement à partir des résultats de cette enquête : elle n'avait pas la prétention d'être exhaustive et beaucoup de sociétés, ayant une grande expérience dans ces domaines, n'ont pas répondu. Néanmoins, on peut penser qu'elle indique des tendances générales :
. le contrôle des soudures sous l'eau de pipelines et de canalisations se fait essentiellement par rayonnements électromagnétiques et, comme déjà dit, essentiellement par gammagraphie. La raison principale est que ce type d'examen est celui prévu par les codes pour les ouvrages à terre. On précisera que, des trois cas signalés de contrôle par ultrasons, au moins un a été rendu nécessaire par l'impossibilité d'un examen complet par gammagraphie, du fait d'un envahissement par l'eau du pipeline en cause, comme cela a été indiqué précédemment;
. le contrôle des réparations de plates formes, par contre, est généralement effectué par ultrasons : la raison principale nous paraît être que, dans ces cas, l'utilisation de la gammagraphie est généralement impossible.

Développements possibles des contrôles par ultrasons. On vient de signaler que diverses sociétés de services cherchent à pallier les inconvénients des méthodes actuelles, principalement liés aux difficultés de communication entre opérateur au fond et autorité en surface et à l'absence de traces permanentes qui puissent être étudiées et servir plus tard de référence. A titre d'exemples, quelques mots seront dits du P-scan et de son développement sous le nom de Corroscan par le Det norske Veritas, puis du DPX développé par Comex Services.

Dans son principe, le P-scan, dû à l'Institut de Soudure danois, permet de

Tableau 1 : Résultats de l'enquête sur les

contrôles de soudures sous l'eau

(Les chiffres indiquent le nombre de cas signalés)

Ouvrages	Types de soudage	Moyens de contrôle utilisés				
		X ou		Magnéto-		
canalisations	dans l'eau	-(a)	-	-	-	-
et	hyperbare	0	14	3	0	(b)
pipelines	atmosphérique	-	-	-	-	-
Plates	dans l'eau	0	0	0	3(c)	12(c)
formes	hyperbare	0	2	6	3	(b)
pétrolières	atmosphérique	0	0	1	1	(b)
Qualification	dans l'eau	0	1	0	0	9
de	hyperbare	2	1	0	0	(b)
procédé	atmosphérique	-	-	-	-	-

(a). Le signe - indique qu'aucune réponse n'a été reçue associant le type d'ouvrage et le type de soudage concernés.

(b). Bien que cela ne soit pas mentionné expressement dans les réponses, il est à peu près certain qu'un examen visuel a été pratiqué dans chaque cas.

(c). On ignore de quelles interventions il est question dans ces cas : il est probable qu'il s'agit de soudures d'éléments accessoires, mais on ne peut exclure qu'il s'agisse de réparations d'éléments structuraux.

représenter le défaut d'une soudure de tube ou de plaque par deux images ultrasonores, sous forme de deux projections rectangulaires, l'une perpendiculairement à la surface, l'autre perpendiculairement à l'épaisseur de l'assemblage. Ces informations sont stockées et affichées et peuvent être restituées à la demande. Elles permettent d'identifier le défaut et de donner sa position et ses dimensions. Une version spéciale de cet équipement a été développée par Det norske Veritas pour les mesures de corrosion sous l'eau de canalisations et de pipelines. Ces mesures sont faites par l'extérieur des tubes, ce qui nécessite donc que la surface soit mise à nu. Un plongeur fixe le dispositif de palpage automatique sur le tube et n'a plus à intervenir, les palpeurs se déplaçant autour du tube, le long d'une chaîne. Toutes les informations sont renvoyées en surface où les conditions d'environnement sont plus propices à leur interprétation qu'en pleine eau. Les compagnies productrices d'hydrocarbures en mer peuvent ainsi disposer d'un moyen d'inspection jusqu'à des profondeurs qui pourront atteindre -300 m. Ce procédé est actuellement utilisé pour le contrôle de la corrosion des tubes ascenseurs en mer du Nord. Mais, comme nous l'a fait remarquer la société de service Intercontrôle, licenciée du P-scan, l'adaptation du procédé aux conditions hyperbares ne verra le jour que lorsqu'il y aura un marché, c'est-à-dire, lorsqu'une demande existera pour le contrôle des raboutages de pipelines. A noter encore que la même société travaille à améliorer, non pas la qualité de détection du P-scan, mais sa précision dans le dimensionnement des défauts, en utilisant des ultra-sons focalisés.

Alors que les appareils précédents donnent une image en modes A et C, le système DPX, en cours de développement, vise à donner, en plus, une mesure dynamique en mode B. L'idée qui est à l'origine du palpeur DPX découle de la constatation suivante : l'orientation des défauts conditionne leur aptitude à être détectés. Donc, il est nécessaire d'utiliser des palpeurs d'incidences diverses pour faire une inspection significative. C'est encore plus indispensable si l'on veut donner une bonne évaluation des dimensions des réflecteurs des ondes ultrasonores. Ceci oblige à des manipulations qui ne sont déjà pas simples à terre et qui deviennent très difficile à mettre en oeuvre dans l'environnement sous la surface de l'eau. Schématiquement, le palpeur DPX est constitué, Fig. 2, d'un émetteur-récepteur envoyant son faisceau sur un réflecteur octogonal, tournant d'un angle de un degré entre chaque train

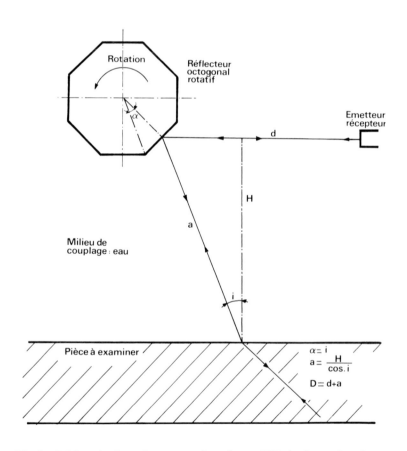

Fig.2. Schéma de fonctionnement du palpeur DPX de Comex Services.

l'ultrasons : à chaque instant donc, la pièce est examinée sous toutes les
incidences de 0 à 90°, par pas de 2,5°.Chaque défaut est donc vu sous l'incidence optimale pour sa détection et pour la mesure de ses dimensions et de
son orientation. Ceci est obtenu par une électronique associée qui envoie
sur l'étage vidéo l'angle d'incidence suivant l'axe Y et le temps entre
l'émission et la réception, c'est-à-dire, la distance du défaut, suivant
l'axe X. L'image de la coupe de l'échantillon balayée par les faisceaux
l'ultrasons est donc reconstituée sur l'écran d'un tube cathodique, en temps
réel ou après mémorisation.

Ces deux exemples, parmi d'autres, montrent que les méthodes d'inspection ultrasonore sous le niveau de l'eau font l'objet de recherches et
de développements constants dans le but d'améliorer leur capacité de détection, leur automatisation et la mémorisation des résultats indispensable pour
assurer leur crédibilité. Elles participent en cela au rapide développement
actuel des méthodes de contrôle non destructif qui doit beaucoup au développement de la micro-électronique. Mais on ne doit pas oublier qu'elles s'at-
taquent à un domaine très spécifique et très exigeant, celui de l'environnement sous-marin.

Autres méthodes de contrôle non destructif.

L'inspection continue des ouvrages et, principalement des plates-formes, est
un souhait formulé par les exploitants dans un double souci de sécurité accrue et, éventuellement, d'économie. Ce besoin est à l'origine de nombreux
développements en cours qui, s'ils ne visent pas spécifiquement les soudures
effectuées sous l'eau, pourront aussi permettre le suivi de leur comportement
dans le temps.

On peut citer trois approches possibles :
- l'utilisation, comme capteurs, de jauges de déformation qui fut la première
  à être mise en oeuvre,
- l'analyse modale des informations fournies par des accéléromètres immergés
  répondant aux vibrations engendrées par des excitateurs posés sur la structure, soit au-dessus, soit au-dessous du niveau de l'eau,
- enfin, l'émission acoustique.

Des développements sur l'application de ces deux dernières méthodes sont en
cours et ont atteint le stade des essais réels sur plates formes en service.

Conclusion.

Le développement du soudage sous l'eau s'est traduit par des applications à
des assemblages structuraux participant à la résistance d'ouvrages exigeant
une sécurité élevée. Ceci n'a pu se faire que par le développement concommi-
tant de moyens de contrôle non destructif adaptés à l'environnement. Encore
actuellement, les méthodes utilisées sont celles du contrôle à l'air libre et
les développements ont porté sur l'adaptation des équipements aux conditions
tout à fait spéciales qui règnent sous l'eau, qu'il s'agisse d'être dans
l'eau ou dans des atmosphères gazeuses sous pression. Deux problèmes essentiels en résultent. D'une part, les opérateurs doivent être qualifiés aussi
bien comme plongeurs que comme contrôleurs. D'autre part, les informations
doivent être transmises en surface à une autorité dûment mandatée pour décider. Ces deux difficultés, entre autres, sont à l'origine des développements
en cours, dont le but final est l'automatisation des méthodes de contrôle.

DEUXIEME PARTIE : COMPORTEMENT EN SERVICE DES SOUDURES EXECUTEES SOUS L'EAU

On sait que la qualité d'une soudure et son adéquation aux sollicitations dont elle est l'objet, en tant que composant d'une structure en service, dépendent d'un nombre considérable de facteurs et de paramètres, parmi lesquels on citera :
- les codes de construction (sollicitations à prendre en compte, calcul des contraintes et coefficients de sécurité),
- le métal de base,
- les produits d'apport,
- les procédé et conditions de soudage,
- le matériel de soudage,
- les opérateurs,
- les méthodes de contrôle non destructif,
- les opérateurs de ces contrôles,
- les codes d'acceptation des soudures.

Dans le cas des soudures exécutées à l'air, une longue expérience antérieure permet de prévoir quels sont les produits et méthodes à mettre en oeuvre pour obtenir un niveau de sûreté des constructions soudées qu'on s'accorde à juger satisfaisant, à condition, toutefois, que les sollicitations prises en compte lors de la conception soient représentatives des sollicitations réelles rencontrées en service. En effet, l'analyse de cas montre que les défaillances de soudures sont dues, le plus souvent à des transgressions des règles de l'art, à des études préliminaires insuffisantes, ou à des abus en service plutôt qu'à des lacunes dans les codes de construction et leur application.

Comme on le sait, il existe de nombreuses méthodes de soudage sous l'eau, qui sont d'ailleurs l'objet de constants développements. Pour notre propos, on peut les grouper en trois catégories, comme les contrôles décrits dans la première partie :
- le soudage dans l'eau,
- le soudage en atmosphère gazeuse sous pression (soudage hyperbare),
- le soudage à l'air à la pression atmosphérique, soit sous la protection d'un batardeau (cofferdam), soit en caisson étanche.

Sauf cette dernière méthode qui - comme l'ont dit Nixon, Allum and Lowe (1982) - remplace les problèmes de soudage par des problèmes de construction, le soudage sous l'eau diffère du soudage à terre par tous les paramètres cités ci-dessus, sauf par les deux premiers. Mais, de plus, le seul fait d'opérer en mer et sous la mer, change radicalement la nature des problèmes par l'énormité des moyens à mettre en oeuvre et par l'influence du milieu sur le comportement des hommes.

Le soudage sous l'eau diffère donc considérablement du soudage à terre et, d'autre part, il s'agit d'une technique en évolution extrêmement rapide. Prévoir le comportement de soudures exécutées sous l'eau revient à prévoir quel type de défaillance est particulier à ces objets nouveaux, au regard des modes de ruine mécaniques, et/ou par corrosion.

On touche là à un problème fondamental, celui de la démarche par laquelle se fait le progrès technique, c'est-à-dire l'insertion de nouvelles techniques dans un contexte économique et social évolutif.

C'est un lieu commun de rappeler que la science vise la connaissance tandis que les techniques sont action, et que la recherche technique comporte nécessairement les trois volets recherche, développement et démonstration, donc expérimentation en vraie grandeur, en vue de rendre acceptable le risque industriel. On constate, enfin, que ce n'est qu'après de multiples applications réelles que vient la reconnaissance, le droit de cité d'une nouvelle technique, sous la forme de normes, de règles et de règlements, à la rédaction desquels participent les trois parties concernées : ceux qui font, ceux qui utilisent et la puissance publique représentant les opinions publique et scientifique.

L'histoire du soudage sous l'eau est un bon exemple de ce processus, puisque c'est une technique déjà ancienne qui n'a fait que récemment l'objet d'une codification et de règlements. Il en résulte que de nombreuses soudures ont été exécutées bien avant que n'existent de tels textes, suivant de nombreuses méthodes et dans des ouvrages très divers.

Il ne fait pas de doute qu'au fur et à mesure que s'étendaient les besoins immédiats ou seulement ressentis, les recherches sur les procédés, les équipements – de plus en plus lourds et complexes – et les propriétés des soudures ont conduit à un développement considérable du champ d'application du soudage sous l'eau dans une étroite collaboration entre les entrepreneurs qui se sont lancés dans cette aventure et leurs clients. Il en est résulté une constante amélioration de la qualité qu'il convient maintenant d'apprécier : qu'il s'agisse de spécifications privées ou de codes et règlements, tous reposent sur la notion de démonstration de l'aptitude à obtenir des propriétés aussi proches que possible de celles des soudures exécutées à terre, pour des assemblages comparables par leur nature et leur service. Les qualifications de procédures de soudage qui en ont résulté ont été de plus en plus exigentes : les conditions d'environnement (pression, nature de l'atmosphère, etc...), les produits d'apport, la nature et les épaisseurs des éléments à souder, etc.., ont dû reproduire, dans les moindres détails, les conditions réelles du travail sous l'eau. De même, on a aussi accru les exigences sur la qualification des opérateurs de soudage comme de contrôle non destructif. Deux buts sont ainsi poursuivis : faire que les soudures sur l'ouvrage réel aient toute chance d'être aussi identiques que possible aux soudures d'essai, et faire que la qualité ainsi atteinte soit suffisante pour garantir un service satisfaisant pendant toute la durée de vie de l'ouvrage concerné. En ce qui concerne ce deuxième objectif, les qualifications de procédure prévoient des essais sur éprouvettes et fixent les résultats minimaux qui doivent être obtenus.

Mais, comment peut-on démontrer qu'il suffit de contrôler quelques uns des multiples paramètres d'exécution d'une soudure pour que celle-ci ait un comportement satisfaisant pendant toute la vie d'un ouvrage particulier ? Avec quelque naïveté, on pourrait essayer de répondre à cette question par un raisonnement théorique. On découvrirait alors que la plupart des caractéristiques des soudures exécutées sous l'eau sont plus ou moins défavorables si on les compare aux soudures exécutées à l'air, qu'il s'agisse de celles dépendant des paramètres humains ou de celles découlant des paramètres physiques. Et nous croyons qu'on en resterait là, car la complexité du problème interdirait de peser quantitativement l'influence de ces différences sur le comportement en service, donc à long terme, des soudures exécutées sous l'eau. Le ferait-on et démontrerait-on que de telles soudures ont des propriétés suffisantes qu'on devrait en conclure qu'il est urgent de déserrer les exigences imposées aux soudures exécutées à l'air. Heureusement, une telle démonstration n'est pas possible en l'état des connaissances et la question de la pertinence des exigences des codes de soud⌐ ` ne sera pas posée.

Reste une approche pragmatique du problème et l'on pense immédiatement à utiliser l'expérience reflètée en principe par la littérature technique. On constate aisément que si le nombre de publications (Masubuchi, 1981) s'accroît d'année en année, elles traitent soit des connaissances de base sur le sujet, soit des procédés, soit d'exemples de soudage sous l'eau. Sauf erreur, aucune ne s'est attachée à dire ce que les soudures sont devenues un, deux, cinq, dix ans après leur fabrication. Que cache ce silence ? Si l'on est pessimiste et sachant que l'industrie n'est guère encline à faire de la publicité sur ses déboires, on peut penser que le vide d'information recouvre des échecs dont il convient de ne pas parler. Si l'on est optimiste et sachant que les peuples heureux n'ont pas d'histoire, on peut penser que les soudures sous l'eau ont un comportement parfait et ne font donc pas parler d'elles.

Nous avons tenté d'en avoir le coeur net et, avec l'aide estimée de collègues de l'IIS, des compagnies pétrolières et d'autres encore, nous nous sommes décidés à faire une enquête d'ampleur limitée.

Dans une lettre circulaire, après avoir indiqué les buts poursuivis, nous demandions que l'on veuille bien nous communiquer, avec garantie d'anonymat pour ceux qui le souhaiteraient, des informations sur le comportement en service réel de soudures exécutées sous l'eau. Sachant d'expérience à quel point ce type d'enquête peut être ennuyeux pour ceux qui sont sollicités, nous accompagnions notre lettre d'un questionnaire, très simplifié et donc très incomplet, auquel il était quasiment possible de répondre en mettant des croix dans des cases prévues à cet effet. Ce document est reproduit à la page suivante.

Nous pensons que ce questionnaire a été relayé et multiplié comme dans une sorte de chaîne du bonheur et qu'ainsi, il a dû atteindre une soixantaine d'organismes et quelques cent cinquante personnes, un peu partout dans le monde. Mais, nous n'avons pas pu, ni voulu être exhaustifs et nous le regrettons pour ceux qui n'ont pas ainsi été touchés et qui auraient voulu répondre.

Le court temps disponible, d'une part, et le poids des activités pétrolières dans le soudage sous l'eau, d'autre part, ont évidemment biaisé cette enquête : des branches industrielles qui utilisent sans doute le soudage sous l'eau n'ont pas été touchées, qu'il s'agisse de la réparation navale, des activités portuaires, des écluses et des ponts ou de la production d'électicité d'origine hydraulique, pour citer quelques exemples.

Le questionnaire

Le questionnaire identifiait seulement trois cas d'application du soudage sous l'eau dans les activités pétrolières :
- les tuyauteries, y compris les tubes ascenseurs de production,
- les pipelines,
  et, dans ces deux cas, une question était posée sur les fluides transportés : huile ou gaz, avec, si possible, indication de la teneur en H_2S,
- enfin, les éléments de plates formes marines.

La profondeur à laquelle la soudure avait été exécutée et la date du travail étaient ensuite demandées.

Une identification sommaire du procédé de soudage était offerte sous forme d'un choix entre soudage dans l'eau, hyperbare ou au sec à la pression atmosphérique.

QUESTIONNAIRE

Ces informations doivent rester anonymes :   oui ☐ - non ☐
(si oui, ne pas répondre aux questions
entre parenthèses)

―――――――

-(Raison Sociale de la Société :   ) (Chantier :   )

- Cas d'application du soudage :   ☐ tuyauterie

gaz ☐   pétrole ☐ H$_2$S % ☐

☐ pipeline

☐ élément de plate-forme fixe

- Profondeur approximative au-dessous du niveau de la mer :   mètre

- Date des travaux :

- Méthode de soudage :

. en pleine eau : .... ☐

. hyperbare : .... ☐

. à la pression atmosphérique .. ☐

- Nature de l'acier :   - Epaisseur :   mm

- Précautions du soudage :

. préchauffage ☐   à   °C

. contrôle de la température
  entre passes .... ☐ à   °C

. traitement thermique de
  relaxation .... ☐ à   °C

- Contrôle non destructif après soudage :

AUCUN ☐   , RADIO ☐   ,   US ☐   , AUTRES ☐   (.......)

- Date de la dernière inspection s'il y en a eu une :

- Méthode d'inspection :

- Résultats :

―――――

Les paramètres métallurgiques essentiels se résumaient à la nuance d'acier e
à son épaisseur.

Les précautions de soudage étaient réduites aux températures de préchauffage
entre passes et lors d'un éventuel traitement thermique de détensionnement.

Suivait une question sur les méthodes de contrôle non destructif utilisée
après soudage.

Enfin, on demandait si, quand et avec quels résultats avaient été inspectée
les soudures préalablement décrites, postérieurement à leur mise en service.

Les destinataires étaient encouragés à donner davantage de détails.

Les réponses au questionnaire

On peut les classer en trois catégories suivant le type d'activité des orga
nismes qui ont envoyé des réponses.

<u>Divers organismes et sociétés faisant des recherches</u> sur les procédés d
soudage sous l'eau, par exemple :
. Thyssen Draht AG,
. Forschungszentrum Geesthacht GKSS,
. Norsk Hydro,
. Nederlandse Aardolie Maatschappij et associés,
. Government Industrial Research Institute (Japon).

Ces réponses, quelque peu en dehors de notre préoccupation immédiate, ont l
mérite de montrer la diversité des organismes qui travaillent à l'amélioration du soudage sous l'eau, chacun s'attachant à résoudre un aspect particulier de cette technique multiforme.

<u>Sociétés faisant des travaux sous la mer</u>. Quatre de ces organismes ont répon-
du à notre demande :
. Comex Services qui pratique divers types d'interventions, telles que l
  soudage dans l'eau, pour des travaux du genre fixation d'anodes de protection cathodique, et le soudage hyperbare pour la construction de pipeline
  et la réparation de plates formes endommagées. Cet entrepreneur nous a envoyé un relevé de 158 soudures hyperbares pour raboutages de pipelines e
  connections de canalisations, s'échelonnant dans le temps entre 1975 e
  1982, pour des profondeurs allant de quelques mètres à -145 m, dans un bo
  nombre de mers du monde.
  D'après ce relevé, 61 soudures ont été effectuées en mer du Nord, 58 ave
  préchauffage et 3 sans préchauffage. Parmi les 97 autres (Méditerranée,
  Golfe de Guinée, Golfe arabique et Golfe du Mexique), seulement 2
  ont subi un préchauffage et les 76 autres, aucun. Cette statistique approximative sur la pratique du préchauffage des soudures de pipelines et d
  risers reflète à l'évidence les exigences des règlements locaux.
  Comex a signalé que, dans un seul cas, la présence d'H_2S dans les fluides
  l'avait obligé à limiter la dureté sous cordon à 248 HV_{10}.
  Aucune des soudures citées n'a subi de traitement thermique de relaxation.
  Comex n'a jamais été informé qu'une des soudures par lui exécutée ait donné
  lieu à un incident en service.

. Taylor Diving and Salvage Co a écrit qu'il n'était pas en mesure de fournir des renseignements détaillés sur le comportement en service de soudures

hyperbares, la raison étant qu'aucun incident ne leur a jamais été signalé
parmi les quelques 250 travaux qu'ils avaient effectués. On conviendra
qu'il s'agit là, brièvement dit, d'un renseignement très important pour
apprécier la qualité du comportement en service des soudures hyperbares.
A noter que la plupart des travaux portaient sur la construction ou la
réparation de pipelines et le reste consistait en réparations de plates
formes pétrolières.

Nakagawa Corrosion Protecting Co, dont l'activité principale est la protection cathodique d'ouvrages marins, pratique le soudage dans l'eau à l'électrode enrobée depuis 1964 et jusqu'à des profondeurs de -30 m, sans préchauffage évidemment. La réponse de cette entreprise ne fait mention
d'aucune défaillance en service de soudures ainsi exécutées.

Sea Con Services a transmis, d'une part, les résultats de neuf tests de
soudage dans l'eau, de deux de soudage hyperbare et d'un en soudage atmosphérique, et, d'autre part, douze cas de soudage dans l'eau sur des éléments de plates formes fixes par des profondeurs variant de quelques mètres
à -78m, s'échelonnant entre 1976 et 1982. Les moyens de contrôle après soudage ont été l'inspection visuelle et/ou la magnétoscopie.

ociétés exploitantes d'ouvrages soudés. Des renseignements ont été communi-
ués essentiellement par des compagnies pétrolières. Ils recoupent presque
inévitablement ceux reçus des sociétés de soudage sous l'eau. Les résultats
ont les suivants :
- Plates formes : neuf travaux de réparation dont :
  deux en soudage atmosphérique avec préchauffage, en mer du Nord, entre
  3 et 5 m d'eau,
  trois en soudage hyperbare avec préchauffage, en mer du Nord, entre 6
  et 12 m d'eau,
  quatre en soudage hyperbare sans préchauffage, dans d'autres mers que
  la mer du Nord, entre 8 et 30 m d'eau.

- Tuyauteries dont tubes ascenseurs (risers) :
  cinq véhiculant de l'huile, dont deux en mer du Nord, avec préchauffage,
  et trois ailleurs, sans préchauffage,
  quatre véhiculant du gaz, tous en mer du Nord, avec préchauffage.

- Pipelines :
  à huile : un en mer du Nord, avec préchauffage et un ailleurs, sans préchauffage,
  à huile plus gaz : deux en mer du Nord avec préchauffage,
  à gaz : deux en mer du Nord avec préchauffage.

Tous ces travaux ont été exécutés entre 1978 et 1982. Bien entendu, on doit
comprendre que, dans le cas des plates formes, il s'agissait de réparations
dues à des avaries et, dans le cas des pipe-lines, de travaux de construction.
Pour les tubes ascenseurs, il peut s'agir de travaux de construction, mais
aussi, dans quelques cas, de réparations.

Dans un seul des questionnaires reçus des compagnies pétrolières, il est fait
état du résultat d'une inspection pratiquée un an après des travaux de soudage de réparation de quatre noeuds fissurés d'une plate forme au niveau -9 m.
L'inspection par magnétoscopie a révélé de nouvelles indications de fissures
dans trois des noeuds. Doit-on en conclure à un échec de soudage hyperbare
dans ce cas ? Il n'est pas possible de trancher, mais on peut émettre l'opinion que l'origine de la fissuration des soudures de réparation est la même

que celle des soudures d'origine exécutées à l'air, ce qui mettrait hors de cause les soudures proprement dites.

Conclusion

Quelque biaisée qu'ait pu être l'enquête à laquelle nous nous sommes livrés, il est très remarquable que tous les travaux de soudage sous l'eau exécutés depuis les années 1960 pour le soudage dans l'eau et dans la seconde moitié des années 1970 pour le soudage hyperbare, soient sans histoire et ce quels qu'aient été les méthodes, les règles et les règlements appliqués.

De plus, on doit remarquer que des différences importantes existent entre les plates formes pétrolières installées dans les diverses mers du monde. Suivant les profondeurs d'eau et la rudessse des environnements marins, les dimensions et, par suite, les épaisseurs mises en oeuvre varient fortement. Ceci est peut-être suffisant pour expliquer pourquoi toutes les soudures sous l'eau effectuées lors de réparations d'ouvrages en mer du Nord l'ont été avec un préchauffage entre 60 et 150°C, le plus souvent à 100°C, alors que pour le même type de travaux dans d'autres régions du monde, aucun préchauffage n'est jugé nécessaire. Mais cette explication perd beaucoup de sa valeur si l'on tente de l'appliquer au soudage des canalisations et pipelines : ils sont construits avec les mêmes aciers, ils ont des dimensions voisines et ils véhiculent des fluides comparables. Or, dans près de 100 % des cas, un préchauffage est effectué en mer du Nord, alors que la situation est inversée ailleurs dans le monde. Puisque, dans un cas comme dans l'autre, il n'est pas signalé d'avarie en service, il y a là un intéressant sujet d'études que nous laisserons à d'autres le soin d'effectuer.

On n'a donc pas pu trouver trace de défaillance en service de soudures exécutées sous l'eau. Nous pensons qu'une telle situation n'a pas été étrangère à l'éclosion, dans les années récentes, de normes, de codes, de recommandations et de règlements. Le soudage sous l'eau, bien qu'en évolution encore rapide, avait fait la preuve qu'il était devenu une technique sûre apte à être utilisée dans des ouvrages exigeant une haute fiabilité : l'écriture de codes sanctionnait cette reconnaissance. Cette idée a été exprimée par Silva (1981) qui concluait son exposé sur les pratiques recommandées par le sous comité D3b de l'AWS, en disant : "Cependant, il était aussi nécessaire de démontrer que le soudage n'était pas seulement une technique à n'utiliser que pour des travaux temporaires ou spéciaux".

CONCLUSIONS GENERALES

Le soudage sous l'eau est une technique qui peut être abordée sous les quatre aspects suivants :
- méthodes disponibles
- propriétés atteintes
- moyens de contrôle
- comportement en service.
Les deux premiers ont été traités par les deux conférenciers précédents. Nous espérons ne pas trahir leurs conclusions en disant que méthodes de soudage et moyens de contrôle, encore en rapide développement, ont permis d'atteindre des propriétés telles que leur champ d'application est devenu infiniment plus vaste qu'il ne l'était il y a une quinzaine d'années. Ce développement est probablement essentiellement dû au rapide essor de la recherche pétrolière en mer et particulièrement, en mers difficiles. Le

oudage exécuté sous l'eau est, en effet, apparu indispensable et compétitif,
oit comme moyen de construction d'ouvrages neufs dans le cas des tuyauteries
t pipelines immergés, soit comme méthode de réparation de pipelines ou de
lates formes. Le besoin a été le moteur de la recherche et du développement
ndustriels.

eut-on dire alors que l'une ou l'autre méthode de soudage sous l'eau peut ac-
uellement permettre de tout souder, quelle que soit la nature de la construc-
ion ou son service ? Certainement non et il nous semble que les industriels
n sont bien conscients, à voir la prudence avec laquelle ils utilisent ces
echniques : à titre de preuve, il ne semble pas qu'elles aient été utili-
ées pour le raboutage de pipelines à gaz acide sans traitement thermique
e relaxation, non plus que pour la réparation à l'identique de soudures de
oeuds de plates formes endommagées par fatigue. C'est notre opinion que
ette prudence des sociétés pétrolières et des sociétés de service, associée
de remarquables développements techniques, est à l'origine du très satis-
aisant comportement en service des soudures exécutées sous l'eau. Il semble
ien que ce même constat ait été fait par d'autres, qui avaient la responsa-
ilité de codifier et de réglementer. Bien entendu, nous sommes prêts à modi-
ier ces jugements si la discussion qui va suivre apporte des éléments nou-
eaux infirmant ceux que nous avons pu recueillir.

REMERCIEMENTS

'auteur voudrait pouvoir remercier tous ceux qui lui ont apporté leur aide
e diverses manières : ils sont malheureusement trop nombreux pour pouvoir
tre tous nommés. Cependant, il ne lui est pas possible de ne pas citer en
remier lieu, l'Institut de Soudure (Paris) et, particulièrement, J.P.
audin, mais aussi ses collègues de l'I.I.S., Bruno L. Alia et U. Girardi et
u groupe Elf Aquitaine, A. Cabiran, W.A. Francis, G. Gainette, R. Guillo,
. Mèche, M. Rabouin, L. Roland, G. Trican, etc... .
es sociétés dont les noms suivent méritent aussi d'être signalées pour leurs
ontributions appréciées :
ouygues Off Shore,
rown and Root,
ureau Veritas,
hevron Petroleum UK,
sso E.P. Norway,
orschungszentrum Geesthacht GKSS,
overnment Industrial Research Institute of Japan,
orrosionscentralen ATV,
itsubishi Heavy Industries,
akagawa Corrosion Protecting Co,
ederlandse Aardolie Maatschappij,
orsk Hydro,
ccidental Petroleum Caledonia,
ennzoil Nederland Co,
hillips Petroleum Company Norway,
lacid International Oil,
ea Con Services,
vejsecentralen,
aylor Diving and Salvage Co,
hyssen Draht AG.

nfin, l'auteur remercie la Société Nationale Elf Aquitaine pour lui avoir
onné les moyens de faire ce travail et l'avoir autorisé à le publier.

BIBLIOGRAPHIE

Ouvrages généraux récents :

. Integrity of Offshore Structures. D. Faulkner, M.J. Cowling and P.A. Frieze
Ed. Papers presented at the Second International Symposium on Integrity of
Offshore Structures, the University, Glasgow, July 1981. Applied Science
Publishers Ltd, London, 1981.
. Underwater Welding of Offshore Platforms and Pipeline. Proceeding of a
Conference, November, 1980, New Orleans, Louisiana. Welding Technology
Series. American Welding Society, Miami, 1981.
. Pénétration sous-marine. Underwater operations and techniques. Conférence
internationale organisée par ATMA et CNEXO, décembre 1982, à Paris. Edi-
tions Technip, Paris.
. Proceedings of the Second International Conference on Offshore Welded
Structures, Session VII, organized by the Welding Institute, London, 16-18
November 1982.

Det norske Veritas, (1976). Rules for the Design and Construction of Submarine Pipelines and Pipelines Risers. Oslo.

Det norske Veritas, (1981). Rules for Submarine Pipelines Systems. Oslo.

Masubuchi, K. (1981). Underwater Factors Affecting Welding Metallurgy. In
Underwater Welding of Offshore Platforms and Pipelines. American Welding
Society, Miami. pp. 81-98.

Nixon, J. H., C. J. Allum and J. M. Lowes (1982). Underwater Welding : a
Review. In Pénétration sous-marine - Underwater operations and techniques.
Editions Technip, Paris. pp. 147-171.

Silva, E.A. (1981). AWS D3b Recommended Practices. In Underwater Welding of
Offshore Platforms and Pipelines. American Welding Society, Miami.
pp. 137-154.

SECTION I

WET AND DRY WELDING/METHODES DE SOUDAGE DANS L'EAU ET A L'ABRI DE L'EAU

IIM-E

PAPERS COMMUNICATIONS

Process Variables and Properties of Underwater Wet Shielded Metal Arc Laboratory Welds

H. Hoffmeister and K. Küster

Laboratorium für Werkstoffkunde und Schweißtechnik, Hochschule der Bundeswehr Hamburg, Federal Republic of Germany

ABSTRACT

Hydrogen pick up and HAZ hardness increase with coating thickness of rutile electrodes in underwater wet SMA welding but decrease with heat input. Multilayer welding of 15 mm thick, 0,14 % C, 0,90 % Mn grade D-steel, provides weld metal CVN-toughness of 35-40 J at 0 °C, applying a total of 14-15 low heat input stringer beads.

KEYWORDS

Wet underwater SMA welding; hydrogen pick up; HAZ hardness; weld metal toughness, multilayer technique.

INRODUCTION

The application of wet SMA welding to fabrication and repair of underwater steel structures and pipelines seems to be hampered by basic metallurgical factors mainly due to high hydrogen pick up and cooling rates producing microstructures and delayed hydrogen assisted cracking. Yet, in the present period of decreasing crude oil prices and profits the availability of alternative flexible and economically working underwater welding procedures should serve to maintain and further develop offshore oil resources. While the aspects of dry hyperbaric welding and wet underwater GMA welding have been discussed frequently in the literature by Schafstall and Schaefer, (1979) quantitative information on process specifics and weld joint quality of SMA wet welding is scarce. In recently published literature Defourny and coworkers (1980) found rutile coatings to provide best arc stability and bead profiles, while iron powder coatings furnished less cold cracking due to relatively low diffusible hydrogen contents, which is in accordance with earlier results of Hoffmeister and coworkers (1979). Since above literature seems to indicate at least partially sufficient weld joint strength and toughness it is the purpose of the present work to deal with aspects of hydrogen pick up and HAZ hardening as well as to provide quantitative information on weld joint properties, in particular, toughness.

The experiments were conducted in a laboratory fresh water basin at 0,4 m
water depth, welding with automatically guided stick electrodes at an in-
clination angle of 75° against plate surface. A power source ESAB LD 600 and
DCSP were applied.

Two series of experiments were carried out. In the first, the effect of
coating thickness on stable welding conditions of a 4 mm core wire diameter
E 5122 RR6 DIN 1913 electrode at 7-10 mm total thickness was established in
bead-on 15 mm thick - St 37-2 plate - welds. For the same range of currents,
voltages and welding speeds as well as coating thicknesses of E 5122 RR6 and,
in addition, for an 8 mm thick E 4332 RR11 180 iron powder electrode, diffusible hydrogen contents according to ISO 3690-1977 and HAZ peak hardnesses
of the St 37-2, 0,13 % C, 0,40 % Mn steel were determined, (Table 1). In the
second series butt welding of a 0,14 % C, 0,90 % Mn GL grade D-steel was
performed, using E 4332 RR11 180 electrodes. Special emphasis was put on the
effect of the number of runs on weld metal toughness. For the purpose of
sound welds, semi-circular edge profiles were machined as indicated in Fig.3.
The number of runs ranged from a total of 4 to 15 stringer beads.

RESULTS AND DISCUSSION

Hydrogen pick up and HAZ hardness

Out of the results of the E 5122 RR6 electrode Fig. 1 shows

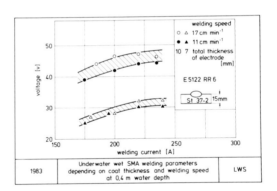

Fig. 1.

stable current and voltage conditions for a total diameter of 7 and 10 mm representing welding speeds of 11 and 17 cm min^{-1} respectively. Increasing
coating thickness as well as higher welding speeds and currents necessitate
a corresponding increase in voltages. At the same time, an increase of cone
depths at the melting tips of the electrodes, was noticed. The wet SMA underwater process encounters high temperature gradients across the coating thickness due to the cooling action of the surrounding water. Cone geometry thus
is stabilized and dependant on welding parameters and thickness. Figure 2a
with increasing heat input reveals a decrease of

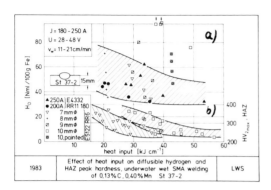

| 1983 | Effect of heat input on diffusible hydrogen and HAZ peak hardness, underwater wet SMA welding of 0.13%C, 0.40%Mn St 37-2 | LWS |

Fig. 2a, b.

diffusible hydrogen contents. In case of the iron powdered electrode E 4332 RR11 180 it reflects the effect of decreasing welding speed at constant currents, the hydrogen contents at 200 A occupying the lower scatter band, clearly below the values for 250 A. From Fig. 1 and 2a, together with cone depth variations, the significance of voltage and thus, arc length for hydrogen pick up can be concluded. According to analysises of the generated gas, 70 - 72 % H_2, 20 % CO and 2 - 3 % CO_2 were characteristic for the E 4332 RR11 180 electrode, independent on welding conditions. Thicker coatings of the E 5122 RR6 electrode, Fig. 2a, especially, at a total diameter of 10 mm, create deep cones and very high diffusible hydrogen contents. Although these may be reduced by protecting the 10 mm electrode coating by a special paint, the hydrogen contents still stay at high levels of 50-80 Nml/100g Fe. According to Fig. 2a, the lowest hydrogen being not below 30 Nml/100g Fe a high risk of delayed cold cracking remains. Yet, increasing heat inputs, Fig. 2b, up to extraordinary high values of 40-60 kJcm^{-1}, at $t_{8/5}$ - cooling times (Hoffmeister and coworkers, 1979) of up to 12 seconds enable HAZ peak hardnesses of the 0,13 % C, 0,40 % Mn St 37-2 steel of as low as 250 HV. Comparing the hardness results of the 7 to the 10 mm total diameter E 5122 RR6 electrodes thick coatings apparently require more arc energy to be melted thus lowering effective heat input and increasing hardness.

Static Strength and Toughness of GL Grade-D-Steel Weld Joints

Static tensile tests revealed failure outside the weld joint in the base plate material of samples without reinforcement provided there were no major weld metal defects, i.e. lack of fusion, which occurred in some cases. CVN test results of a 6-bead weld joint, Fig. 3, for a test temperature of 0 °C, show a

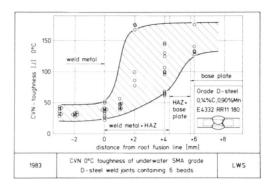

Fig. 3.

sharp drop of toughness down to 30-40 J within the weld metal, while the HAZ approaches values of the unaffected base plate material of around 150 J. There fore, weld metal toughness apparently represents a major problem due to high proportions of coarse grained martensite and grain boundary ferrite in the as welded microstructure. Moreover, according to Table 1, manganese contents in the weld metal burned down to as low as 0,43 % which according to Bosward and John (1978) is likely to produce unfavourable microstructures, in particular, higher grain sizes of the weld heat refined sections. For increasing number of stringer beads in the multilayer technique decreasing heat input per run according to Fig. 4

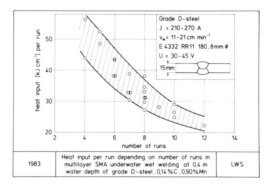

Fig. 4.

had to be applied, incorporating a higher risk of hydrogen pick up and higher hardness in the as welded state. As a result of the multilayer technique, peak hardnesses of 420 HV2 in the near surface section of the HAZ were not influenced by the number of runs while minimum HAZ hardness in the reheat treated root areas revealed a slight increase from 220 at 4 runs to 250 HV2 at 12 runs. There was no marked effect of the number of runs on weld metal maximum and minimum hardness, varying between 260 in the as welded metal and 205 in the reheat treated fine grained areas. The as welded microstructure is

composed of mostly martensite and aligned ferrite-carbide (upper bainite)
within the primary grains, together with grain boundary ferrite (Fig. 5a).
Reheat treating of microstructures by the following layers reveal Fig. 5b
and 5c, showing smaller grains of still martensite

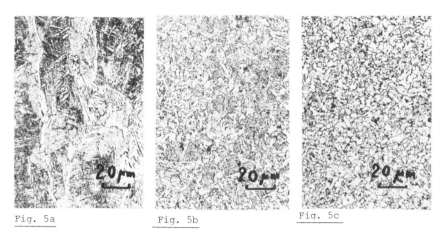

Fig. 5a   Fig. 5b   Fig. 5c

with aligned ferrite-carbides in Fig. 5b and fine grained ferrite-perlite in
Fig. 5c. CVN toughness of the complex weld metal microstructure at O °C,
Fig. 6, shows a clear increase, with number of runs, starting at a range of
between 15 and 30 J at 4 stringer beads and reaching 42-50 J at 15 beads.

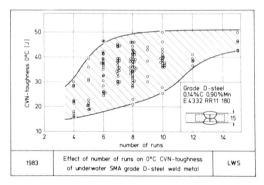

Fig. 6.

The reduction of coarse grained martensite and grain boundary ferrite and
the subsequent formation of fine grained perlite/ferrite or even fine grained
martensite/aligned ferrite-carbides due to the multilayer technique produces
increased toughness of the weld metal, which may satisfy specified requirements. Yet, this can be achieved only at welding conditions, Fig. 2, which
most likely introduce high amounts of hydrogen and delayed cold cracking
into the weld metal and HAZ. The existence of cracking in both areas was
proved by microscopic examination though not yet quantified with respect to
effect of number of runs.

CONCLUSIONS

1. Increasing coating thickness of underwater rutile SMA electrodes results in accellerated hydrogen pick up and higher HAZ peak hardness.
2. By increased heat input hydrogen may be reduced only slightly, therefore, delayed cold cracking will still be a risk.
3. HAZ hardness may be reduced to specified levels of the 0,13 % C, 0,40 % Mn steel by high heat inputs.
4. Inspite of considerable hydrogen pick up and thus cold cracking at low heat inputs, the application of multilayer small stringer beads to butt welding of a 15 mm grade GL-D-steel results in comparatively good weld metal CVN toughness at 0 $^{\circ}$C. This was achieved at, due to burn out, low Manganese contents.

REFERENCES

Bosward, J.G., and R. John (1978) Intern. Conf. Trends in Steels and Consumables for Welding. The Welding Institute, London, paper 14

Defourny, J., A. Bragard and H. Lamberigts (1980) Convention no. 7210-KA/2/201 Comm. Communautés Européennes, Rapport final

Hoffmeister H., K. Küster, P. Nölle and M. Werner (1979) DVS-Bericht 57, p. 91/97

Schafstall, H.G. and R. Schaefer (1979) Schweißen u. Schneiden, 31, 9 p. 374/81 5, p. 186,92

TABLE 1 : Chemical composition of plates and weld metal

Wt.-%

	C	Si	Mn	P	S	Ni	Mo	Al	Nb	Ti
St 37 - 2	.13	.28	.38	.011	.049	.08	.02	.002	.002	.002
GL grade D	.14	.20	.90	.006	.009	.10	.09	.056	.002	.003
Weld Metal-grade D	.07	.14	.43	.026	.011	.07	.03	.002	.002	.006

ACKNOWLEDGEMENTS

The experiments were carried out by cand. ing. E. Röper, cand. ing. A. Lausberg and cand. ing. Röpenack as part of their graduation work.

Process Variables and Properties of Wet Underwater Gas Metal Arc Laboratory and Sea Welds of Medium Strength Steels

H. Hoffmeister and K. Küster

Laboratorium für Werkstoffkunde und Schweißtechnik, Hochschule der Bundeswehr Hamburg, Federal Republic of Germany

ABSTRACT

Wet underwater GMA welding by assistance of a gas cavity stabilizing water jet was investigated with respect to critical gas flow for the reduction of diffusible hydrogen and to the effect of multilayer technique on CVN toughness. Diffusible hydrogen contents as low as 2-4 Nml/100gFe can be achieved, depending on gas flow and jet conditions. In butt welding of St E 36 increasing the number of stringer beads leads to CVN toughnesses at -20 $^{\circ}$C of up to 50-70 J in the weld metal using a Mn-Ni flux cored filler wire. Too small stringer beads yet encounter the danger of hydrogen assisted cold cracking.

KEYWORDS

Underwater GMA wet welding; hydrogen control ; water curtain; multilayer technique; toughness of medium strength steel weld metal.

INTRODUCTION

Wet underwater gas metal arc welding has been shown to establish crack free weld joints provided the arc metal droplet transfer is protected from the access of hydrogen by, among other means, a high velocity conical water jet (Hamasaki and Sakibara, 1976). For future applications weld joint quality requirements for medium strength offshore steels as for instance St E 36 are to be met. Previous work of the authors (Hoffmeister and Schafstall,1982), for St E 36 and X60, has still indicated comparatively low levels of weld metal toughness as well as high HAZ hardness due to quenching by the water jet after welding. Moreover, the tolerances to water cavity instabilities with respect to hydrogen pick up should be precisely known. It is the purpose of the present work to establish stable parameter ranges for the water jet protected underwater FCA MAG welding of St E 36 using constant water jet geometry and a flux cored filler wire. In particular, the effect of shielding gas flow rate and heat input on diffusible hydrogen contents together with and without water jet as well as the influence of multilayer technique on weld metal toughness and HAZ hardness were to be established and to be compared to earlier results.

EXPERIMENTAL PROCEDURE

The experiments were conducted in a fresh water basin at a water depth of 0,5 m using a CP power source. Flux cored seamless filler wire with a diameter of 1,2 mm was used providing a weld metal composition mentioned in Table 1. For the hydrogen determination by the mercury method according to JSO 3690-1977 different shielding gases were applied according to Fig. 1 and 2. Two test series on St 37-2 were carried out, one with increasing gas flow rate of the 50 mm diameter, 0,5 mm wide jet, see Fig. 1, at constant heat inputs and water jet flow rates. The second was run at increasing heat inputs, keeping gas- and water flow rates constant. Since the welding speed was also kept constant at 21 cm·min^{-1} the welding currents and voltages ranged between 210 and 280 A resp. 23 and 32 V. For the multilayer, flat position butt welds 330x150x20 mm St E 36 plates (Table 1) were prepared with double-V edges, using opening angles of 60° for tests with up to a total of 10-12 stringer beads. For tests with up to 12-18 total passes wider angles of 90 and 100° were applied. CO_2 was used as a shielding gas for up to 8 runs, i.e. 4 runs on each side, while, for less spattering, Ar+8%O_2, Ar+5%O_2+5%CO_2 and Ar+18%CO_2 were used for the higher numbers and thus smaller sections of the beads. The gas flow rate was constant for each type of shielding gas. At increasing the number of runs it was found necessary to reduce heat input applying welding speeds of 10-24 cm min^{-1} at voltages of 22-34 V and currents of 150-330 A. The water jet flow was kept constant at 33 l min^{-1}.

RESULTS AND DISCUSSION

Diffusible Hydrogen Contents

For shielding gases CO_2 and Ar+8%O_2 the increase of the gas flow rate at constant water jet flow and heat input causes a sharp decrease in diffusible hydrogen, Fig. 1. At a

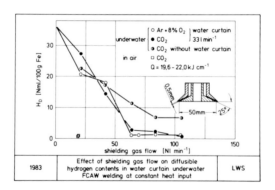

Fig. 1.

critical flow rate of about 55-60 Nl min^{-1} hydrogen contents as low as 2-4 Nml/100g Fe are established and kept at this level also at higher flow rates. Parallel jet stability tests without welding but observing water cavity conditions by means of acrylic glas plates as described by Shinada and coworkers, (1978) revealed critical flow rates of about 80-100 Nl min^{-1} for a stable, "waterfree" cavity. Heating of the shielding gas therefore, may serve as an additional means for stabilizing the fluid water cavity thus providing a "dry" spot at lower gas flow rates than without welding. The beneficial effect of the water jet stabilisation of the shielding gas cavity can readily been seen by comparison to the hydrogen contents of the test runs without water jet. Though, decreasing hydrogen is also achieved with increasing shielding gas flow, the final levels remain as high as about 7-8 Nml/100g Fe, due to frequent access of water vapor to the arc. Figure 2 reveals a slight decrease of diffusible hydrogen with

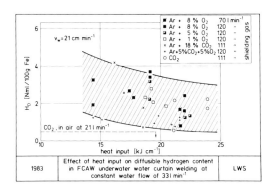

| 1983 | Effect of heat input on diffusible hydrogen content in FCAW underwater water curtain welding at constant water flow of 33 l min^{-1} | LWS |

Fig. 2.

increasing heat inputs, collecting the results for different shielding gases in a common scatter band. Though in general, the small increase of diffusible hydrogen with decreasing heat inputs does not seem to encounter too high contents, individual results of the shielding gases indicate a stronger increase as for instance in the Ar+18%CO_2 tests. Concerning possible hydrogen access to the arc the welding process has to deal not only with the stability of the gas cavity but also with the thin rest layer of water at the plate surface. This should be removed by the heat of the arc and sucked off by the shielding gas stream, a process, which, depending on weld parameters, may govern hydrogen pick up of the weld metal droplets expecially at high welding speeds and low heat inputs. However, low enough hydrogen levels are achieved in order to avoid hydrogen assisted cracking of offshore medium strength steels, provided the process works at stable cavity conditions.

Effect of multilayer Welding on Hardness and Toughness of St E 36 butt Welds.

Figure 3 shows the decrease in heat inputs per run

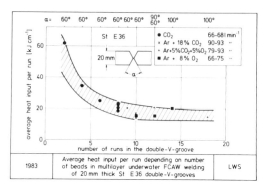

| 1983 | Average heat input per run depending on number of beads in multilayer underwater FCAW welding of 20 mm thick St E 36 double-V-grooves | LWS |

Fig. 3.

at increasing number of stringer beads in the double-V-grooves. For the weld metal, this implies less plate material fusion which can be drawn from the average contents of the center of the V-grooves, <u>Fig. 4</u>

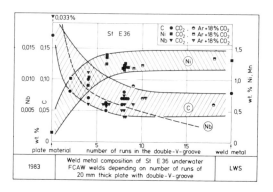

| 1983 | Weld metal composition of St E 36 underwater FCAW welds depending on number of runs of 20 mm thick plate with double-V-groove | LWS |

Fig. 4.

The contents of C and Nb, fused from the plates, are decreasing with in-creasing number of runs, while, at the same time, higher Ni contents from the deposited metal are resulting. The scatter bands are including the C- and Ni-contents of both the plate material and the deposited metal which was deter-mined by a separate bead-on plate test.

The macrostructure of the weld metal showed decreasing areas of coarse graine as welded metal at increasing number of runs as recently discussed by Dawson (1982). The microstructure in the as welded sections was grain boundary fer-rite with mostly martensite and little ferrite with aligned carbides in the primary grains. Grain boundary ferrite with slightly less martensite/ferrite with aligned carbides were found in the reaustenitized and quenched coarse grained areas. Within the fine grained HAZ of the weld metal, mostly polygo-nal ferrite with ferrite/carbide aggregates were found. With increasing num-ber of runs, the amount of grain boundary ferrite decreased in the as welded and coarse grained reaustenitized zones, which, in accordance to recent re-sults by Taylor and Evans (1982) may be attributed to increasing Ni-contents,

Fig. 5a, b.

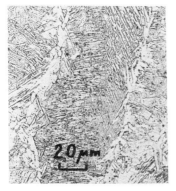 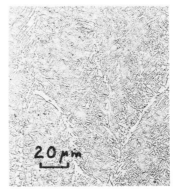

Fig. 5a: As welded, 2 runs, CO_2 Fig. 5b: As welded, 14 runs, $Ar+8\%O_2$

Looking at <u>Fig. 6</u> the hardness HV_1-values of the as welded areas

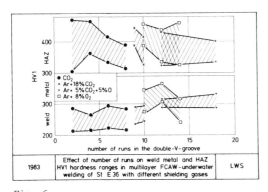

| 1983 | Effect of number of runs on weld metal and HAZ HV1 hardness ranges in multilayer FCAW-underwater welding of St E 36 with different shielding gases | LWS |

Fig. 6.

range between 260 and 290 for the CO_2-welds, but between 290 and 380 for the Ar-gas mixtures, indicating accellerated cooling and higher martensite contents as compared to CO_2. Weld-temperature-time measurements at fusion lines of bead on plate welds (Hoffmeister and coworkers, 1979) indicated $t_{8/5-1}$ cooling times of not more than 6-7 sec at heat inputs of up to 40 $kJ cm^{-1}$, even with CO_2 as a shielding gas. In both the as welded and the weld metal reaustenitized and quenched zones there seems to be no effect of number of runs or heat input. Yet, hardnesses HV_1 of the HAZ near to the surface are decreasing with increasing number of runs, for the CO_2 welds starting at 470 with 2 runs and ending at 390 with 8 runs. Though, for steels with lower carbon contents than the present plate, only smaller hardness alterations are possible, this indicates the possibility of decreasing hardness values to specified requirements. However, at a higher number of passes and by use of Ar-gas mixtures higher hardnesses are resulting in the HAZ's. <u>Moreover, in the case of the 18 run-weld with Ar-5%O_2-5%CO_2 severe HAZ cold cracking was</u>

observed in several cases. Therefore, a limit seems to exist for increasing
the number of runs with respect to quench hardening and delayed hydrogen
cracking, the latter being due to insufficient protection of the arc against
the access of hydrogen by too low heat generation as indicated above. However
as previously shown by Satoh and coworkers (1981) weld thermal cycles in
water jet underwater welding are dependant on the welding velocity to water
jet-diameter relation which means that the above mentioned limits apply to
the respective constant conditions of the present investigation.
The clearly beneficial effect of increasing the number of passes and thus
refining weld metal structure on CVN toughness at -20°C can be seen from
Fig. 7. Starting with CVN values between 10 and

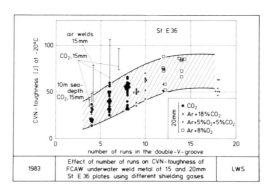

| 1983 | Effect of number of runs on CVN-toughness of FCAW underwater weld metal of 15 and 20mm St E 36 plates using different shielding gases | LWS |

Fig. 7.

40 J for the 4-run weld, toughnesses as high as between 50 and 90 J are
achieved in the 12-14 run welds. The effect of the refining of the weld metal by multilayer welding is even more pronounced in the groove sides welded
first and thus reaustenitized partly by the other side's runs, the former
occupying the upper portion of the CVN-scatter band. SEM investigations show
predominantly cleavage fractures in the as welded metal at low numbers of
runs, Fig. 8a, while in the same samples, grain boundary fracture and even
dimple fracture occurred in the coarse grained reheated areas together with
cleavage fracture. At a high number of runs, Fig. 8b almost exclusively
ductile dimple fracture occurred regardless of which section of the weld
metal was engaged by the fracture path. The increase in toughness therefore,
is based on the redistribution and grain refinement of microstructural constituents, mainly ferrite and martensite, as well as on the decrease of C
and increase of Ni contents with increasing number of beads. Earlier results
of the authors (Hoffmeister, Küster and Fußhöller, 1982) using the same
welding process are included in Fig. 7. For the same or

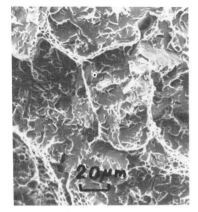

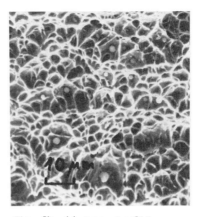

Fig. 8a: 4 runs, CO_2 Fig. 8b: 14 runs, $Ar+8\%O_2$

similar plate chemical composition, see Table 1, but plate thickness of 15 mm, higher CVN-values are resulting. Regarding the effect of number of runs they are almost in line with results of welds made in air. In the 4 run welds of the 10 m seadepth welding test (Hoffmeister, Küster and Schafstall, 1982) the scatter band of the present CVN values is occupied. It may be concluded that, at smaller plate thicknesses, the refining capacity of the multilayer technique is more pronounced due to a slower heat transfer to the surrounding water at increased surface boiling.

CONCLUSIONS

For the application of water curtain assisted wet underwater GMA welding to medium strength steels the following conclusions can be drawn:
1. Diffusible hydrogen contents of the weld metals can be reduced to such low levels, that hydrogen assisted cold cracking will not be likely provided a stable shielding gas cavity exists.
2. Increasing the number of stringer beads in but welding of St E 36 with a Mn-Ni-filler wire will increase toughness of the weld metal by weld reheat refining of microstructures as well as by increased Ni-contents and thus, more favourable as welded microstructures.
3. Multilayer technique seems to be limited by welding conditions, which, in underwater wet GMA welding inspite of a stable cavity produce too rapid cooling and too small dry weld groove areas, thus leading again to hydrogen assisted cold cracking.

REFERENCES

Dawson, G.W. and P. Judson (1982). Sec. Intern. Conf. Offshore Weld. Structures, paper 2, The Welding Institute, London
Hamasaki, M., J. Sakibara and Y. Arata (1976). Metals Constr. 8, 3, p. 108/112
Hoffmeister, H., K. Küster, P. Nölle and M. Werner (1979). DVS Ber. 57 p. 91/97
Hoffmeister, H., K. Küster and K.J. Fußhöller (1982). Schw. Schn. 34 p. 315/19

Hoffmeister, H., K. Küster and H.G. Schafstall (1982). Sec. Intern.
  Conf. Offshore Weld. Structures, paper 17, The Welding Institute, London
Satoh, K., M. Tamuar, K. Kawakami, K. Shinada, T. Ohame and Y. Manabe
  (1981). IIW doc. XII-B-17-81
Shinada, K., Y. Nishio and H. Wada (1978). IIW doc. XII-B-237-78
Taylor, D.S. and G.M. Evans (1982). Sec. Intern. Conf.
  Offshore Weld. Structures, paper 29, The Welding Institute, London

TABLE 1: Chemical composition of plates and deposited
          Weld Metal

Wt.-%

thickness	C	Si	Mn	P	S	Ni	Al	Nb
20 mm	.17	.40	1.30	.011	.006	.14	.057	.033
15 mm	.17	.40	1.44	.016	.010	.02	.036	.029
Weld metal	.08	.33	1.46	.012	.011	1.29	.009	.003

ACKNOWLEDGEMENT

Parts of the experiments were carried out by cand. Ing. B. Paulus
and cand. Ing. U. Schwab as part of their graduation work.

L'avenir des différents procédés pour le soudage en condition hyperbare

J. Fuin et JP. Gaudin

Institut de Soudure, Paris, France

RESUME

La situation actuelle du soudage en condition hyperbare amène à se poser un certain nombre de questions quant à l'avenir de ce moyen d'assemblage en grande profondeur.

Après un bref rappel de l'état des connaissances dans le domaine du soudage en condition hyperbare et des moyens dont dispose l'Institut de Soudure pour étudier les problèmes qu'il pose, cette communication propose les bases d'une réflexion sur l'utilisation des procédés de soudage pour des applications à de plus grandes profondeurs que celles jusqu'alors pratiquées et du point de vue de leur automatisation pour le soudage des canalisations.

MOTS CLEFS

Soudage hyperbare ; automatisation ; fissuration à froid ; évolution caractéristiques mécaniques ; influence pression ; caisson de simulation ; profondeur limite.

INTRODUCTION

Le soudage en condition hyperbare est une technique couramment employée depuis 1960 pour le raccordement des canalisations sous-marines et qui tend à se développer pour la réparation des structures immergées. Or, l'évolution de cette technique n'est pas à l'image de celle que connaissent les technologies offshore en général. Ceci tient à la multiplicité des paramètres régissant cette technique d'assemblage, à l'empirisme dont souffre la connaissance des phénomènes physico-chimiques liés aux procédés utilisés et au coût des expérimentations en atmosphère sous pression nécessaires à toute évolution.

Le présent exposé se propose de retracer l'évolution des résultats obtenus avec les techniques les plus employées dans ce domaine et envisage leur avenir dans le cas d'une automatisation ou d'une utilisation à de plus grandes profondeurs.

RAPPEL DES CONNAISSANCES EN SOUDAGE EN CONDITION HYPERBARE

La dynamique de FRIGG (avant 1980)

Dans les années 1970, l'intérêt des compagnies pétrolières pour les ressources de la Mer du Nord et la volonté de construire un gazoduc pour l'exploitation du champ de Frigg allaient donner son élan au soudage utilisé comme moyen d'assemblage de canalisations sous-marines. Malheureusement, cette volonté a abouti, à cause d'investissements tardifs dans la recherche, au choix et à la mise au point d'un procédé de soudage pour sa faculté à être mis en oeuvre le plus rapidement possible.

Les études entreprises à l'Institut de Soudure dans le cadre de cette mise au point ont permis, par l'utilisation d'une procédure de soudage avec électrodes enrobées, d'évaluer certains effets de la pression sur la soudabilité opératoire et métallurgique. Il a été ainsi possible de poser le problème, nouveau en soudage, de l'influence de la pression sur les équilibres physico-chimiques durant l'élaboration de la zone fondue (accroissement, dans le métal déposé, des éléments à faible réactivité avec le laitier tel le carbone, et diminution d'autres espèces telles que le manganèse, le silicium, etc...), de l'évolution de la solubilité de l'hydrogène, produit de décomposition de l'enrobage, et de ses conséquences sur les risques de fissuration qui rendent nécessaire le préchauffage des pièces dès les plus faibles profondeurs d'interventions (voir Fig. 1).

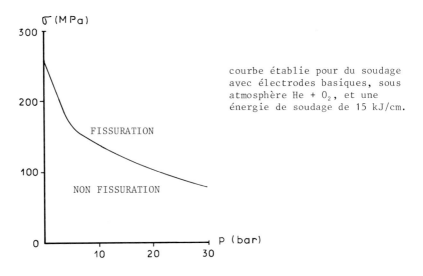

courbe établie pour du soudage avec électrodes basiques, sous atmosphère He + O_2, et une énergie de soudage de 15 kJ/cm.

Fig. 1   Essais de soudabilité par la méthode des implants, Détermination de la limite de non fissuration.

Grâce à ce programme de recherche à court terme, des caractéristiques mécaniques acceptables vis à vis des règlements en vigueur ont été obtenues
lors des qualifications de modes opératoires, mais compte tenu des résultats, les perspectives d'utilisation de ce procédé à de plus grandes profondeurs ou pour des températures de fonctionnement plus faibles, étaient
peu encourageantes.

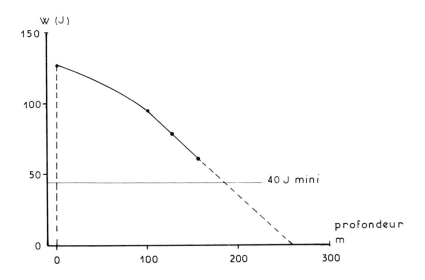

Fig. 2   Résultats d'essais de résiliences obtenus sur qualifi-
          cation de mode opératoire avec électrodes enrobées.
          (essais CHARPY V à -10°C)

L'extrapolation des résultats d'essais de résilience effectués dans le
métal fondu indique   (Fig. 2) que la profondeur limite pouvant être
atteinte, si la température de fonctionnement était identique à celle
retenue dans le cas de FRIGG (0°C), se situerait aux environs de 200 m.

Il paraissait intéressant de poursuivre l'action entreprise afin de répondre aux éventualités d'explorations plus profondes. C'est pourquoi
l'Institut de Soudure, avec le soutien d'une compagnie pétrolière, a
décidé de se doter d'un moyen d'essai capable d'expérimenter le soudage
hyperbare dans des conditions proches de la réalité.
C'est pour celà qu'a été construit l'ensemble de simulation HYPLAB 3300
qui répond aux 11 critères principaux d'une reproduction la plus fidèle
possible des conditions de soudage en chambre hyperbare (Fig. 3).

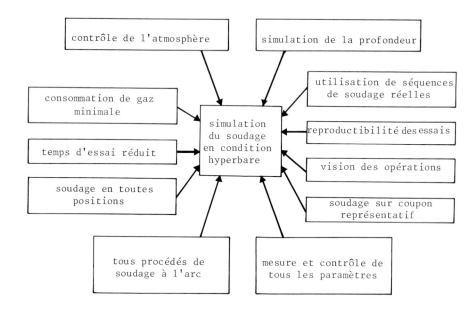

Fig. 3   Objectifs principaux à atteindre pour simuler en laboratoire le soudage en condition hyperbare.

Le laboratoire de soudage HYPLAB 3300 de l'Institut de Soudure (Fig. 4) a
été ainsi construit.
Il permet dans un volume d'environ 1 m³ d'effectuer une préqualification de
mode opératoire de soudage sur éprouvette de 300 x 400 x 50 mm jusqu'à la
profondeur de 1000 m pour tous les procédés de soudage à l'arc.

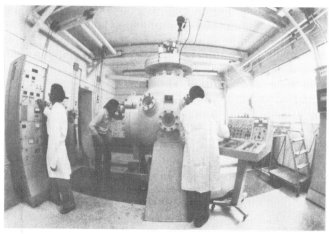

Fig. 4   Laboratoire HYPLAB 3300.

Après FRIGG (depuis 1980)

Les procédés mis à la disposition des soudeurs sont à priori nombreux.
Cependant, les contraintes de l'intervention en chambre hyperbare habitée
limitent considérablement l'éventail des techniques utilisables dans ces
conditions. Ainsi, des procédés comme le soudage plasma, le soudage par
friction ou par induction n'ont pas dépassé le stade de l'étude prélimi-
naire en laboratoire à faible profondeur et n'ont jamais été utilisés sur
des canalisations de la taille de celles installées actuellement.

Le soudage avec électrodes enrobées reste incontestablement le plus ancien
mais le plus utilisé de tous les procédés de soudage en condition hyperbare ; il bénéficie d'une grande simplicité de mise en oeuvre mais nécessite en revanche un personnel hautement qualifié. Aussi, des études ont-
elles été menées afin d'améliorer la maniabilité des électrodes. Des progrès substantiels ont été obtenus quant aux caractéristiques mécaniques du
métal fondu grâce à la mise au point de nouveaux enrobages ; ainsi, pour
une température de fonctionnement identique à celle de FRIGG, a-t-il été possible de qualifier ce procédé jusqu'à la profondeur de 320 m. Cette amélioration a été obtenue par approches successives, sans une étude fondamentale
des équilibres physico-chimiques qui régissent les échanges chimiques durant
l'élaboration de la zone fondue. Cette profondeur de 320 m semble être proche de la limite d'utilisation de cette technique, d'autant plus qu'au-delà
apparaissent d'autres limites, d'ordre physiologique celles-là.

Le procédé de soudage TIG est demeuré un procédé utilisé exclusivement pour
l'exécution de la première passe : homogénéité de composition chimique du
métal fondu, insensibilité à la fissuration par l'hydrogène, etc... Mais les
meilleurs résultats sont obtenus avec de l'argon comme gaz de protection, ce
qui oblige les soudeurs à revêtir un masque durant toute l'opération. Dans
des conditions similaires au soudage avec électrodes enrobées, la qualification de mode opératoire a été réalisée à 320 m ; toutefois, à cette profondeur, il a été nécessaire d'utiliser une source de courant particulière
(à haute fréquence) pour éviter les instabilités de l'arc. Le problème de
l'amorçage de l'arc sans contact n'a toujours pas été résolu et le risque
de pollution du bain de fusion par le tungstène de l'électrode existe toujours. Ce procédé fait toutefois l'objet d'études avancées dans certains
pays.

Le soudage MIG avec fil plein ou fil fourré, bien que peu utilisé en opération, fait l'objet de nombreuses études afin de dominer les conditions
de transfert du métal dans l'arc et optimiser ainsi l'énergie dissipée dans
l'arc et dans la pièce. Ce procédé, en cours d'évolution, est celui qui
offre le plus de possibilités d'avenir, aussi bien en soudage manuel qu'en
soudage automatique, car il peut être non polluant pour le soudeur et il
permet d'obtenir les caractéristiques mécaniques requises par ajustement
des différents paramètres. Des essais menés à l'Institut de Soudure pour le
compte de la COMEX ont permis de qualifier ce procédé jusqu'à des profondeurs de 600 m.

La Figure 4 illustre les résultats obtenus pour ces procédés durant des
qualifications de mode opératoire effectuées soit dans le caisson HYPLAB 3300,
soit en soudage réel.

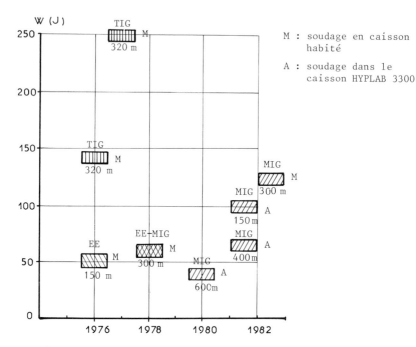

Fig. 4 Evolution des caractéristiques obtenues sur
assemblages depuis 1976.
(essais CHARPY V à -10°C)

Il ressort de cette analyse que le soudage en condition hyperbare est envi-
sageable jusqu'à des profondeurs de l'ordre de 600 m, et peut être même au-
delà. Ces perspectives sont toutefois indissociables d'une automatisation
intégrale des procédés.

VERS UNE UTILISATION DU SOUDAGE A PLUS GRANDE PROFONDEUR

Evoquer une telle possibilité amène obligatoirement à réfléchir sur les
problèmes susceptibles d'être rencontrés dans ces conditions et de tenter
de proposer un certain nombre d'orientations des recherches pour atteindre
cet objectif.

A l'instar des essais effectués aux différentes profondeurs au cours de
ces dernières années, il faut s'attendre, compte tenu de l'accroissement
de la conductibilité thermique du gaz avec la pression, à voir s'amplifier
les phénomènes liés à cette caractéristique ; ainsi, par exemple, le confinement de l'arc, sa tension et les difficultés de son amorçage seront
accrus.
Pour minimiser l'effet de ces facteurs sur l'élaboration de la zone fondue,
il sera nécessaire de modifier un certain nombre de paramètres définissant
le procédé. Pour cela, il faut considérer le procédé de soudage comme un
système constitué d'éléments interdépendants ; tout procédé comprend ainsi
un élément "source d'énergie" relié à un élément "élaborateur de la zone
fondue".

Bien que le procédé semble n'être composé que de deux éléments principaux distincts, il ne faut pas négliger, en soudage hyperbare, l'organe de liaison : les cables électriques, qui jouent un rôle important dans la défi-nition par exemple de la constante de temps qui régit les conditions de transfert du métal dans l'arc en soudage MIG.

Ainsi, toute modification de l'"élaborateur de la zone fondue" devra être examinée du point de vue de ses implications tant sur sa propre chaine d'éléments, que sur la production de la zone fondue ou sur la source d'é-nergie.

Prenons l'exemple d'un maillon de l'élément "élaborateur de zone fondue" en soudage MIG : le gaz de protection.

Son choix définit un certain nombre de paramètres qui peuvent avoir une influence sur l'ensemble de la chaine, à savoir, pour :

la conductibilité thermique : - la forme de l'arc (confinement),
- le débit de gaz (renforcement du confinement, vitesse de refroidissement de la zone fondue)
- la tenue de la buse à l'échauffement (mouvement de convection),
- la répartition de la température dans la partie libre du fil d'apport (coefficient de dépôt),
- etc...

le potentiel d'ionisation : - la tension de l'arc,
- l'énergie dissipée dans l'arc,
- les caractéristiques de la source d'énergie,
- etc...

l'activité chimique : - les interactions avec la zone fondue,
- les interactions avec le métal durant son transfert,
- etc...

Cette notion de chaine est mise en évidence lorsque l'on examine par exemple la tension de l'arc ; dans la liste ci-dessus, il est indiqué qu'elle est liée au potentiel d'ionisation mais une augmentation de la conductibilité thermique du gaz confine l'arc et provoque aussi son accroissement ; il en est de même de l'activité chimique qui peut faire apparaître, durant le transfert des gouttes dans l'arc, des espèces qui, avec un potentiel d'ionisation plus faible, peuvent tendre à diminuer la tension d'arc.

Ce concept de chaine n'est pas nouveau, il intervient de la même façon à pression atmosphérique, mais compte tenu du fait que la pression reste constante , il ne revêt pas autant d'importance. En soudage en con-dition hyperbare, par contre la conception d'une nouvelle méthode de sou-dage devra obligatoirement s'en inspirer.

VERS L'AUTOMATISATION DU SOUDAGE EN CONDITION HYPERBARE

Deux objectifs successifs sont définis pour l'automatisation du soudage en condition hyperbare : la mécanisation des procédés de soudage avec inter-vention d'un opérateur pour des profondeurs accessibles à la plongée humaine, une automatisation complète pour les profondeurs plus importantes.

135

Le deuxième objectif pose le problème de l'automatisation de l'intégralité des opérations avant, pendant et après soudage.

L'automatisation d'un procédé de soudage en condition hyperbare passera là encore par la définition d'une "chaine" propre à l'objectif à atteindre. Après avoir fait une liste des paramètres opérationnels pouvant être rencontrés, il sera nécessaire d'évaluer leur incidence vis-à-vis du procédé et d'établir inversement qu'elle est la faculté du procédé à dominer toute variation de ces paramètres. Ainsi, grâce à la définition du domaine de soudabilité opératoire, dont les limites sont obtenues à partir de critères d'acceptabilité de forme de cordon de soudure, de conditions de transfert du métal dans l'arc, d'énergie de soudage, etc..., une relation sera établie entre les paramètres opérationnels et les paramètres plus particuliers au procédé de soudage.

Des essais systématiques aboutiront à la production de graphes représentant ces différents domaines et relations (voir Fig. 5 et 6) dont l'intersection définira un domaine de plus grande fiabilité de la procédure.

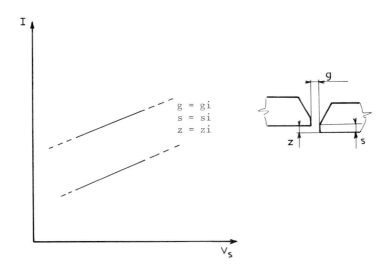

Fig. 5   Domaine de soudabilité opératoire.

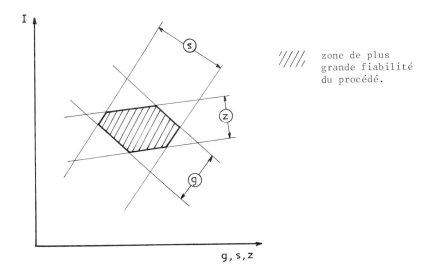

Fig. 6   Relation entre paramètres opérationnels et paramètres
de soudage.

Ces relations pourront ainsi être introduites dans un système informatisé
qui par reconnaissance de la forme du chanfrein et la définition des paramètres de position de la torche, pourra attribuer des paramètres électriques
de soudage au système en tout point de sa trajectoire sur le tube.

CONCLUSION

Ce rapide survol des techniques de soudage utilisées en condition hyperbare
indique que les limites de profondeur de chacun des procédés ne sont pas
encore atteintes et qu'un examen systématique des éléments ou paramètres
définissant chaque procédé permettra, par la connaissance de leurs évolu-
tions et de leur interdépendance, d'atteindre des profondeurs supérieures à
celles actuellement obtenues et de définir l'automate de demain pour le
soudage en condition hyperbare.

Underwater Dry TIG Welding Using Wire Brush Nozzle

M. Hamasaki and J. Sakakibara

Machinery & Metal Department, Government Industrial Research Institute, Shikoku,
Takamatsu City, Kagawa, Japan

ABSTRACT

In an underwater dry welding in a deep water, mixed gas of helium and oxygen
is used as the environmental atmospheric gas and to reduce the fume and the
spatter, the most suitable welding method is TIG welding. However the helium
gas as the shielding gas has such drawbacks as tungsten electrode is eroded
seriously and arc-starting is difficult in high pressure, while argon gas is
superior to helium gas in the said characteristic but argon is about twice
as narcostic as nitrogen, thus a special wire brush nozzle which consist of
dual nozzle was developed: Argon was flowed out from a inner nozzle and the
gas was inhaled by a vacuum pump from a circular gap between the inner nozzle
and the outer nozzle having wire brush which prevented to leak the argon to
the environmental atmosphere. With this nozzle as the result that inhalation
ratio is selected 2.0 under the current of 200A, satisfactory welds could be
obtained even in a water depth of 200m and welding could be done comparatively
in long time under the condition that the shielding gas and the fume did not
leak to the environmental atmosphere.

KEYWORDS

underwater welding; underwater dry welding; underwater TIG welding; high
pressure; wire brush nozzle; fume

INTRODUCTION

In an underwater dry welding, a work to be welded is covered with a submerged
chamber, of which bottom is open to a sea, displacing a water in the chamber
with an environmental atmospheric gas, after that a diver/welder works in it.
As the environmental gas for the submerged chamber in general, air is used.
But with increasing of a water depth, nitrogen and oxygen gases which form
the air cause narcotic and oxygen toxicity. The mixed gas of helium and oxygen is therefore used for a deep water and oxygen percentage is reduced as a
water depth increases.
As the underwater dry welding method in a deep water, TIG welding, MIG/MAG
welding and short-circuiting arc welding are adopted, but from a point of view

to reduce the spatter and the fume, the most suitable welding is TIG method, in which an inert gas as a shielding gas is required, because a tungsten ele rode is oxidized by the oxygen in the environmental gas. Helium, one of the inert gas, has such drawbacks as the tungsten electrode is eroded seriously and arc starting is difficult in high pressure, while argon gas is superior to helium gas in the said characteristics, but argon is about twice as narco tic as nitrogen. To solve these problems and obtain satisfactory welds, a special wire brush nozzle which consists of a dual structure having a diverg type wire brush was developed, and the experiment was carried out.

EXPERIMENTAL APPARATUS

Structure of Special Nozzle and Measuring Method of Leak Gas

Fig.1 shows a special wire brush nozzle which consists of dual nozzle and schematic drawing of measuring method of leaked argon gas to the environment atmosphere. Argon shielding gas flows out from a inner nozzle in the same manner as the conventional TIG welding and the argon gas is absorbed by a vacuum pump from a circular gap between the inner nozzle and the outer nozzl having a divergent type wire brush which prevents to escape the argon gas to the environmental atmosphere. The wire brush is made up of 0.15 mm diam.
stainless steel wires and stainless steel cross wire, which surrounds the space between the outer nozzle tip and the work. Even if the nozzle-work-distance changes, covering effect is always kept, since the tip of the wire brush which has spring action is kept in contact with the work. The distance h between outer nozzle and inner nozzle is changeable. The wire brush nozzle can be taken away to know covering effect. Fig.2 shows this special wire brush nozzle in arcing.

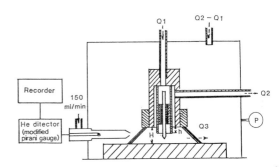

Fig.1. Schematic drawing of wire brush nozzle and measuring method of leaking gas

Shielding gas flow rate is Q_1, inhalation rate absorbed by pump is Q_2, leaking gas rate is Q_3. If the leaking gas through the wire brush nozzle is 0, Q_2 is equall to Q_1, but in actual welding, some content escape to the environmental atmosphere. Q_2 must be therefore selected larger to the Q_1 lest the shielding gas flow out to the environmental atmosphere and Q_2-Q_1 must be supplied from out of the chamber to keep the pressure constant, as the gas

Fig.2. Appearance of wire brush nozzle

absorbed by the pump is argon shielding gas and a part of the environmental gas. Q₃ was calculated by the leaked gas content measured with a modified Pirani meter, which can measure a little helium in argon, but not a little argon in helium. Thus the experiment was achieved in helium gas as a shielding gas and argon as an atmospheric gas, but the actual welding was achieved in argon gas as the shielding gas and helium as the atmospheric gas. The leaking gas through the wire brush nozzle is more plentiful in helium than argon, and Q₃ measured in this method will be therefore much more than that in actual welding, in other word, leaking gas in actual welding is less than Q₃ shown in here and this Q₃ is safty side value for the diver/welder.

Measuring Method of Fume

Fig.3 shows measuring method of the fume; the fume generated in arcing in a constant time was flowed away with the curtain water which was flowed out from cylindrical nozzle. This water was filtered through a paper and weight of the remained fume on the paper was measured.

Welding Apparatus Used in This Experiment

Two high pressure chambers were emploied in this experiment. One, 1.5m long, 0.35m diam., was used for researching of arc phenomena and electrode erosion, while another, 1.2m long, 1.0m diam., shown in Fig.4 was used for actual welding using wire brush nozzle. Flow rate of the shielding gas was regulated 15-100Nl/min corresponding to the environmental pressure. Welding was carried out moving the work straight with a fixed TIG torch and arc welder used was 300A DC with drooping characteristic and was selected straight polarity. A high frequency ionizer for non-touching arc start is twice in power compared with that for the conventional TIG welding. The TIG torch has a mechanism that electrode-work distance is set up about 0.5mm initially and pulled up to 2-3mm after the arc is generated.

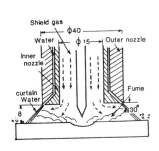

Fig.3. Measuring method of fume

Fig.4. High pressure chamber used in this experiment

EXPERIMENTAL RESULTS AND DISCUSSION

Arc Characteristic in High Pressure

The experiments of the arc characteristics in high pressure of both argon and helium gas were carried out. Fig.5 shows the relation between apparent arc length and arc voltage in both argon and helium gas under 100A current condition. The arc characteristics of 200A and 300A current conditions were also studied on parts of the experiments. From these results, it was cleared that arc voltage and arc gradient increased as current increased and those values were higher in helium gas than argon gas, and the greater the pressure and the current increased, the wider the bead and the deeper the penetration became. Fig.6 shows arc shapes. The arc in helium gas forms bell type at low pressure, but stiff cone type of which cone angle decreases with increasing of pressure.

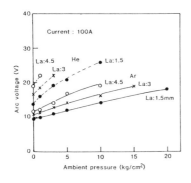

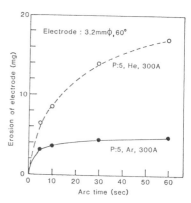

Fig.5. Relation among apparent arc length, pressure and arc voltage under Ar and He shield gases

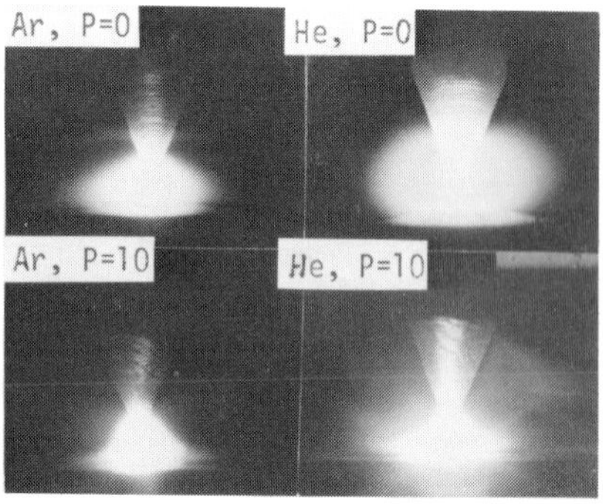

Fig.6. Arc shapes in Ar and He gases

Erosion of Tungsten Electrode

Fig.7 shows the relation among pressure, current and erosion of tungsten electrode in both argon and helium gases. Argon gas is superior to helium gas concerning to the erosion of the electrode, which increases with increasing of both pressure and current. Fig.8 shows the effect of arcing time on the erosion

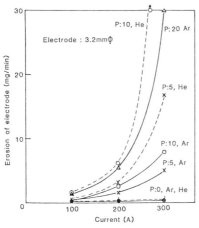

Fig.7. Effect of current and pressure on erosion of electrode

Fig.8. Effect of arcing time on erosion of electrode

ion of the electrode under a condition of 5 kg/cm^2 gauge pressure and 300A current. In helium gas, the erosion rate decreases with increasing of arcing time, but the erosion has proceeded even after 1 minute, while in argon gas erosion has not proceeded after about 20 seconds. Judging from these experiments, almost erosion of the electrode occures at arc generation directly. Fig.9 shows the relation between the erosion after 1 minute arcing and pressure in both argon and helium gases under 300A current with indirect water cooled electrode of 3.2mm diam. and direct water cooled electrode of 6mm diam.. The erosion of the electrode in helium gas was greater than that in argon gas and increased as the pressure became high. So as the effect of water cooling, 6mm diam. direct water cooled electrode is less eroded compared with 3.2mm diam. indirect water cooled electrode. Fig.10 shows the electrode tips of 3.2mm diam. and 6mm diam. which were eroded in 1 minute under a condition of 300A current in both argon and helium gas of 0 kg/cm^{2}G (1 atmospheric pressure) and 10 kg/cm^{2}G. The electrode is eroded 5mg under a condition of 300A current even the 6mm diam. direct water cooled electrode, and higher the pressure becomes, the greater the electrode is eroded. So the TIG welding of 300A current can not be used in actually, thus the current was reduced to 200A. The electrode erosion after 10 minutes arcing indicated 1.3mg and 2.6 mg in argon gas of 5 kg/cm^{2}G and 10 kg/cm^{2}G respectively. Judging from the results, we think it is possible to apply in actual welding under this condition, since the electrode grinding is not necessary so many times.

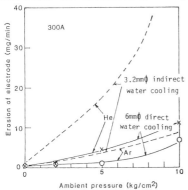

Fig.9. Comparison of erosion with 3.2mm and 6mm diam. electrodes

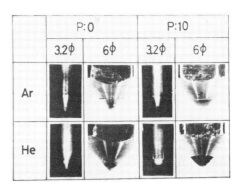

Fig.10. Appearances of eroded tungsten electrodes

Fume in High Pressure

Fig.11 shows the effects of current, kinds of gases and pressure on the fume generation. In helium gas, the fume shows a tendency to increase with increasing of the current and the pressure, while in argon gas, it is not much affected with the pressure and the current. Fig.12 shows the fume on the both sides of bead-on-plate welds in both argon and helium gas of 5 kg/cm^{2}G. The fume is much greater in helium than argon gas. W in this figure shows

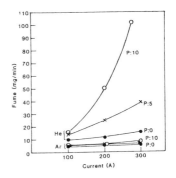

Fig.11. Effects of shield gas, pressure and current on fume

143

Fig.12. Appearance of fume adhered
on work

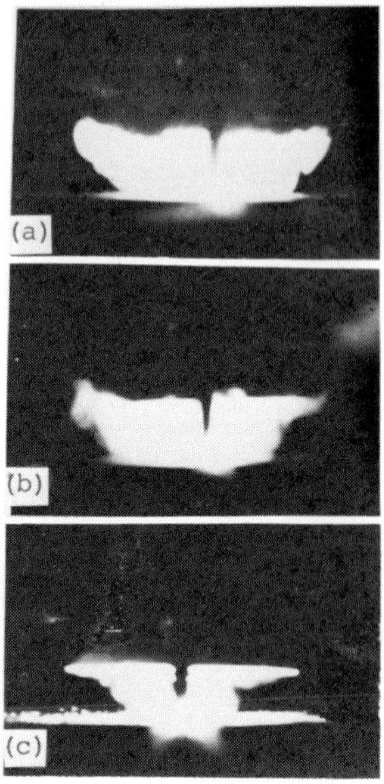

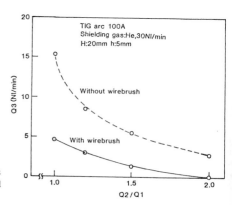

Fig.13. Diffusing fume observed
from side direction

width of adhered fume. In actual welding,
the much fume diffuses in the environmental atmosphere except for adhesion on the
work. Fig.13.shows the diffusing fume
which is generated under a 200A current in
helium gas of 5 kg/cm^{2}G. Under over these conditions, clear observation becomes difficult. Fig.13(a) is the same as conventional TIG welding; or without inhalation. (b) and (c) is inhalation ratio of 1.0 and 2.0 under that th
other conditions are the same as (a) respectively. In (b) and (c), it can be
observed that the fume is absorbed into a nozzle. The wire brush is taken
away from the nozzle lest it disturb to observe the diffusing fume. If the
special wire brush nozzle is used and welding is carried out in argon gas, th
fume can be more reduced.

Inhalation Ratio and Leaking Gas

Fig.14 shows the relation between
the inhalation ratio Q_2/Q_1 and the
leaking gas rate Q_3 under the arcing
proceed. The conditions are current
of 100A, shielding gas flow rate of
30Nl/min and h of 5mm. The effect
of the wire brush is remarkable.
When the wire brush nozzle is used,
leaking gas Q_3 is 0 under the Q_2/Q_1
of 2.0, but no arcing condition, Q_3
is 0 under the Q_2/Q_1 of 1.5. This is
because the shielding gas is expanded
by the arc to increase the volume.

Fig.14. Effect of inhalation ratio
on leaking gas in actual
arcing

144

Welding Result at Pressure of 20kg/cm²G

An actual welding experiment, in which argon gas was used for the shielding gas in the environmental atmosphere of helium, was carried out at a pressure of 5kg/cm²G, but almost experiments were carried out in air, because the use of helium gas as the environmental gas is costly. However no difference was recognized between air and helium so far as observing arc voltage, adhered fume and bead appearance. Fig.15 shows surface bead appearance, bottom bead appearance, X-ray inspection result and cross section of butt welds made with 3.2mm thick killed mild steel under argon pressure of 0 and 20 kg/cm²G. The bead appearance shows satisfactory even at the pressure of 20 kg/cm²G, still more at the pressure of 0kg/cm²G. As arc voltage becomes high and penetration depth increases with increasing of water depth, welding speed is 10cm/min and 16cm/min at the pressure of 0 kg/cm² and 20 kg/cm²G respectively under the same welding current.
Fig.16 shows the results of surface and root bending test of 180° without no crack. In tensile testing, no fracture occured even in welds made at the pressure of 20 kg/cm²G.

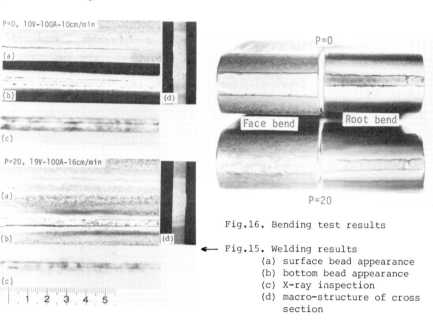

P=0, 10V-100A-10cm/min

(a)

(b)   (d)

(c)

P=20, 19V-100A-16cm/min

(a)

(b)   (d)

(c)

1 2 3 4 5

P=0

Face bend   Root bend

P=20

Fig.16. Bending test results

← Fig.15. Welding results
(a) surface bead appearance
(b) bottom bead appearance
(c) X-ray inspection
(d) macro-structure of cross section

CONCLUSION

As the underwater dry welding in a deep water, TIG welding was carried out in helium gas environmental atmosphere using argon shielding gas and a special wire brush nozzle which consists of dual nozzle having wire brush. The results obtained were as the follows.
(1) As the regard of the fume and the electrode erosion, argon gas was superior to helium gas as the shielding gas
(2) The fume and the electrode erosion increased with increasing of current and shielding gas pressure
(3) As the regard of the electrode erosion, 6mm diam. water cooled electrode was less eroded than 3.2mm diam. indirect water cooled electrode

(4) In the case that inhalation ratio of 2.0 was selected, TIG welding could be done lest argon gas flow out to the environmental atmosphere

(5) If the current was selected in 200A, satisfactory welds could be obtained and welding was carried out in long time even in water depth of about 200m

Preliminary Experiment on Improvement of Underwater Wet Plasma Welds Using Filler Metals

S. Fukushima, T. Fukushima and J. Kinugawa

Welding Div., National Research Institute for Metals, 3-12, 2-chome, Nakameguro, Meguro-ku, Tokyo 153, Japan

ABSTRACT

The possibility of improving the ductility and notch toughness of underwater wet plasma welds was investigated by controlling their chemical compositions using filler metals.
The welding was carried out in fresh water under the pressurized condition of 9.8bar(gauge).
Results of this experiment are summarized as follows.
(1) The ductility of a single pass double-Vee butt joint was expected to be improved by using either of the two kinds of filler metals, one of them had a basic composition of 0.013m% carbon, 0.081m% silicon, 1.05m% manganese and iron as the balance -- base filler metal -- and the other had 0.59m% titanium added to the above-mentioned base filler metal.
(2) Addition of niobium alone or molybdenum with boron to the base filler metal contributed less to improve the ductility of the weld metal.
(3) It is necessary to reconsider the chemical composition of filler metal for a filled pass in multi-pass welding, because the dilution of the deposited metal with the base metal in a filled pass differs from that in a root pass.

KEYWORDS

Underwater plasma welding; underwater wet welds; filler metals; titanium addition; niobium addition; molybdenum-boron addition; extra-low-carbon weld metal; weld metal strength; weld metal ductility; notch toughness.

INTRODUCTION

In underwater wet welding processes, the ductility and notch toughness of the weld metal tend to deteriorate because of rapid cooling, hydrogen absorption and oxidation of it induced by surrounding water.
The purpose of this study is to search for the possibility of improving the ductility and notch toughness of the weld metals by controlling their chemical composition using filler metals in underwater wet plasma welding. In the first step of this experiment, single pass bead-in-groove welding on small test plates was carried out in fresh water under the pressurized condition of 9.8

147

bar(gauge) which corresponded to a depth of 100m, and amount of diffusible hy drogen, changes of chemical compositions and tensile properties of the weld metals were measured.
Following this, double-Vee groove welds were made using the selected welding conditions and their tensile properties and notch toughness were confirmed.

BASE METAL

A carbon-manganese hot rolled steel plate for welded structure (JIS SM41A) wa annealed at $600^{O}C$ for 2 hours in order to expel hydrogen. This base plate ha a thickness of 19mm. Rectangular prisms, 135mm long, 25mm wide and 18mm hig were cut out from that plate and were machined the groove shown in Fig. 1(A) at their center. Test plates used for making the welded joints were machine to the groove shapes shown in Fig. 1(B) and (C) from the above-mentioned ste plate in the condition as received.

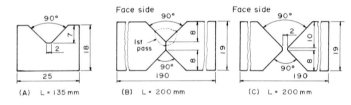

(A) L = 135 mm   (B) L = 200 mm   (C) L = 200 mm

Fig. 1. Shapes of groove. (L denotes the length of test plate.)

FILLER METALS

According to the result (Fukushima, 1977) of tension test for all weld metal produced by melt-run welding under various water pressures, a tensile strengt of about 510MPa and an elongation of about 19% (gauge length of 20mm) were o tained for the weld metal containing 0.04m% carbon, 0.09m% silicon and 0.5m% manganese; and, the test showed that the elongation of weld metals decreased with the increase in the above-mentioned elements while their tensile strengt increased.
In order to obtain the weld metal in the above-mentioned constituent, a base composition of Fe-Si-Mn filler metal was calculated assuming that the dilutio of filler metal is 50%, the recovery of carbon is 60%, the recovery of silic is about 70% and the recovery of manganese is 85%. Hereafter, the filler met al will be referred to as the "base filler metal". Titanium and niobium whic will prevent under-bead cracking in ordinary steels were added to the base filler metal. Furthermore, molybdenum with boron, which will decrease cold cracking susceptibility in steel, was added to the base filler metal. This trial is based on the prediction that the ductility of the weld metals would be improved if the weld discontinuities such as microfissures did not exist in them.
Electrolytic iron were melted with an addition of respective amounts of metal lic silicon, metallic manganese, metallic titanium, ferro-niobium (60m% in N metallic molybdenum, ferro-boron (22m% in B) and casted in a form of 3kg ing in a vacuum. Subsequently, these ingots were hot forged into rods, descaled by turning and finished into wire with a diameter of 1.6mm by cold drawing wi the annealing process being carried out in a vacuum in the middle of the pro Chemical compositions of the filler metals are shown in Table 2.

WELDING CONDITIONS

The welding was carried out in fresh water under the pressurized condition of 9.8bar(gauge) corresponding to the water depth of 100m, where the welding was done in the stable state. Weld beads were deposited by forehand welding and an aiming point of the filler metal was the point of 5mm just under the orifice of the nozzle. The welding conditions are shown in Table 1.

TABLE 1 Welding Conditions

Welding environment	In fresh water under a pressurized condition of 9.8bar(gauge)
Welding process	Plasma arc welding with plasma jet
Welding position	Flat
Flow rate of plasma gas	2.5 1/min (Argon) at 9.8bar(gauge)
Plasma jet current	100A
Plasma arc current	150A
Apparent arc length	20.5mm (including electrode setback of 10.5 mm)
Welding speed	5cm/min
Feeding rate of filler metal	3.2 grams per 1cm in weld length

TABLE 2 Chemical Compositions of Weld Metals, Filler Metals and a Base Metals

Filler metal		Element m%								
		C	Si	Mn	P	S	Ti	Nb	Mo	B
No.1	F.M.	0.013	0.081	1.05	0.002	0.005	--	--	--	--
	W.M.	0.031	0.073	0.61	0.012	0.013				
No.2	F.M.	0.01	0.094	1.20	0.002	0.005	0.20	--	--	--
	W.M.	0.06	0.030	0.41	0.009	0.012	0.008			
No.3	F.M.	0.01	0.11	1.16	0.002	0.005	0.37	--	--	--
	W.M.	0.07	0.046	0.51	0.010	0.012	0.008			
No.4	F.M.	0.02	0.11	1.14	0.002	0.005	0.59	--	--	--
	W.M.	0.08	0.070	0.57	0.010	0.013	0.016			
No.5	F.M.	0.04	0.11	1.13	0.002	0.005	0.80	--	--	--
	W.M.	0.07	0.075	0.60	0.010	0.012	0.026			
No.6	F.M.	0.03	0.11	1.11	0.002	0.005	1.24	--	--	--
	W.M.	0.08	0.085	0.57	0.010	0.013	0.060			
No.7	F.M.	0.024	0.11	1.10	0.002	0.005	--	0.19	--	--
	W.M.	0.051	0.057	0.50	0.012	0.013		0.03		
No.8	F.M.	0.011	0.12	1.10	0.003	0.005	--	0.44	--	--
	W.M.	0.043	0.076	0.53	0.012	0.013		0.09		
No.9	F.M.	0.018	0.14	1.14	0.004	0.005	--	0.84	--	--
	W.M.	0.046	0.098	0.61	0.011	0.013		0.25		
No.10	F.M.	0.013	0.14	1.16	0.005	0.005	--	1.28	--	--
	W.M.	0.047	0.102	0.56	0.011	0.012		0.39		
No.11	F.M.	0.014	0.10	1.14	0.003	0.005	--	--	1.06	0.008
	W.M.	0.046	0.053	0.60	0.012	0.013			0.64	0.002
B.M.		0.14	0.23	0.92	0.016	0.019	--	--	--	--

Note: Weld metals were produced by single pass bead-in-groove welding with each filler metal.

EXPERIMENT, EXPERIMENTAL RESULTS AND THEIR CONSIDERATIONS

Chemical Compositions of Weld Metals

The four single pass weld beads were deposited in each groove, shown in Fig.
1(A), through a length of 130mm using the welding conditions, shown in Table
1, and the same filler metal. Choosing three beads among them, samples for
chemical analysis were taken from the weld metals in the range of 40mm from
the end of the weld. Chemical compositions of weld metals, filler metals and
a base metal are shown in Table 2.

Mechanical Properties of All Weld Metals

All weld metal test specimens for tension having the dimensions shown in Fig.
2(A) were cut out from the residual parts of the welds which were submitted t
take samples for chemical analysis, and the thicknesses of these specimens we
finished in the range of 4mm to 6mm owing to the unequal crosssections of the
weld metals.
A tension test was carried out under a tension speed of 1mm/min. The cross-
head of the tension testing machine was stopped for a moment to measure the
gauge length at the point of the maximun tension load by the dividers, and ar
uniform elongation was calculated from the measured value obtained by the mar
ner described above. Prior to the tension test, cracks were checked by the d
penetrant test, and no crack was detected in any of the specimens.

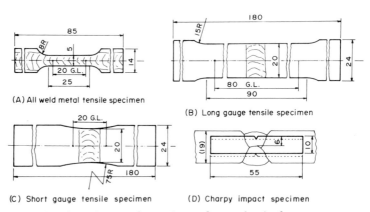

(A) All weld metal tensile specimen

(B) Long gauge tensile specimen

(C) Short gauge tensile specimen

(D) Charpy impact specimen

Fig. 2.   Shapes of specimens for mechanical tests.

Tensile strength, uniform elongation (gauge length of 20mm) and reduction in
area of each weld metal are shown in Table 3 as the mean value, the maximum
and the minimum values for three test specimens.
The tensile specimens for weld metals No.1 and No.4 were necked down to a con
siderable extent in their crosssections.
Based on the results, filler metals No.1, No.4, No.7 and No.11 were selected
from each group of the filler metals with consideration given to the mean val
and the amount of scatter in the measured values in the reduction in area of
each weld metals.

Micro-hardnesses of the weld metals (VHN 9.8N), which were measured in the direction of plate's thickness from the weld bead surface, are in the range of 150 to 270 for all of weld metals, because their carbon contents are extremely low.

TABLE 3 Tensile Properties of Weld Metals

Filler metal		Tensile strength (MPa) Max.- Min./Mean	Uniform elongation (%) Max. - Min. / Mean	reduction in area (%) Max. - Min. / Mean
No.1	Base	510 - 445 / 446	12.5 - 8.5 / 10.7	65.3 - 48.8 / 57.4
No.2	0.00Ti	539 - 499 / 511	11.0 - 10.0 / 10.4	47.6 - 23.3 / 35.0
No.3	0.37Ti	543 - 533 / 537	10.5 - 8.0 / 9.5	35.1 - 25.6 / 28.8
No.4	0.59Ti	577 - 534 / 560	11.0 - 9.0 / 9.7	44.2 - 41.7 / 42.5
No.5	0.80Ti	602 - 499 / 564	9.5 - 9.0 / 9.3	31.1 - 10.4 / 23.6
No.6	1.24Ti	630 - 547 / 589	7.5 - 5.5 / 6.7	42.3 - 10.6 / 28.4
No.7	0.19Nb	603 - 543 / 582	7.0 - 6.0 / 6.5	45.6 - 33.6 / 38.2
No.8	0.44Nb	647 - 490 / 557	6.5 - 4.0 / 5.2	55.6 - 22.3 / 34.0
No.9	0.84Nb	710 - 690 / 703	5.5 - 5.0 / 5.2	27.7 - 21.2 / 25.4
No.10	1.28Nb	757 - 672 / 708	5.5 - 5.5 / 5.5	24.1 - 11.4 / 16.1
No.11	1.06Mo -0.008B	604 - 527 / 566	8.5 - 8.0 / 8.3	50.0 - 6.4 / 30.2
Base metal		446 - 445 / 446	38.0 - 36.5 / 37.5	62.2 - 61.0 / 61.5

Shape of tensile specimen : Fig. 2(A)

Microstructures of Single Pass Weld Metals

A fair amount of massive ferrite can be seen in the microstructure of the No.1 weld metal when compared with the others, because its carbon content is extremely low, no more than 0.03m%. In the weld metals No.4, No.7 and No.11 in which a third element such as titanium, and so forth is added to the No.1 weld metal, ferrite precipitates at grain boundaries of the parent austenite and its size becomes large in the order of No.4, No.7 and No.11. In the No.4 weld metal containing titanium, a fine-grained structure was observed when compared with the others and the precipitates in its grains seem to be more than those of the others because of its slightly higher content of carbon, 0.07m%.
All of the nonmetallic inclusions are spheres and the one in No.11 weld metal was slightly larger than the others. Furthermore, the amount of nonmetallic inclusions in all weld metals were fairly large and this fact shows that the oxidation of the weld metal in water was fairly severe.

Amount of Diffusible Hydrogen in Single Pass Weld Metals

In order to know the approximate volume of diffusible hydrogen in the weld metals, measurements were carried out using the "increasing method" -- essentially the same as the method by Onishi (1967) -- as the first step.
The collecting conditions are as follows:
1) Collecting liquid: about 300ml of glycerin,
2) Collecting temperature: $40^{\circ}C$,
3) Time before the beginning of collection: 30min after the welding was finished, in order to keep the time for taking the specimens out from the pressure chamber,
4) Collecting duration: more than 200 hours, this is the time when an increment

in the volume of diffusible hydrogen is scarely recognized in its evolution
curves.

Amounts of diffusible hydrogen were in the range of 14ml to 21ml per 100 gram
of weld metal at $0^{\circ}C$ and 1.013bar(abs.).

Mechanical Properties of Welded Joints (by Single Pass Welding on Both Sides)

Using each of the filler metals No.1, No.4, No.7 and No.11 selected in the previous experiment, double-Vee butt joints were welded. Their groove shape was
shown in Fig. 1(B) and tack welds were carried out on both ends of the test
plates. The welding conditions were the same as shown in Table 1 except that
the plasma arc current and the feeding rate of the filler metal were set at
125A and 2.6 grams per 1 cm weld length in the root pass of the face side.
No backings and sturdy restraint jigs were employed. In welding the backside
the groove face was cleaned using only a wire brush and back chipping was not
employed. The test plates after welding were left in the room for five days
or more in order to remove diffusible hydrogen. Rectangular plates having a
width of 25mm were cut out from the above-mentioned plates in the direction n
mal to the weld line. These coupons were machined and ground their surfaces
until no concave part existed at the toe of the weld and were finished like t
tensile specimens shown in Fig. 2(B) and (C). The final thicknesses of the
test specimens were in the range of 12mm to 13mm. Specimens for bending test
had their laterals finished to a constant width of 24mm and their surfaces we
treated in the same way as described above.

All of long transverse tensile specimens shown in Fig. 2(B) were broken at ba
metal except for one. Furthermore, the specimen broken at the weld metal had
almost the same strength as that of the base metal.

Accordingly, it may be considered that the strength of the welded joint is
equal to that of the base metal.

The tensile strength and elongation (at initial gauge length of 20mm includin
the weld metal part of about 10mm) of four kinds of weld metals No.1, No.4, N
7 and No.11 were measured by using the short transverse tensile specimens
shown in Fig. 2(C) which were made for examining the weld metal strength at
the joint. The surface elongation was measured in a roller bending test with
an inner radius of 25mm, where the weld metal was put in the middle part of t
gauge length, 20mm. The fractured point (the start of unloading) was set at
the point where crack initiation was clearly recognized by observing visually
the surface on tension side during loading. Measurements of elongation were
carried out after unloading. In all welded joints produced with each filler
metal, there were specimens which were able to bend nearly 180 degrees, however, fine cracks except for the aforementioned one were found at the boundary
or intersection of the columnar grains in the observation of the surface on
tension side after testing.

In order to examine the notch toughness of the weld metal, Charpy impact tests
were carried out at $0^{\circ}C$ using the 2mm Vee notch specimens shown in Fig. 2(D)
which were cut out from the parts shown in the same figure as a general rule.
In each of weld metals, the absorbed energy of specimens having a notch on fac
side pass was higher than that of specimens having a notch on backside pass.
This fact is considered because the root of the notch on face side pass was
located at the HAZ of backside pass due to the lack of weld metal thickness in
it, when dimensions of the specimens were divided from the bottom of backside
weld metal. This fact suggests that the notch toughness of the weld metal ma
be improved by reheating it after the final pass.

Concerning each of the filler metals, above-mentioned results are summarized
as follows.

As regards the filler metal No.1, the welded joints produced with this filler
metal have a mean tensile strength of 503MPa and a mean elongation of 24.3%.

This value of elongation was the highest among the filler metals used. In roller bending test, these joints had a mean surface elongation of 19% in the face bend and 20.3% in the root bend, and they had a mean Charpy absorbed energy of 41J in the specimens having a notch on their face side and 20J on their backside. These values are second to those obtained by the filler metal No.4. Concerning the filler metal No.4, the welded joints had a mean tensile strength of 499MPa and this was the lowest value among the filler metals used. However, this was about 49MPa higher than that of 446MPa in the base metal. The mean elongation of these joints was 19.2% and was ranked the second place. In the roller bending test, they had a mean surface elongation of 21% in the face bend and 24.5% in the root bend, and these values were the highest among the filler metals used. In Charpy impact test, the mean absorbed energy of these joints was 48J in the specimens having a notch on their face side and 22J on their backside, and these values were also the highest.

Concerning filler metals No.7 and No.11, the welded joints produced with the respective filler metals had a mean tensile strength of 518MPa and 507MPa, respectively, and they were superior to the aforementioned values, however, their mean elongations were as low as 17% and 10.5%, respectively. In the roller bending test, their mean surface elongations were 16.5% and 13.5% in the face bend and 11.5% and 13.5% in the root bend, and these values did not reach those of the former two. The mean values of their Charpy absorbed energy were 28J and 25J in the specimens having a notch on their face side and 15J and 16J on their backside and these values were also lower.

From the above results, filler metals No.1 and No.4 were selected for multi-pass welding in the following experiment. In addition, it is conjectured that the ductility of the welded joint can not be improved easily as the elongation of the base metal was 59.2% in the short transverse tensile specimen and 36.0% in the roller bend specimen.

Mechanical Properties of Welded Joints (by Multi-pass Welding on Both Sides)

In the previous experiment, some of joints had incomplete root penetration, and there were some troubles to obtain full penetration stably. So, in this experiment, the groove shape shown in Fig. 1(C) was adopted and joints were made by two-layer-three-pass welding on both sides and copper backing was only used in the root pass. Filler metals No.1 and No.4 were used. The welding conditions used in this experiment were shown in the aforementioned Table 1 except that the plasma arc current and feeding rate of the filler metal at the second layer on the backside were set at 125A and 2.6 grams per 1cm weld length and the apparent arc length was decreased 2mm in the welding of the second layer on both sides. Slag was removed by wire brushing in each pass after draining off the water. The specimens for the mechanical tests were taken from the welded joints which were left in the room for seven days or more in order to remove diffusible hydrogen. The shapes of specimens were the same as in the previous section. The thickness of specimens was about 16mm owing to the removal of an amount of angular distorsion by machining.

Multi-pass welded joints produced with filler metals No.1 and No.4 had a mean tensile strength of 421MPa and 441MPa, respectively, and all of tensile fractures occurred at the weld metal. No the parts, which were clearly judged to be a lack of fusion, existed in the fractured surface in tension. Elongation of these joints was not so good as that of the joints obtained by single pass welding on both sides. The reason why the tensile strength of the multi-pass welded joints was lower, 59-78MPa in this case, compared with that of the above-mentioned joints was because the dilution of the deposited metal with the base metal decreased with an increase in the number of passes and this phenomenon resulted in a decrease in the carbon content in the weld metal. This can be understood from the fact that the hardness of the weld metal obtained by multi-

pass welding was lower, about 10 in microhardness(VHN 9.8N), compared with
that of the weld metal obtained by single pass welding on backside (final pas
Observing the crosssections of the weld metals microscopically, some of the
microcracks were detected on the second-pass weld metals on the backside, how
ever, they scarcely existed in the others.
Charpy absorbed energy of multi-pass weld metals was in the range of 99J to
24J and these values were scattered considerably due to a change in the location of the notches.
Considering dilution of the deposited metal with the base metal, it is conclu
ed that the above-mentioned results show the necessity of a further investiga
tion into the chemical composition of the filler metal used for filled passes
after the root pass.

CONCLUSIONS

The possibility of improving the mechanical properties of underwater wet plas
welds, especially their ductility and notch toughness, was investigated by co
trolling their chemical composition using trial filler metals. These filler
metals were extra-low-carbon Fe-Si-Mn filler metal containing 0.013m% carbon,
0.081m% silicon and 1.05m% manganese -- the base filler metal -- and the fill
metals in which titanium, noibium and molybdenum with boron were independentl
added to the base filler metal.
Main results obtained in this study are summarized as follows.
(1) Two kinds of filler metals, one of them was the base filler metal and the
    other was the filler metal which had 0.59m% titanium added to the base fil
    er metal, suggested the possibility of improving the ductility of a single
    pass double-Vee butt joint.
(2) Addition of niobium alone or molybdenum with boron to the weld metal con-
    tributed less to improve its ductility.
(3) It is necessary to reconsider the chemical composition of filler metal fo
    a filled pass in multi-pass welding.

REFERENCES

Fukushima, S., T. Fukushima, and J. Kinugawa (1977). Underwater wet plasma
    welding in pressurized water. Trans. National Research Institute for Meta
    19, 133.
Onishi, I., and co-workers (1967). Foundamental research of the behavior of
    hydrogen in steel (part 1). J. of the Japan Welding Society, 36, 1126.

The Effect of Pressure on the TIG Welding Process

O. Dijk and G. den Ouden

Department of Metallurgy, Delft University of Technology, Delft, The Netherlands

ABSTRACT

In this paper the results are reported of a study dealing with TIG welding under pressure (up to 21 bar). Information is given about the influence of pressure on the behaviour of the arc and the properties of the weld.

KEYWORDS

TIG welding; pressure dependence; hyperbaric welding; arc behaviour; weld shape.

INTRODUCTION

During the last years there has been a growing interest in underwater welding, especially in the offshore industry, where joining techniques under water become more and more important.
The technique most frequently employed up till now is wet welding, manual arc welding in a water environment using specially developed coated stick electrodes. The method has the advantage that it is rather simple and that no sophisticated equipment is required. One of the disadvantages of this method, however, is that the weld obtained has poor mechanical properties. This is mainly due to the hydrogen absorbed during the welding process, which together with the fast cooling rate and the presence of local stresses can lead to cold cracking. A different approach to underwater welding is hyperbaric welding. In this case welding is carried out in a gas filled chamber which encloses the workpiece to be welded. Since the chamber is in open contact with the surrounding water, the gas pressure must balance the pressure of the water.
It is evident that the pressure in the chamber affects the behaviour of the arc and the weld pool and hence, influences the properties of the weld itself. In this paper the results are reported of a study dealing with some of these effects in the case of TIG welding. The study was a continuation of Neptunus, a Dutch research project on underwater welding (see Kapteijn and Den Ouden, 1982).

EQUIPMENT AND EXPERIMENTAL PROCEDURE

The welding apparatus used in the experiment was composed of a specially
designed water-cooled TIG torch, a workpiece table and a guiding system. With
this guiding system torch and table could be moved with variable speed in
three directions perpendicular to each other.
Welding was carried out in the pressure vessel shown in Fig. 1 at pressures
between 1 bar and 21 bar. The welding operation could be controlled from outside the vessel, windows made it possible to observe the actual welding. A
DC unit in the constant current mode was used as power source. During welding
current and voltage were simultaneously recorded.
The gases used were technically pure (99,99%) argon and helium. During welding the gas pressure could be maintained at a nearly constant level by means
of adjusting valves in the vessel. Welding was carried out under static conditions, i.e. without a forced gas flow along the arc. Before filling with
gas, the vessel was evacuated.
Unless otherwise stated bead-on-plate weld runs were made on Fe 510 steel
plates of dimensions 330 x 150 x 25 mm.

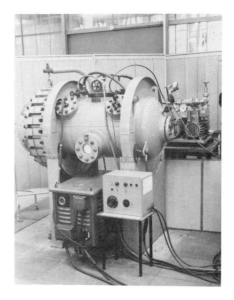

Fig. 1.   Test facility for hyperbaric
welding. Inner diameter 0.6 m.

RESULTS AND DISCUSSION

Arc Voltage versus Arc Length

Measurements of the arc voltage V as a function of the arc length l were
carried out for different arc currents and different gas pressures. The
results show that V(l) increases markedly with pressure.

As an example, Fig. 2 gives a plot of V versus l for argon for an arc current of 150 A and three values of the gas pressure (1, 11 and 21 bar). All obtained V(l) curves have a shape similar to the curves shown in Fig. 2: they are concave at small arc length (below about 5 mm) and become linear when the arc length is increased. The deviation from linearity below 5 mm is due to the fact that at arc lengths below that value changes in arc cross-section, and hence in current density, become significant.
The V(l) curves of helium were found to lie well above the corresponding curves of argon, the slope of the linear part being much higher.

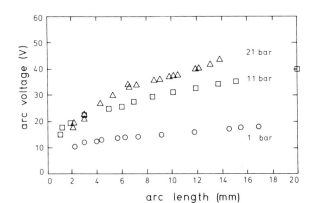

Fig. 2. Arc voltage as a function of arc length for
three pressures (argon, 150 A).

The Electric Field Strength

Values of the electric field strength E were obtained from the slope of the various V(l) curves. It appears that E depends on the gas pressure p. Theoritical work (Weizel and Rompe, 1949) indicates that E and p are related by an expression of the form $E \sim p^{\alpha}$, α being a constant depending on the type of gas. It was found that the present results can indeed be described by a relation of this kind. By plotting E as a function of p on a double-logarithmic scale, straight lines were obtained from the slope of which α could be calculated. Values of about 0.45 and 0.30 were obtained for argon and helium respectively. These values seem to be rather insensitive to arc length and do not differ much from those reported by other investigators (Allum, 1982; Bauder, 1976). It must be remembered that the present results were obtained under static conditions, that is without a forced gas flow around the arc. In order to check the influence of such a gas flow some additional tests were carried out. From the results of these tests it may be concluded that a forced gas flow around the arc raises the electric field strength but does not influence the value of α.

The Voltage Drop at Zero Arc Length

An important feature of the arc welding process is the energy transferred per unit time from the arc to the electrode and the workpiece. This energy is mainly determined by the voltage drop in the anode and cathode regions. Whereas these voltages are difficult to measure separately, their sum can easily be obtained by extrapolation of the V(l) curves to zero arc length. This has been done for all V(l) curves measured. Unfortunately, the obtained results are found to be rather inaccurate due to the concave shape of the curves at small arc length. In order to obtain more reliable values, an experiment was carried out in which the electrode was moved with constant speed (0.15 mm/sec) towards a tungsten workpiece till contact was made, while simultaneously the voltage was recorded. From the recording the voltage drop at zero arc length could easily be obtained. See, for instance, Fig. 3. Using this method the voltage drop at zero arc length was measured as a function of pressure. As is shown in Fig. 4 the voltage drop slightly decreases with increasing pressure.

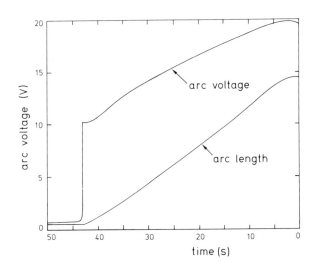

Fig. 3.   Recording of arc voltage during movement of electrode towards workpiece (argon, 6 bar).

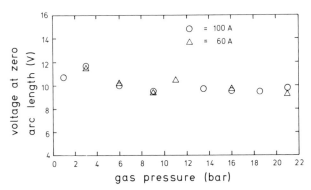

Fig. 4.   Arc voltage at zero arc length as measured for several pressures (argon).

The Current-Voltage Characteristic

Current-voltage characteristics were measured for a number of arc lengths and gas pressures using a tungsten plate as workpiece. Some of the results for argon are given in Fig. 5. The measured characteristics are all similar in shape: they show an initial decrease, reach a minimum and increase again for larger current. Increase of pressure raises the characteristics towards higher voltage while the minimum is shifted to lower current. For an argon arc of 5.5 mm, for instance, the minimum voltage at 1 bar is 13,5 V occuring for 170 A, while at 21 bar these values are 26 V and 40 A respectively.

From Fig. 6 it can be seen that the behaviour of helium is similar to that of argon, be it that the characteristics of helium are shifted towards higher values of voltage and current.

Considering Figs. 5 and 6 it must be remembered that the stability of the arc depends, apart from other factors, on the relative positions of the arc characteristic and the characteristic of the power source. More specifically, the arc becomes unstable when both characteristics partly coincide or lie close together over a certain length. This situation can easily occur at higher pressure, especially in the case of helium.

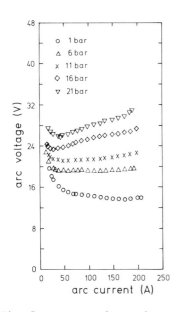

Fig. 5 . Current-voltage characteristics of argon for different pressures (arc length 5.5 mm).

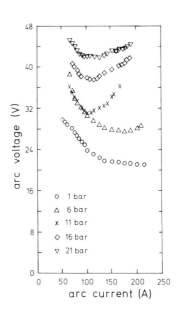

Fig. 6 . Current-voltage characteristics of helium for different pressures (arc length 5.0 mm).

Shape and Properties of the Weld

In order to investigate the effect of pressure on the shape of the weld, cross-sections were made of bead-on-plate welding runs, produced under different pressure conditions. These cross-sections were polished and etched, thus making it possible to measure width, penetration depth and nugget area. From these measurements it can be concluded that penetration depth and nugget

area increase with pressure, whereas the width becomes constant after a small initial increase.
From the nugget area the arc melting efficiency (defined as the energy used for actual melting, divided by the total arc energy) was calculated for various situations. It appears that the arc melting efficiency is more or less independent of pressure, whereas it decreases with increasing arc length, especially for low pressures.
In order to get an idea of the mechanical properties which may be expected of TIG welds produced under pressure, square-groove butt joints were made between 3 mm thick Fe 510 steel plates in argon under various pressure conditions. Strength and Charpy V-notch toughness were measured, using specimens which were taken from the welds as indicated in Fig. 7. The results of these measurements are summarized in Table 1. It appears that over the whole pressure range the weld metal is somewhat stronger and more brittle than the base metal (higher tensile strength , lower elongation, higher brittle-ductile transition temperature). More important, however, is the conclusion that the mechanical properties are not (or only slightly) affected by pressure. This conclusion is in line with recently published work carried out elsewhere (Knagenhjelm, 1982).

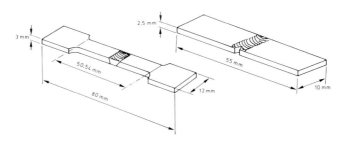

Fig. 7. Specimens for measuring strength and Charpy V-notch toughness.

TABLE 1   Mechanical Properties of TIG Welds produced in Argon under different Pressure Conditions.

pressure (bar)	tensile strength (N/mm^2)	elongation (%)	Charpy V-notch value (joule)*				
			-70^0C	-50^0C	-30^0C	-10^0C	20^0C
1	584	15	23.9	24.2	39.7	51.6	61.0
6	586	15	15.5	26.5	29.4	50.0	52.9
11	587	15	20.0	23.9	38.7	45.5	50.7
16	584	15	14.8	24.2	25.2	45.2	48.1
21	575	13	19.0	25.2	38.7	47.7	48.0
base material	577	20	77.1	72.6	73.9	76.1	74.5

*Obtained from measured values by normalizing to 10 x 10 mm specimen size.

CONCLUSIONS

On the basis of the results presented in this paper the following conclusions can be drawn.

1. The pressure dependence of the electric field strength in the TIG arc can be described by the relation $E \sim p^{\alpha}$, α having a value of 0.45 for argon and 0.30 for helium.
2. A forced gas flow around the arc raises the electric field strength, but does not influence the value of α.
3. The voltage drop at zero arc length (the sum of anode fall and cathode fall) slightly decreases with increasing pressure.
4. Arc instability occurring at higher pressure, especially in the case of helium, may partly be due to shape and location of the V-I arc characteristic.
5. Penetration depth and nugget area increase with pressure, weld width remaining constant after a small initial increase.
6. With TIG welding under pressure welds can be obtained having good mechanical properties.

REFERENCES

Allum, C.J. (1982). TIG's underwater role: present and future. Welding and Met. Fabr., 124-132.
Bauder, U.H. (1976). Properties of the high pressure and plasma. Appl. Phys., 9, 105-115.
Kapteijn, J., and G. den Ouden (1982). Neptunus, a Dutch research project on underwater welding. Meerestechnik, 13, 139-144.
Knagenhjelm, H.O. (1982). Development of hyperbaric TIG welding to 500 m water-depth. Proceedings Int. Conf. Underwater Operations and Techniques, Paris, pp. 173-199.
Weizel, W., and R. Rompe (1949). Theorie elektrischer Lichtbogen und Funken. Barth Verlag, Leipzig, p. 27.

Underwater Welding with Fluxed-cored Wire

S. H. van den Brink and M. P. Sipkes

Metaalinstituut TNO, P.O. Box 541, 7300 AM Apeldoorn, The Netherlands

ABSTRACT

Wet underwater welding with flux-cored wire was investigated. Basic wires and special developed stainless steel wires were used in welding V-joints in Fe 510. Slagforming compounds appeared to be of prime importance for good results.

KEYWORDS

Underwater welding, flux-cored wire, stainless steel cored wire.

INTRODUCTION

There is a widely held belief that GMA welding offers a number of specific advantages over welding with coated electrodes, such advantages being a continuous supply of filler wire, easier shielding of the arc area, in some cases better vision as a result of low fume production, a higher weight of deposited metal and good opportunities for process automation. However, from a metallurgical standpoint, it is by no means favourable that the weld should cool down extremely fast as a result of the presence of water. The consequence of such rapid cooling is that there is more likelihood of hardening and cracking occurring. A further factor to be borne in mind is that, if welding is being carried out in an aquatic environment, it is more or less inevitable that a large amount of hydrogen will be absorbed into the weld pool. This only further increases the possibility of cracking occurring in or next to the weld.

Apart from the features which enable welding to be carried out in dry conditions, not a lot is known with great certainty of the possibilities offered by the process. There does seem to be more or less general agreement that welding with a solid wire does not give the weld those qualities which are regarded as a minimum requirement for corrective welding. The welds are porous and brittle and contain large quantities of hydrogen. In the light of the above, it was decided within the framework of the Neptunus project, a Dutch project on underwater welding, to focus attention specifically on welding with a cored filler wire.

This decision was based on experiments on which publications have already appeared, in many cases drawn up by Russian researchers. According to their results, hardness is reasonable and hydrogen concentration is relatively low, two factors which reduce the likelihood of cracking.

The research programma started out by using cored wires as are also used when welding in dry conditions. At a later stage, however, use was made of experimental wires.

METHODS OF RESEARCH

During the laboratory experiments with GMA welding, the following points were given consideration:
- the power source
- polarity
- commercial basic-type cored wires
- the choice of shielding gas
- stainless steel cored wires
- welding qualification

Power State and Positioning

In order to ensure that there was sufficient range in the open circuit voltage, and that welding could be carried out at sufficiently high voltage, where necessary two power sources were employed, connected in series. The units used were a Philips PZ 2330 and a Cloos D.C. unit (both with horizontal characteristic). The welding torch was suspended from a rail and could be moved horizontally through a range of speeds. In all cases, welding was in the downhand position. The workpiece was positioned on a rack and placed in a container filled with fresh water. Earth terminals were welded on to the rack at a number of points in order to prevent any instability from occurring in the arc as a result of faulty earthing.

Materials and Workpieces

Those trial welds which were necessary in order to arrive at the correct settings were carried out by performing fillet welds on Fe 400 steel at a thickness of 10 mm. The welds were in Vee grooves (Fig. 1), and there were also a number of Vee grooves welded using Fe 510.

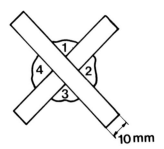

Fig. 1. Geometry of testspecimen

Polarity

Most of the experiments were carried out with the electrode positive. Changing the polarity did not produce any improvements in results as far as penetration and the appearance of the weld were concerned.

Types of Wires

A list of the types of wires used is given in Table 1.

TABLE 1   Composition of welding wires

Type		Ø mm	Analysis						Weight % slagform- ers	Remarks
			C	Mn	Si	Cr	Ni	Mo		
a	E-70-T-1 (basic)	1.6	0.08	1.4	0.7					
b	E-70-T-5 (basic)	1.6	0.05	1.7	0.5					
c	E-70-T-1 (basic)	1.6	0.08	1.0	0.7					
d	AISI 308 LSi hydrogen nr. 4316 (solid wire)	1.2			1.2	19	10			Commercial wires
e	AWS E 308-LT-1-2 (cored)	1.6	0.04	1.2	1.0	19	10	–		
f**	G-745 (cored)	1.6	0.05	2.5	0.5	15	8.5	1.5	8	** experimental wires Information: Dutch Welding Institute (NIL)
g**	G-757 (cored)	1.6	0.05	2.5	0.5	15	8.5	1.5	8	
h**	G-761 (cored)	1.6	0.05	2.5	0.5	15	8.5	1.5	9	
i**	G-810 (E309Mo) (cored)	2.4	0.05	1.2	0.6	21.7	12.0	2.3		
l**	G-838 (cored)								9	

BASIC-TYPE CORED WIRES

Setting the Parameters

In order to determine the working range with regard to the parameters, test welds were carried out using E 70-T-1 cored wire with Ar + 10% O_2. Figure 2 shows the results obtained, with the pertinent weld grades. Lines with a constant open circuit voltage are also indicated.

A choice had to be made regarding the shielding gas. Although there are advantages in using argon with a high percentage of oxygen (resulting in a low hydrogen content in the weld), it does present the concomitant danger of explosions occurring when the mixture is used at great depth. Since a cored wire including slagforming constituents in its core was used as the filler wire, a point which is of some importance is whether or not the slag formed can offer the weld pool sufficient protection under water, even if no shielding gas is used.

The cored wires a, b and c, as indicated in Table 1, are suitable for use with the normal grades of steel. As far as weldability is concerned, there are practically no differences between the three types. For the purposes of the research, fillet welds were made and wire a (basic) was used as the filler wire. Welding was done with argon, with argon + 10% oxygen and without any shielding gas.

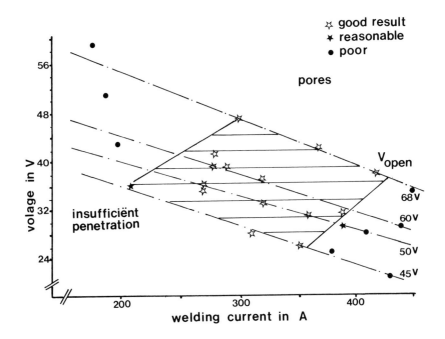

Fig. 2. Operational envelope for underwater welding with flux-cored wire

In order to study the geometry of the weld runs, weld beads were deposited under water using wire on a 10 mm thick, flat sheet. Using a constant welding speed (18 cm/min) and with the welding machine set in all cases at 260-270A, 43-44V, the following weld beads were deposited:
a) without any shielding gas
b) using argon as the shielding gas
c) using argon + 10% O_2 as the shielding gas

Figure 3 includes a number of cross-sections of each of the weld beads referred to above. It came to light in these experiments that is is not the type and quantity of shielding gas that is the determinant for the quality of the weld, but a continuous layer of slag. The rate of travel must be adapted to the quantity of slagforming constituents present in the core of the wire. Penetration is greatest when no shielding gas is used. It would appear that the use of argon results in the weld showing a somewhat higher degree of hardness.

Mechanical Properties

In order to be able to compare the mechanical properties of various types of basic cored wire, a 60° Vee groove was welded without a root gap in 25 mm thick Fe 510 sheeting. Argon + 10% oxygen was used as the shielding gas. Charpy impact tests and side bend tests were carried out on samples from the welding sheeting in order to determine the mechanical properties. These properties are shown in Table 2.

Shielding gas / Hardness
Ar + 10% O_2
Hv10: 205

Ar
Hv10: 200

Shielding gas/
hardness
none
Hvl0: 200

Fig. 3. Underwater weldruns for various shielding gases

TABLE 2 Results of Mechanical Tests: V-joint 60°, 25 mm Plate Thickness

Type of wire	Exp. nr.	Charpy-V temp. 0°C			Side bending test			Hardness HV 10
		Nr.	Joules	Mean	Nr.	D/d	Angle	
E 70-T-5 type a	35	35A	39					207-196 not annealed
		B	63					
		C	29	48.1	35H	4	10-20°^{*)}	
V-joint filled with 5 runs (no shielding gas)		D	80		35I	4	10-20°^{*)}	196-190 annealed
		E	54					
		F	24					
E 70-T-5	36	36A	22^{**)}					weld not annealed
		B	28^{**)}					190-182-185-186-185
		C	45	50.5^{***)}	36H	4	10-20°^{*)}	
		D	28^{*)}		36I	4	10-20°^{*)}	163-159-198-195-185
V-joint filled with 11 runs (no shielding gas)		E	47					
		F	49					
		G	52					weld annealed

*)   Bending properties are poor, angle is difficult to discern
**)   Fracture surface shows defects, pores and slag inclusions
***)   Lower values are not included in mean

168

RESEARCH INTO STAINLESS STEEL CORED WIRES

From a metallurgical standpoint, the use of stainless steel cored wire has
some appeal, in that the solubility of hydrogen is greater in the weld metal.
The hydrogen will be less prone to diffusing away, as a result of which
there will be less likelihood of cracking occurring in brittle areas.

Initially, commercial wires were used (referred to in Table 1 as types d and
e) with a nominal composition of 19% Cr and 10% Ni, followed by a number of
cored wires made specially for the research programme containing large
amounts of slagforming constituents in the core.

Both the solid and the cored wire d) and e) produced bad results (welding
with the solid wire being in the first instance with pure argon and then
with argon + 10% oxygen). It is interesting to note that the water in the
container remained exceptionally clean during welding. It was ascertained
that ther commercial cored stainless steel wires with a diameter of 1.6 mm
contained hardly any slagforming constituents in the flux of the core. The
reason for this is that, with a diameter of 1.6 mm, it is only possible to
fill 36% of the total volume. This proportion is then used in the main for
alloying elements and to a minimal degree for slagforming constituents.
Where there is a diameter of 2.4 mm, the percentage rises to 45% of the
volume, with the result that a greater proportion of slagforming constituents can be added to these types of wires.

In the case of these wires, 8 to 9% by weight of the flux was used for slagforming constituents.

The best result was obtained with type 1. The core of this type of welding
wire is more or less the same as that in types f, g and h. The only difference is that extra alloying elements were added as a means of keeping the
welding metal austenitic following fusion with the base metal.

Weldability is good if the heat input is no less than 19 and no more than
25 kJ/cm per weld. There is, however, the qualification that the only possibilities for welding are fillet welds and root passes in Vee grooves. The
hard, solid slag formed apparently only excercises its protective function
in respect of the weld metal if the slag is held firmly between the sides
of the seam. Welding the cap pass of a Vee groove and welding on smooth
sheeting do not proceed successfully, since the slag formed offers no protection.

The hardness of the weld metal is on average 230 (expressed in HV 10). It
was possible to bend the side bars through an angle of about 90° before the
first signs of cracking were observed.

CONCLUSIONS

1. The commercial cored basic wires produce reasonable welding results. In
   view of their low hydrogen contents, basic wires are to be recommended
   in preference to rutile wires.
2. Bending properties show considerable improvement for basic wires if there
   is a reduction in the quantity of hydrogen as a result of annealing.
3. Whether or not gases can provide sufficient protection is questionable.
   A better solution is to allow slagforming constituents in the wire to
   take on the protective role. It should be added that, in this case, welding should be done with a stable arc.

4. Stainless steel wires can also be used to good effect in underwater welding. A better level of protection is attained if extra slagforming constituents are added to the core of the wire.
5. The composition of the stainless wire has to be such, that the weld metal is austenitic after fusion and contains as much ferrite as possible.
6. The unalloyed types of cored wire can be used over a rather wide range of currents, voltage and welding speed. The stainless steel wires are suitable only for a very limited range.

ACKNOWLEDGEMENT

The authors would like to express their gratitude to the Dutch Welding Institute (NIL) which initiated and supervised the Neptunus underwater welding project. Thanks also to Mr. F. Schat who carried out the experimental work.

Effect of Pressure on Arcs

C. J. Allum

School of Industrial Science, Cranfield Institute of Technology, Cranfield, Bedford MK43 0AL, UK

ABSTRACT

The effect of pressure (1-1000 bars) on arcs is reviewed. Special attention is given to arcs used in hyperbaric welding. Emphasis has been placed on arc stability, electrical characteristics and power dissipation in high pressure columns.

KEYWORDS

Arcs, welding, subsea welding, high pressure arcs, electric fields, arc stability.

INTRODUCTION

Arcs at pressures above one atmosphere have been the subject of investigation for a wide range of industries encompassing the exploration of both outer space and subsea environments. Only relatively recently has significant interest been shown in deep water applications and these form the central interest of this paper.

Gaseous conductors are highly susceptible to the influence of pressure. Generally it is found that the major influences are on arc stability and power dissipation in the column. It is a pre-requisite for welding that an arc is stable and with suitable heating characteristics and so the influence of pressure on these properties is of great interest for underwater welding applications. In this paper the behaviour of a variety of arcing arrangements at pressures of up to 1000 bar and currents as high as 50,000A is reviewed hence allowing the influence of pressure on welding arcs to be viewed in a more general perspective.

ARCS AT HIGH PRESSURE

Creation of High Pressure Plasma

High pressure plasma may be created in a variety of ways the most obvious of which is to increase the pressure of the region in which an arc burns. This may be associated with a more or less uniform pressure enclosure, as in hyperbaric welding, or the environment may exhibit a pressure gradient. An example of the latter occurs in a gas blast circuit breaker where a gas stream is exhausted from a high pressure tank into the atmosphere so extingu shing an arc. In very low pressure environments high local plasma pressure may be created by stagnation flow effects, e.g. the ablation shield of a space craft re-entering the earth's atmosphere (Bauder, 1968). High vapour pressures may possibly be created in vacuum arcs by electrode material. However, considerable uncertainties in the magnitude of this effect exist (Guile, 1971, points to estimates of between 2000 atm and 10^{-3} torr for mercury arcs). Arcs operating under normal ambient conditions may experience significant axial and radial pressure gradients due to self magnetic field effects. These are greatest in the most constricted arc zone and may generate high plasma streaming velocities, and influence metal detachment and weld pool motion. For typical TIG welding arcs the pressure near the cathode has been measured using a manometer (Schoeck, 1967) as typically 10^{-2} bar. Theoretical predictions indicate that the maximum excess pressure (Δp) due to this effect should increase with current as

$$\Delta p \simeq \frac{\mu_0 I J}{4\pi}$$

where I is arc current and J the current density at the cathode, μ_0 is the permeability of free space. Calculations by Jones and Co-workers (1982) based on photographic observations of J indicate that $\Delta p \sim 10$ atmospheres for currents of order 40,000A, as found in circuit breakers. However, measurements of Δp at these currents are markedly less, but still significant, arguably (Jones and Co-workers) because of a pressure reduction caused by plasma swirl at very high currents.

Experiments with Arcs at High Pressures

Applications employing high pressure plasmas are wide ranging and have covered re-entry simulation, arc heaters for wind tunnels, arc light sources nuclear fusion, and dry underwater welding and cutting, circuit breakers and rocket engines for space flights have also been mentioned (Petrenko and Co-workers, 1969). Interest has also been shown in magnetically driven arcs at high pressures and in high pressure welding arcs for applications other than underwater. Indeed the first reported investigation of welding arcs in dry high pressure environments (Dynackenko and Olshanskii, 1968) was aimed at manipulating fusion characteristics and not for underwater applications. Accompanying these investigations, have been analytical investigations into the transport properties of high pressure plasma, mainly argon (Drellishak, Knopp, Bauder, 1968; Devoto, 1973; Emmons, 1963, Kannappan and Bose, 1977).

The range of arcing parameters considered include currents of 1-65,000A (Petrenko and Co-workers), steady and pulsed modes, arc lengths of 1 to 100mm and pressures of 1 to 1000 atmospheres (Borovik and Co-worker, 1967) in gases such as argon, helium and hydrogen. Long arcs at high pressures generally require stabilisation. In some situations an arc is stabilised by self induced gas flows (e.g. free arcs such as welding arcs). In other situation the arc must be stabilised by means of a highly conducting wall (wall stabili sed), and applied gas flow or magnetic field. Axial gas flows may stabilise arc behaviour by means of a thermal pinch effect when the arc burns in an orifice. For tangential flows or flows with swirl (vortex stabilisation) a positive radial pressure gradient is established tending to locate an arc on the axis of the flow. Axial magnetic fields can have a similar effect by creating plasma swirl within an arc. The physical mechanism of wall stabilisation is associated with the inability of magnetic flux to penetrate

a perfectly conducting wall. Magnetic pressure then forces an arc away
from a wall and towards the centre of the discharge tube and the angular
velocity of the swirling gas flow (Marlotte, Cann, Harder, 1968). In general
a large number of factors will combine to determine the stability of a given
system. Some general observations concerning stability will now be given.

Russian workers (see below) have performed extensive experiments with long
arcs (80mm) at pressures of up to 1000 atmospheres with a view to producing
high arc temperatures at relatively low power inputs. Arc stabilisation
was generally achieved by vortex effect either by rotating a screen within
the pressure chamber (the screen axis of rotation being common with the arc
axis) or by an impeller mounted at one end of the discharge chamber. It is
found (Borovik, Mitin, Knyasev, 1962) that with hydrogen arcs at 2 bars the
maximum arc length could be increased from 7-8mm with no rotation, to 30mm
with rotation (steady current of 150A, 120V arc in a 85mm diameter chamber
with rotational speeds of 2500 rpm). However, hydrogen proved insufficiently
stable for further investigation. With helium arcs, 40mm arc lengths could
be drawn at pressures of up to 30 atmospheres without rotation. This could
be increased to at least 80mm with rotation. Subsequent work by these
investigators in argon (Mitin, Knyasev and Petrenko, 1964) found that rotational speeds greater than 2000-2500 rpm were required (4000-6000 rpm was
used) to fix the cathode spot and stabilise arc behaviour over the parameter
range 10-150A, 0-80mm, 3-100 atmospheres. The requirement of having to
increase rotational speed with pressure has also previously been noted by
Fortzik (see Finkelnburg and Maecker, 1962). Ultra high pressure (1000 Atm)
experiments were performed by Borovik and Co-workers (1967) in argon using
pulsed currents (10^{-3}s pulses) with peaks of 50,000A.

The stability of high pressure arc heaters has been examined under contracts
for the U.S. Air Force (e.g. Marlotte, Cann, Harden). These workers note
that while the problems of local wall heating may be overcome with gas
vortex flows, other problems may arise, i.e. re-circulation may occur leading to breakdown into secondary flows and local burning of critical arc
heater components. Axial flows and magnetic fields have also been investigated by the above workers. However, under some conditions magnetic
fields may generate cork-screw shaped arcs. For axial flows at high pressures the mass flow rate is often high enough for the Reynolds number to be
high and the flow turbulent. This may then disturb the arc (Frind, 1964,
1966). The phenomena of intermittent turbulence may also occur. This is
associated with a very high local Reynolds number near the gas injection
point and a subsequent axial decrease to below the critical value (due to
down stream gas heating and the related increase in the coefficient of
viscosity).

Considerable data exists concerning the influence of pressure on electrical
characteristics. This data is often expressed in terms of the pressure
dependence of mean electric field strength (E) in a column and represented by

$$E \propto p^n$$

where n is a positive number (generally such that $1 > n > 0$) depending on
arcing conditions, parameters and pressure. Guile (1972) has reviewed the
literature on E and so only a general account will be given here. The above
representation of the influence is concerned only with the arc column.
However the literature suggest (Guilé, 1972) that the sum of the fall voltages
is largely unaffected by pressure and hence any change in voltage is associated with the column. The value of n mirrors energy exchange processes
occurring within the plasma. These vary significantly in mechanism depending on the arcing arrangement employed, i.e.

    I wall stabilised arcs

    II   arcs in axial or cross gas flows (e.g. vortex stabilised or
          magnetically driven)
    III   arcs with significant self generated convection (e.g. gas blas
          circuit breakers and welding arcs)

Bauder (1969) has performed experiments on Class I type arcs and reports
$n \simeq 0.5$ at 150A in argon and $n \simeq 1$ at 30A (for 1-50 atmospheres). This lo
current dependence is among the highest to be found in the literature and i
somewhat surprising as conduction mechanisms are expected to dominate and
these are little influenced by pressure. Smits (1939) who examined low
current arcs in nitrogen (1-7A) found $n \simeq 0.16$ for the range 1-1000 bars.
Magnetically driven arcs (between parallel rails in argon) are reported by
Guile (1972) where $n \simeq 0.6$ near 2 bars and falls to $n \simeq 0.22$ between 20 and
50 bars. This has been explained in terms of a decrease in arc velocity
(of transverse movement) with pressure causing a reduction in heat convecte
away by the cross flow. The results for gas vortex stabilised arcs (see
Russian work previously discussed) have been given by these workers as
$n \simeq 0.25$ for argon, $n \simeq 0.4$ for helium and $n \simeq 0.33$ for hydrogen for pulsed
currents of 40,000A and pressures of up to 80 bars. For steady current ar
of 100A the same workers have found $n \simeq 1$ in argon and $n \simeq 0.66$ in helium a
pressures up to 60 bars. Clearly the impeller characteristics will contribute to the observed behaviour. It is interesting to note (Borovik et al
1967) that at very high pressures (100-1000 bars) the field saturates in
argon and helium, i.e. $n \simeq 0$ (400,000A, pulsed currents). This has been
explained by these authors in terms of a transition from a radiative
dominated loss mechanism at lower pressures to nearly complete re-absorbtio
of radiation within the arc at high pressures. Type III arcs have been
considered by Lowke (1979) using simple physical models. He predicts that
buoyancy generated flows at low currents ($\sim$ 10A) should generate $n \simeq 0.25$
while for self magnetic field generated flows $n \simeq 0.25$ to 0.38 depending on
cathodic current density behaviour. Welding arcs generally fall into
Class III and these will be considered separately in a following section.
Before considering welding arcs some attention will be given to other
physical changes associated with the influence of pressure.

Many workers have reported that arc diameter decreases with increasing pres
sure. Few systematic experimental results are available but a number of
theoretical predictions exist. Blix and Guile (1965) quote the work of
Lord who considered a magnetically driven arc in a high pressure environmen
and obtained $R \propto p^{-0.15}$. For Class III arcs Lowke has predicted $R \propto p^{-0.2}$
if buoyancy forces dominate and between $R \propto p^{-0.13}$ and $p^{-0.20}$ for high curr
behaviour.

Measurements of arc temperature have been made by Bauder (1969) for 125A
Cascade arcs in argon at 10 and 15 atmospheres. He finds temperature to
be little influenced by pressure falling from 10,200K to 9,600K over this
range. However, enhanced radiative transfer appears to have considerably
flattened the profile (compared to ambient). Estimates of arc temperature
made by Borovik and Co-workers (1967) are as high as 40,000K in pulsed argo
arcs (40,000A) at 1000 atm. This is attrbiuted to re-absorbtion of radiation in the column. Theoretical models of Ramakrishnan, Stokes and Lowke
(1978) and Lowke (1979) indicate that the temperature of a low current (10A
Class III arc in air increased with pressure but may decrease at high curre
($\sim$1000A).

               WELDING ARCS AT HIGH PRESSURE

The arc welding processes in common off-shore use are SMAW, TIG and FCAW.
Laboratory experiments or trials have also been undertaken with MIG,
(Matsunawa and Nishiguchi, 1979) plasma and plasma-MIG welding as well as

plasma cutting (Waldie, 1980). The deepest reported operation is 80 atm. a figure significantly less than for non welding situations. TIG is the only process for which experiments have been systematically performed and little scientific information is available for other processes.

Tungsten Arcs

Argon TIG arcs have been operated under laboratory conditions at pressures of up to 80 bars (Allum, 1982a). Helium arcs have been considered to substantially lower pressures largely due to electrode erosion (Allum 1982b). To avoid this problem (and for physiological reasons) argon/helium mixtures are presently receiving attention (Knagenhjelm, 1982). This combination also generates more efficient weldment melting than pure argon (see Allum, 1982b)

As with other arc forms so far reported, instabilities can be problematic at high pressures. The least stable arc gas is helium. With this gas electrode erosion can generate a self pulsating arc form which changes colour periodically (Allum, 1982b). Experiments using argon arcs seeded with low amounts of hydrogen (1% to 5%) are also found to be less stable than pure argon arcs (Whitcroft, 1982). Pure argon arcs may exhibit instabilities associated with local cathodic melting and 'floppy' arc behaviour. (Stalker and Salter, 1975). Experiments with argon arcs on cooled copper anodes have shown (Allum, Pinfold, Nixon, 1980) that high shielding gas mass flow rates (i.e. high Reynolds numbers and hence turbulence) can disrupt arc operation. However, to maintain a given shielding efficiency the mass flow rate must increase with pressure. This creates a potential conflict between the demands of arc stability and shielding. Instabilities are not confined to the arc column but may also occur at the anode where, above 30 atmospheres erratic weld pool behaviour has been observed (Allum and Apps, 1982). One type of instability often reported operationally is due to pipe magnetism. Laboratory tests (Allum and Apps, 1982) have found that a given arcing condition (arc current, length etc.) becomes increasingly susceptible to a given magnetic field with pressure. This is thought to be associated with a change in arc column flow and current distributions with pressure.

A number of methods for stabilising arcs have been proposed. Knagenhjelm (1980) reports that the use of very high frequency current (16KHz) has a stabilising influence. This has been associated with an increase in 'arc pressure' due to the skin effect in arc current travels through the outer reaches of the arc (Allum, 1982c). Stabilisation may also be produced (Allum 1982c) by arc constriction as in plasma welding. Here the arc burns in an imposed gas flow. Other problems such as orifice erosion may however arise. In the case of weld pool instabilities axial magnetic fields have been used (Allum, 1982b, 1982c; Whitcroft, 1982). These may be used to straighten the path of a weld bead, manipulate the fusion characteristic and align the arc. No detailed understanding of weld bead instabilities exists. However, models have been proposed for arc instabilities (Allum 1982b, 1982c). These are based on the generation of unsteady conditions in the outer arc reaches by unsteady motions due to turbulent shields or buoyancy forces which may become significant at high pressures. Such behaviour is then argued to create time dependent current density and magnetic field distributions and hence influence bulk motion of an arc through the $\underline{J} \wedge \underline{B}$ force. Stabilisation might then be achieved by influencing conditions at the arc boundary (as in n-f power or arc constriction).

Electrical characteristics of arcs have been reported by a number of workers. Experiments with arcs on copper anodes (Allum and Apps) indicate that the electrode fall voltages (V_o) are insensitive to pressure. A number of workers have reported an increase in V_o for steel anodes (Levin, 1969). However recent work with electrostatic probes (Allum, 1982b) would indicate

175

that V_o is not influenced by pressure on steel and that the reported effects
are due to difficulties in extrapolations associating arc length with electro
separation settings.   Changes in arc voltage are therefore largely associate
with the column.   Values of n reported (see $E \alpha p^n$) range from 0.25 for long
arc length estimates by Dynachenko and Olshanskii (argon arcs) to 0.4 to 0.6
depending on arc gas, shield gas flow etc. (see TNO Project Neptunus and
Allum, 1982c).   Behaviour is generally described by a square root law which
enables the voltage of an argon arc to be estimated from

$$V = 8 + 0.8 \cdot \ell \sqrt{p}$$

where ℓ is the arc length in mm and p the pressure in bars.   It is interesting to note that a $\pm$ 1mm change in arc length under normal ambient conditions
changes V by $\pm$ 0.8V.   At 160m of sea water this would become $\pm$ 3.2V.   Clearl
some appreciation of the implications of such behaviour is required for proce
dural purposes (e.g. significance of process variables, how does heat input
change? etc.).   Arc models have been developed (see Allum, 1982b, 1982d) to
predict V.   These indicate that for argon arcs convection is the dominant
heat transfer mechanism at high pressures with radiation and conduction
increasing and decreasing in importance respectively.   For example, measurements of radiative output show that about 18% of the arc column power appears
in this form at 1 bar for argon.   The corresponding figure at 42 bars is
26% (Allum 1982d).   Column field strength is predicted to behave according
to $n \simeq 0.33$ to $n \simeq 0.50$ depending on the relative significance of radiation.
Higher relative values of n may occur for arcs in turbulent shield flows, as
has been observed ($n \simeq 0.59$, Edmonds and Co-workers, 1979).

Accompanying these changes in electrical behaviour and stability are changes
in arc appearance.   It is observed that the arc column contracts, and in the
case of helium arcs, intense plasma jetting action replaces the diffuse ball
shaped 1 atmosphere appearance.   Photographic and electrical measurements
of argon TIG arc dimensions indicate that $R \propto p^{-0.17}$ to $p^{-0.25}$ with the
stronger dependence being appropriate in turbulent shielding flows (Allum,
Pinfold and Nixon).   Such behaviour is exactly that expected on theoretical
grounds (Allum, 1982d).   In plasma welding the simple pressure dependence
of arc radius given above is complicated by variable plasma forming gas mass
flow conditions with pressure (Allum, 1982b).

Arc temperature estimates (Allum, 1982b) based on the behaviour of electrical
conductivity in Ohms law suggest that arc temperature falls with pressure
(for argon arcs between 1 and 14 bars).   However, more detailed consideration
based on measurements of axial distributions in electric field strength and
radiation, show that pressure has virtually no influence on the axial dependence of radially averaged temperature (Allum, 1983).   The same experimental
data has also been used to estimate plasma velocities.   It is found that the
streaming velocity falls with pressure.   This may be understood in terms of
the pressure insensitivity of momentum generation mechanisms (see Allum, 1983
i.e. flow velocity must then fall as density increases.

The column changes reported above have a significant effect on behaviour at
the anode.   Both the total heat and distribution of anodic heating are
influenced by pressure.   Increases in anodic heating have been associated
with an increase in the column contribution (Matsumawa and Nishiguchi;
Edmonds and Co-workers, 1979; Allum, 1982b), i.e. the total electronic
contribution is insensitive to pressure.   However, the intensity of electro
heating increases markedly with pressure.   Measurements (Allum, Pinfold, Nix
of current density on cooled copper anodes show an increase by a factor of 4
over the range 1-14 bars in argon (typically 10 A/mm^2 at 1 bar to 40 A/mm^2
at 14 bars for I = 100A and $\ell \simeq 3.2mm$).   Calorimetric work at high pressur
(up to 40 bars) has revealed significant differences between heat transfer o
steel and cooled copper anodes (Matsumawa and Nishiguchi; Allum, 1982b).

For both materials the heat transfer efficiency (η) falls with pressure but this behaviour is stronger on steel (e.g. typically η(Cu) falls from 80% at to 60% at 40 bars while the corresponding fall in η(Fe) is 75% to as low as 40%).

No attempt is made here to give a comprehensive survey of bead characteristics resulting from behaviour described above. However, it is noted that for argon arcs the bead width is largely unaffected by pressure but penetration increases until the weld bead aspect ratio (W/D) is about 2. This results in an increase in fusion area and generally higher melting efficiencies (Allum, 1982b) than observed under normal ambient conditions. For helium arcs more complicated melting behaviour occurs e.g. bead width increases with pressure, beads may have W/D ratios of as low as 0.9, fusion area is strongly coupled to arc length, keyhole action may occur and melting efficiencies as high as 40% have been observed (Allum, 1982b).

Consumable Arc Systems

The operation of consumable systems is generally adversley affected by pressure. In both SMAW and MIG it is found that smaller gauge consumables are required for hyperbaric use than for surface applications. It is also observed that the preferred polarity (Matsunawa and Nishiguchi) may change with pressure, i.e. electrode negative MIG is preferred to electrode positive beyond about 7 bars. Consumable arcs have a substantially higher fall voltage than TIG arcs and so most power generation remains at the electrodes with increasing pressure. This indicates that process heat transfer efficiencies should be little influenced by pressure. Calorimetric tests on SMAW have confirmed this (Allum, 1982e). Arc column field strengths for MIG have been obtained by Guile (1972) from data presented by Burrill and Levin (1970). It is deduced that at 12 bars $0.53 > n > 0.42$ where the upper limit is for an 0.8mm wire diameter and the lower for a 1.6mm wire.

Electrode burn off rate often behaves in a complex manner with increasing pressure. For electrode positive solid wire MIG and FCAW the burn off is substantially independent of pressure (Matsunawa and Nishiguchi; Allum, 1982e). However in electrode negative MIG the burn off rate falls such that the value at 7 bars can be less than half the normal ambient value (Matsunawa and Nishiguchi). For SMAW in the electrode positive mode, increases in burn off of 30% have been reported over the first 4 bars (Allum, 1982e). This has been associated with a redistribution in arc heating such that more power appears at the rod and less at the plate. In electrode negative MIG welding behaviour can be such that the burn off initially increases reaching a peak between 2 and 3 bars before falling to a level below that observed under normal ambient conditions.

ACKNOWLEDGMENT

The author would like to thank Mr. J.F. Lancaster and IIW Study Group 212 colleagues for discussions concerned with this paper.

REFERENCES

Allum, C.J. (1982a). Offshore Research Focus, 33.
Allum, C.J. (1982b). Ph.D. Thesis, Cranfield Institute of Technology, Bedford MK43 OAL, England.
Allum, C.J. (1982c). Welding and Metal Fabrication, April.
Allum, C.J. (1982d). Proceedings of 7th International Conference on Gas Discharges and Their Applications, 377-380.
Allum, C.J. (1982e). American Gas Association, Report No. PR-147-135, Sept.
Allum, C.J. J. Phys.D.Appl.Phys. 14, 1041-59.
Allum, C.J. (1983). Paper submitted to J. Phys.D.Appl.Phys.

Allum, C.J. and Apps, R.L. (1982). Offshore Bergen Conference.

Allum, C.J., Pinfold, B.E. and Nixon, J.H. (1980). American Welding Journal Vol.59, No.7.

Bauder, U. (1968). J. Appl.Phys., Vol.39, No.1, 148-152.

Bauder, U. (1969). Proceedings of International Conference on Phenomena in Ionised Gases, Bucharest, 3-10.

Blix, E.D., Guile, A.E. (1965). Brit. J.Appl. Phys., 16, 857-864.

Borovik, E.S. and Co-workers. (1967). Soviet Physics-Technical Physics, Vol.12, No.4, 502-506.

Borovik, E.S., Mitin, R.V. and Knyasev, Yu.R. (1962). Soviet Physics-Technical Physics, Vol.6, No.11, 968-973.

Burrill, E.L. and Levin, M.L. (1970). Proceedings of International on Gas Discharges, London.

Devoto, R.S., (1973). Phys. of Fluids, Vol.16, No.5.

Drellishak, K.S., Knopp. C.F. and Cambel, A.B. (1962). Northwestern Univ. Report No. NU-GDL A-3-63, Illinois.

Dynachenko, V.V. and Olshanskii, A.N. (1968). Svar. Proiz, 11, 3-6.

Edmonds and Co-workers. (1979). Proceedings of W.I. International Conference on Arc Physics and Weld Pool Behaviour, Paper 46

Emmons, H. (1963). Modern Developments in Heat Transfer, Edited by W.Ibele Academic Press, New York, 401-478.

Finkelnburg, W and Maecker, H. (1962). U.S. Airforce Report No. ARL 62-302, and Handbuch der Physik, XXII, 254-444.

Frind, G. (1964). U.S. Airforce Report No. ARL 64-148.

Frind, G. (1966). U.S. Airforce Report No. ARL 66-0073.

Guile, A.E. (1971). Proceedings IEE, 118 (9R), IEE Reviews, 1131-1154.

Guile, A.E. (1972). Proceedings of 2nd International Conference on Gas Discharges, London.

Jones, G.R., Leclerc, J.L. and Smith, M.R. (1982). Proceedings 7th International Conference on Gas Discharges and Their Applications, 73-75

Kannappan, D., and Bose, T.K. (1977). Phys. of Fluids, Vol.20, No.10 1668-1673.

Knagenhjelm, O. (1982). ATMA Conference, Paris.

Knagenhjelm, O. (1980). IIW Doc. CREAU-34-80.

Levin, M.L. (1969). International Conference on Gas Discharges, Bucharest.

Lowke, J.J. (1979). J. Phys.D.Appl.Phys, 12, 1873-1885

Marlotte, G.L., Cann, G.L. and Harder, R.L., (1968). U.S. Airforce Report No. ARL 68-0049, March

Matsunawa, A and Nishiguchi, K., (1979). Proceedings of W.I. International Conference on Arc Physics and Weld Pool Behaviour, Paper 8.

Mitin, R.V., Knyaser, Yu.R. and Petrenko, V.I., (1964). Soviet Physics-Technical Physics, Vol.9, No.2, 267-270

Petrenko, V.I. and Co-workers, (1969). Proceedings of International Conference on Phenomena in Ionised Gases, Bucharest, 311.

Ramakrishnan, S., Stokes, A.D. and Lowke, J.J., (1978). J. Phys.D.Appl.Phys, 11, 2267-2279

Schoeck, P.A., (1963). Modern Developments in Heat Transfer, Edited by W.Ibele Academic Press, New York, 353-400.

Smits, C.G., (1939). Phys. Rev. 55, 561-567.

Stalker, A.W. and Salter, G.R. (1975) W.I. Report No. RR-SMT-R-7504, April.

Waldie, B., (1980). Proceedings of International Conference on Gas Discharges and Their Applications.

Whitcroft, C. (1982). M.Sc. Thesis, School of Industrial Science, Cranfield Institute of Technology, Bedford, MK43 OAL, England.

Power Sources for a Wet Underwater Mechanized Electric Welding

V. K. Lebedev, Yu. A. Usilevskii and V. Ya. Kononenko

E. O. Paton Welding Institute of the Academy of Sciences of the Ukrainian SSR, USSR

ABSTRACT

The requirements to the power sources for mechanized wet welding are considered. Power source-welding circuit-semi-automatic machine-arc-weld system and its peculiarities are studied. The posibility of self-controlling the welding arc when burning at various depths is investigated. The process of self-controlling the arc with abrupt disturbance along its length and also the experimental dependencies of disturbance setup time on external welding circuit resistance value are studied.

KEYWORDS

Wet underwater welding, mechanization, power source, arc, welding circuit, arc self-adjustment.

The performance of the power source - welding circuit - semi-automatic machine - arc-weld system in a wet underwater mechanized welding has a number of peculiar features as compared to the process of the open atmosphere mechanized welding. The main of them are being considered in the present paper.

Studies, carried out at the E.O.Paton Welding Institute of the Ukrainian SSR Academy of Sciences, showed that in the underwater mechanized welding with the self-shielding electrode wire a continuous electrode melting and a spray metal transfer, not accompanied by arc gap short-circuiting, are observed (Fig.1). Due to this, the variation in arc length during the welding process also occurs continuously, without short-circuitings. Therefore, it may be concluded that self-adjustment of the arc in its pure form takes place in this kind of underwater welding. Under such conditions the stability of the underwater mechanized welding process and intensity of self-adjustment of the underwater arc with real disturbances essentially depends

on the paremeters of power source and welding circuit, the
determining factor of the welding process stability being the
rigidity of external characteristic of the power source.

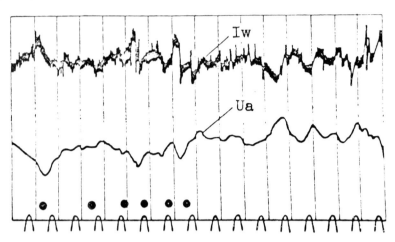

Fig.1. Typical oscillogram of arc current and voltage in underwater mechanized self-shielding wire welding.

The experience of underwater jobs and experimental studies
showed that the process of the underwater mechanized welding
is more stable when using the power sources with rigid, falling and rising external characteristics. Besides, the application of power sources with sloping external characteristics
is also rational due to their technical-economical advantages.
These advantages consist in that the dimensions and mass of
these power sources, determined by the product of U_{oc} by
I_{nom}, are minimum with other conditions being equal. Here one
of the main requirements made of the underwater welding electric equipment is met, i.e. minimum mass-dimensional characteristics of the equipment used are provided.

It should be noted that when evaluating the rigidity of the
power source external characteristic it is necessary to determine the voltage at the arc, but not at the power source terminals. Otherwise, the output voltage of the power source
should be calculated, considering the total resistance of the
underwater welding electric circuit $R_{ex.cir.}$ consisting of
a number of resistances connected in series (Fig.2): resistance of the stationary welding cable $R_{stat.cable}$ (from power
source up to the control cabinet of the semi-automatic machine), resistance of the underwater flexible cable $R_{flex.cable}$
(from control cabinet up to the semi-automatic machine holder)
resistance of the holder cable R_{holder}, resistance of the
flexible hose R_{hose} from the current conductor up to the contact tip, resistance of the "tip-electrode" contact R_c, resistance of the electrode stickout R_g. It should be noted that
the location of the welding arc power sources on the floating
drilling platforms, surface ships or barges, that ensure the

diving jobs, does not allow to bring the power source nearer
to the place of performance of the underwater electric welding
jobs. Therefore, when welding at a depth, for instance, in
welding off-shore pipelines, located on the seabed or in repair of the drilling equipment, the length of the external
welding circuit from the power source down to the underwater
semi-automatic welding machine reaches considerable sizes. It
leads to more essential voltage drop in the welding circuit
and to the sharp reduction of the rigidity of external characteristics of the arc supply sources. In the wet underwater
mechanized welding the semi-automatic machines with a constant
electrode wire feed speed, independent of the current and the
arc voltage, are mainly used. Hence, the reduction of the rigidity of the power source external characteristics leads to
the decrease of the intensity of the arc self-adjustment process.

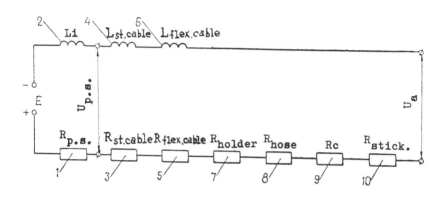

Fig.2. Scheme of welding circuit for the case when the underwater mechanized welding is performed from the floating
drilling platforms:
1-equivalent resistance of power source; 2-inductance
of power source; 3-resistance of stationary welding
cable; 4-inductance of stationary welding cable;
5-resistance of underwater flexible cable; 6-inductance
of underwater flexible cable; 7-resistance of holder
cable; 8-resistance of holder flexible hose; 9-resis-
tance of "tip-electrode" contact; 10-resistance of
electrode stickout. $U_{p.s.}$ - voltage at power source terminals; U_a - voltage of underwater welding arc.

At the same time, the process of the wet underwater mechanized
welding is characterized by frequent disturbances both in the
welding arc length and in the value of the electrode wire stickout, which are specified by unstable equilibrium of the body
of the diver-welder, its negligible negative buoyancy and low
stability. Besides, these disturbances more often occur in ti-

me and they are of a larger absolute value than in welding in conventional atmospheric conditions. As a result of the effect of the mentioned disturbances the deviations of the underwater welding condition from the assigned one become longer, thus leading to the deterioration of the quality of the underwater welded joints.

The above-mentioned features of the wet underwater mechanized welding process show that the fast response of the arc self-adjustment system in correction of typical disturbances is of a special importance in wet underwater welding at large depths for ensuring the stability of the welding process and improving the quality of the welded joints. Therefore, the main object of the present work is to study the peculiarities of operation of the system power source - underwater welding arc and development of measures of improving the intensity of the arc self-adjustment process in performing the welding jobs at the depths of the continental shelf.

When the disturbances occur in the length of the underwater arc Δl_{a_0} in the system power source - welding circuit - arc the transition process of correcting this disturbance takes place, due to the arc self-adjustment phenomenon. The differential equation of this transition process has a form of

$$\frac{L_{ss}}{k_{sc}} \cdot \frac{d^2 \Delta l_a}{dt^2} + \left(\frac{L_{ss} k_{sv} k_a + k_s}{k_{sc}} \right) \frac{d \Delta l_a}{dt} + \left(k_s \frac{k_{sv}}{k_{sc}} + 1 \right) k_a \Delta l_a = 0 \qquad (1)$$

where K_a is the field intensity in the arc column, V/m;

$k_s = \partial U_a / \partial I_a - \partial U_{ss} / \partial I_a$ is the coefficient of stability of power source - arc system.

k_{sc} and k_{sv} are the coefficients of self-adjustment of arc by current (m/s A) and by voltage (m/s V);

L_{ss} is the total inductance of the supply system, GN.

Let us write the characteristic equation for expression (1)

$$T_1^2 p^2 + T_2 p + 1 = 0 \qquad (2)$$

where T_1 and T_2 are respectively equal to

$$T_1 = \sqrt{L_{ss} / k_a (k_s k_{sv} + k_{sc})} \qquad (3)$$

$$T_2 = \frac{L_{ss} k_{sv} k_a + k_s}{k_a (k_s k_{sv} + k_{sc})} \qquad (4)$$

Let us find the roots of the characteristic equation

$$p_{1,2} = \frac{-T_2 \pm \sqrt{T_2^2 - 4 T_1^2}}{2 T_1^2}$$

At $T_2^2 > 4 T_1^2$, the process of processing of disturbance Δl_{a_0} is aperiodical and the solution of equation (1) has a form of

$$\Delta l_a (t) = C_1 e^{-t/T_a} + C_2 e^{-t/T_b}$$

where $T_a = -1/p_1$, $T_b = -1/p_2$

After determining and substitution of integration constants C_1 and C_2 the solution of equation (1) has the form of

$$\Delta l_a (t) = \Delta l_{a_0} \left(\frac{T_a}{T_a - T_b} e^{-t/T_a} + \frac{T_b}{T_a - T_b} e^{-t/T_b} \right) \qquad (5)$$

It is known that the system of self-adjustment will be stable i.e. the occurred deviation of the arc length Δl_a will decrease with time if the roots of the characteristic equation (2) have a negative real part. Since $T_1^2 > 0$, then the condition of the system stability can be written as follows:

$$\frac{L_{ss} k_{sv} k_a + k_s}{k_a (k_s k_{sv} + k_{sc})} > 0 \qquad (6)$$

To determine the stability boundaries of the system of underwater welding arc self-adjustment at the increase of depth of arc immersion we studied the fulfilment of the condition of stability (6) and calculated the real value of time constants T_a and T_b for different depths of underwater welding and the values of inductances of underwater welding cables suitable for these depths.

The results of calculations (Table) showed that with an increase of depth of the underwater welding performance the system stability coefficient K_s increases due to the rise of the power source external characteristic steepness, the intensity of arc self-adjustment being simultaneously decreased and the total time t_{corr} of disturbances correction in the arc increased. The rise of the arc column field intensity K_a intensifies to a certain extend the process of the underwater arc self-adjustment, the former increasing with the depth of the arc immersion into the water. However, in spite of this, the enlargement of the steepness of the power source external characteristic sloping occurs more intensively than the increase of the field intensity in the arc column. As a result, with the increase in the depth of the immersion the rate of the transition processes running in the system reduces and the total time of disturbance correction in the arc length rises.

TABLE Estimated data on intensity of arc self-adjustment at different depths in wet underwater mechanized welding

H, m	$R_{ex.cir}$,	U_{oc}	Steepness of external characteristics of the power source $\bar{\gamma} = \partial U_{ss}/\partial I_a$ V/A		Coeffi-cient of stabili-ty of power source-arc system, K_s V/A	Intensi-ty of arc co-lumn field, $K_a = E_c \cdot 10^{-3}$ V/m	Time of correc-ting the distur-bances in the arc length t_{corr}, S
			at termi-nals of po-wer source	at arc			
0	0.049	39–41	−0.01	−0.059	0.059	4.19	0.292
40	0.108	62–65	−0.01	−0.118	0.118	5.47	0.429
100	0.167	80–84	−0.03	−0.197	0.207	6.25	0.619
160	0.226	100–104	−0.04	−0.266	0.306	6.70	0.811
220	0.285	118–120	−0.06	−0.345	0.435	7.05	1.013

According to the estimated data (see Table) the intensity of
the underwater arc self-adjustment process is almost 2.5 times reduced with the depth variation from 40 down to 220 m.
For practical applications the rate of correction of disturbances in the arc length is only provided for the depth of
down to 120...130 m. At the depth from 130 down to 160 m the
efficiency of the self-adjustment is considerably reduced and
the transition processes are excessively delayed. In this connection the disturbances of the process stability may occur
in welding at the depths of more than 160 m using the semi-
automatic machines with a constant feed speed.

The experimental checking of the obtained estimated data and
study of the effect of the transition processes duration on
the quality of the welded joints were carried
out in the experimental installation (Fig.3), that allowed to
reproduce the disturbances in the arc length resulting from
holder deviation from the workpiece welded during the underwater mechanized welding performed by the diver-welder. By
using the experimental installation the effect of two parameters of the supply system on the process of the arc self-adjustment was studied, such as: the steepness of the power

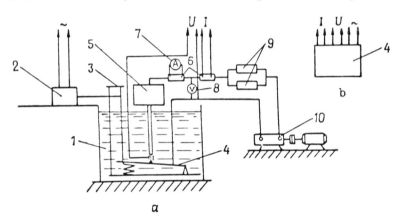

Fig.3. Elementary diagram of experimental equipment (a) and
       H-115 type oscillograph connection for arc current and
       voltage recording (b)
       1-water tank; 2-electric relay ЭЛ-41v-22; 3-device for
       imitation of staggered disturbances along the underwa-
       ter welding arc length; 4-steel plate, on which the un-
       derwater bead deposition is performed; 5-TC-17M tractor
       with an extended nozzle; 6-shunt; 7-ammeter; 8-voltme-
       ter; 9-ballast reostats РБ -300; 10-welding converter
       for underwater mechanized welding; 11-oscillograph
       H-115.

source external characteristic $\gamma = \partial U_{ss} / \partial I_a$, which is proportional to the resistance of the external welding circuit
Rex.cir. and open-circuit voltage U_{oc} of the power source. Here, the possibility of changing any of these two parameters was
provided, the other one being unchangeable.

The experimental welding was carried out at rigid characteristics of the power source with the following values of U_{oc} : 33,40,50,60,70,80 V. The variation of the power source external characteristic steepness was ensured by connection of various steps of resistance of the ballast reostats into the welding circuit, the total resistance of the welding circuit $R_{ex.cir.}$ being varied from 0 down to 0.23 ohm. The value of the reproduced disturbance in the arc length was set before each experiment and could vary from 1 up to 7 mm. In total, more than 80 experiments were carried out on the study of transition processes of disturbances correction at various values of U_{oc} $R_{ex.cir.}$ and Δl_a . The typical oscillogram of the transition process in correction of the abrupt disturbance along the arc length is given in Fig.4.

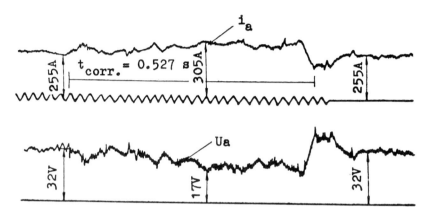

Fig.4. Typical oscillogram of transition process in correction of staggered disturbances along the underwater arc length by the system of self-adjustment. The value of disturbances Δl_a = 3.5 mm, U_{oc} = 80 V, $R_{ex.cir.}$ = 0.22 ohm.

The experiments showed that with some limited value of $R_{ex.cir.}$ resistance in the welding circuit separate pores of 0.4...0.6mm size appear on the macrosections of the deposited metal, that correspond to the moment of disturbance correction. Therefore, a number of pores on the mentioned macrosections was assumed as a criterium for evaluating the allowable durations of deviations of the arc voltage and current during correcting of the abrupt disturbance in the arc length.

The selected criterium of the weld density allowed by the results of experiments and experience of off-shore welding to set the limited value of resistance of the external welding circuit $R_{ex.cir.lim.}$ for each value of U_{oc} . In exceeding this value the transition process of the disturbance correction becomes inadmissibly delayed, thus considerably increasing the weld metal porosity. Relationships $t_{corr}=f(R_{ex.cir.})$ for various values of U_{oc} were plotted, taking into account the selected criterium of quality and on the basis of experiments and results of oscillogram processing.(Fig.5).

The investigations showed that the steepness of the external characteristics of the power source, the field intensity in the arc column (E_c) and the values of coefficients of arc self adjustment in current (K_{sc}) and voltage (K_{sv}) affect the rate of transition processes in the self-adjustment system during the mechanized underwater welding in the same way as in the open air welding. However, the peculiarity of the self-adjustment process in the underwater welding is that the resistance of the external welding circuit, in deep welding, in particular, is so high and, therefore, its effect on the reduction of the arc self-adjustment intensity is so considerable that the effect of E_c, K_{sv} and K_{sc} on the process of the self-adjustment of the arc burning under the water is less noticeable than in air welding. Hence, the most effective method of increasing the fast response of the arc self-adjustment and improving the weld quality in welding at large depths is the reduction of the resistance of the external welding circuit. It can be achieved by one of the following methods:
-parallel connection of two or more cables into the direct and return welding leads;
-application of power sources with a controlled rigidity of their external characteristic;
-application of immersible arc power sources, installed on the seabed near the place of the performance of the underwater welding jobs;
-accomplishment of underwater welding from power sources installed on the submarine machines designed for accomplishment of errection and repair jobs in construction and service of the off-shore drilling platforms.

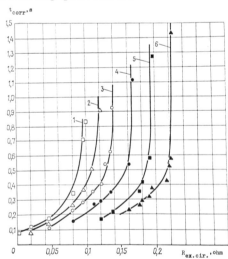

Fig.5. Experimental relationships between the time of correcti on of Δl_a disturbances along the underwater arc length and value of resistance of external welding circuit Rex.cir.
1-Uoc =33V; 2-Uoc=40V; 3-Uoc=50V; 4-Uoc=60V; 5-Uoc=70V; 6-Uoc=80V.

The semi-automatic machines, based on the principle of automatic control of the electrode wire feed speed with feed-backs by current and arc voltage, can be also used for the underwater welding beginning from 160...170 m depths to increase the fast response of the arc adjustment system.

CONCLUSIONS

1. In the underwater mechanized welding with semi-automatic machines having a constant wire feed speed and arc supply from the non-immersible welding current sources, the rate of correction of the typical disturbances in the length of the underwater arc is considerably reduced with the increase of depth of the welding jobs accomplishment.

2. The increase of the intensity of the underwater arc self-adjustment and the resulting improvement of the underwater joint quality can be achieved by using the immersible power sources, semi-automatic machines based on the principle of the arc voltage automatic control; application of welding generators with the compound series winding, by accomplishment of welding jobs with the submarine machines equipped with manipulators and arc supply from the power system of the machine.

SECTION II

INSPECTION AND PERFORMANCE / CONTROLE ET COMPORTEMENT

Remote-controlled Underwater Welding in the Dry

D. van der Torre and M. P. Sipkes

Metaalinstituut TNO, P.O. Box 541, 7300 AM Apeldoorn, The Netherlands

ABSTRACT

As operation depths for all off-shore work will grow in the near future a welding system has been developed with which remote controlled welding of pipes can be carried out.
The system makes use of the pulsed TIG-welding process with cold wire feed. The welding is observed by a (patented) CCTV-system. Naturally for a remotely controlled system all torch movements are fully mechanized and independent of the torch the wire feed-nozzle can be moved in two perpendicular directions. The system has been qualified in the laboratory and tested in open water.
As it is known, that TIG-electrodes suffer from erosion when they are working under higher pressures, tests have also been undertaken with the plasma-welding process.

KEYWORDS

Remote-controlled, underwater welding, TIG welding, plasma welding, mechanization, automation.

INTRODUCTION

Underwater welding can be carried out manually under wet or dry conditions. In both cases, however, the weld quality is often poor.
The presence of water causes many major problems in the welding process. For instance, water cools the heated area of the weld very quickly, causing hardening, and increasing the danger of coldcracking. The finished weld has usually a high porosity as well. In the welding process itself, the diver/welder has a difficult job since he suffers from lack of visibility underwater and pressure at working depths has a detrimental effect on his handling of the welding equipment.
As a result, manual welding under dry or wet conditions on depths of more than 100 meter is hardly likely to lead to an acceptable weld quality. By overcoming these difficulties with wet and dry welding, remote-controlled welding may prove an appropriate solution.

After three years of research, TNO has developed a remote-controlled underwater welding system. Welding is controlled by an operator in a ship anchored over the pipe. The welding equipment itself is contained in an habitat placed over the pipe. Pulsed TIG-welding with cold wire feed is used as
the welding process.

General

The remote-control system comprises of four main parts: controle console,
observation unit; welding head and welding manipulator. The control console
allows the operator to perform all necessary manipulations of the welding
equipment. The observation unit consists of a closed-circuit TV camera and
monitor system under the control of the operator and is used for observing
the welding and inspecting the weld afterwards. The CCTV-system has patented
(1) optical elements and filters (Fig. 1).

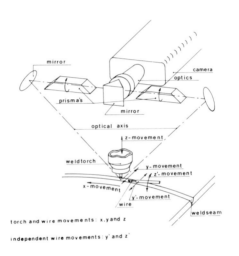

Fig. 1. Arrangement of CCTV camera system and
the weld torch

The welding head consists of the torch, the wire feed and the necessary
motors, gear boxes and transmissions. The manipulator provides for the
orbital movement. The CCTV camera and torch are of modular construction, so
they are easily attached to the manipulator unit. The equipment is designed
to withstand pressures at depths down to 200 meters.

The Modular System

As pointed out above the system consists of modular parts, which can be installed easily. This modular system is chosen for several reasons:

---

(1) Patent application nr. O.A. 82.01673. Netherlands filed 22 April 1982.

1. The several parts are drip-tight, but not water-tight. So these parts shall be placed after the weldhabitat is emptied from water.
2. The fully water-tight manipulator allows a certain range of diameters. Over a certain diameter another manipulator has to be used.

In Fig. 2 the basis is shown. This is formed by the manipulator, which allows diameters of 6 to 12 inches. On this manipulator a spring loaded base plate is placed. On this plate the welding head with wire feed unit and observation system can be installed. The manipulator is the heaviest part and the whole with cables included is supposed to stay in the habitat. The manipulator consists of two U-formed parts: a stationary one that clamps around the pipe and a rotating one on which the welding head is mounted.

Fig. 2. Manipulator with spring loaded base plate

The complete remote-controlled welding system is shown in Fig. 3.

The Control System

This is built-up of a number of units. On a console the following functions are present:
a. The control of the x, y and z movement of the torch and y' and z' movement of the wire feed tip
b. The read-out of the torch position
c. The most important functions for the welding cycle, i.e. start, weld-current, downslope-on, stop, wire feed speed
d. The controls of the camerasystem as diaphragm, focusing and filters.

The control equipment can be installed in a freight-container to make the system mobile. Figure 4 gives an example of such a set-up.

The Welding Performance

The operator uses the information on the monitor screen and the arc voltage plotted on a recorder to control the welding. An example of the information as the operator gets from the screen is shown in Fig. 5. This is a split image of the weld zone. It is necessary to observe the front and the rear of the weld-pool: the front in order to control the wire feed and the rear to control penetration during the root-pass and fusion during the filler passes. On the screen a read-out can be projected of the average current, the orbital position (x), the torch position towards the edge of the bevel (y) and to the surface of the weld (z).

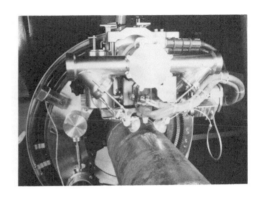

Fig. 3. Remote-controlled welding system

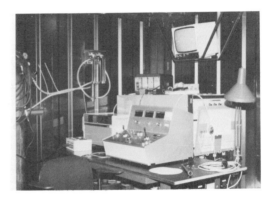

Fig. 4. Control equipment with CCTV system

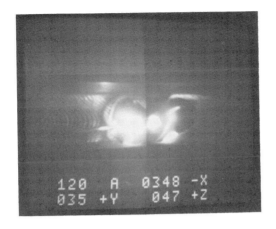

Fig. 5. Image of the weld-pool viewed on the
monitor screen

THE WELD PARAMETERS

All weld parameters are preset before starting. The welding speed is kept at
a constant value throughout the whole welding sequence. Only the current,
and the positions of torch and wire are changed when necessary. The system
allows a high-low of the surface of both pipe ends to a maximum of 2 mm.
The gap can vary from 0 to 1 mm. The used welding parameters for the welding
of a pipe grade X-52 steel according to API 1104 are given in Table 1.

TABLE 1   Weld Parameters

Bevel	: U-shaped (taper 10°)
TIG electrode	: W + 2% ThO_2
Diameter of electrode	: 3.2 mm
Diameter of wire	: 0.8 mm
Wire speed	: 1000-1300 mm/min.
Weldspeed	: 113 mm/min.
Shielding gas	: He + 20%Ar, 7 1/min.
Backing gas	: Ar 10 1/min.
Pulse frequency	: 1.7 Hz
Current ratio high/low	: 50/50%
Current low	: 35A
Current high	: 195-280A
Arc length	: 4 mm

The dimensions were diameter 220 mm and wall-thickness 12.7 mm. The weld
tests were performed at pressures of 100 (atmospheric pressure) and 640 kPa,
the latter being the highest permissible pressure in the chamber used for
the testing. To obtain an idea of the influence of the gas argon as well
as nitrogen was used.

THE RESULTS OF THE INVESTIGATION

From the welded pipes samples were taken for a mechanical and microscopical
investigation. Every pipe was also investigated by NDT. X-ray as well as US
testing were carried out.

Non Destructive Testing (NDT)

The results of the NDT are given in Table 2.

TABLE 2   Non Destructive Testing

Chamber		NDT	
Pressure kPa	type of gas	X-ray IIW-code	US
100	Ar	1	no
100	N_2	1	indica-
640	Ar	1Aa-2Aa	tions
640	N_2	2Aa-3Aa	

Increasing pressure gives more porosity and this is more outspoken when
nitrogen is used.

Mechanical Testing

Mechanical testing results are given in Table 3. Pressure has no influence
on the results of the tensile testing, although nitrogen as a pressure gas
seems to influence the ductility. This can be noticed from the hardness
testing.

TABLE 3   Results of the Mechanical Testing of the Welds

Type of mechanical testing	100 kPa	640 kPa	Remarks
Tensile strength in MPa	645	635	
Bending test, D = 3t	good	2x good 1x 60°	pore from surface
Hardness HV 20 kgf	210	261-277 218-235	base material welded in nitrogen welded in argon
Charpy-V in J, testtemp. 0^oC	112 38	42 21 -27	1st delivery 2nd delivery

Quite a large scatter in mechanical properties was noticed of the two
different charges of pipes used, especially for impact testing. A guarantee
for a certain level of impact value is not given by the supplier according
to API 1104. A drop in impact value for increasing pressures can be noticed.
The faster cooling of the weld seems to be responsible for that. For this

and other reasons preheating will be necessary. Successful tests have been carried out with the arc as a preheating source.

Microscopical Investigation

From the welds made at a pressure of 640 kPa with Ar as chambergas sections were taken for microscopical investigation. To obtain an idea of the influence of the weld position the sections were taken at 3, 6, 9 and 12 o'clock. The results are given in Fig. 6.

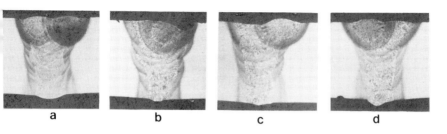

Fig. 6.   Cross-section of weld; position (a) 3 o'clock, (b) 6 o'clock, (c) 9 o'clock, and (d) 12 0'clock.

The microstructure itself showed no defects other than those noticed from the X-ray testing.

TESTING OFF-SHORE

To test the equipment in practice it was put aboard a ship especially constructed to carry out work with a weld habitat. This test was performed by a diving firm, the owner of the ship, and TNO. The location of the test was the estuary of the Eastern Scheldt (Zeeland). The maximum water depth there is about 35 m. Because of tidal currents the working conditions were rather hard. Only during dead tide (currents of less than 0.5 m/min) was it possible to use the habitat. The control equipment was installed in a container, Fig. 4.
In October 1981 the habitat was transported to the Eastern Scheldt and brought aboard, Fig. 7. During three days a number of tests was undertaken. It turned out that all systems worked satisfactorily. As a result of the tests a number of improvements had to be made, they were, however, marginal in relation to the total effort in this enterprise. After these improvements it should be possible to use the equipment for welding jobs under water.

PLASMA WELDING

Due to the use of higher pressures elaborate erosion of the TIG electrode will take place. Trials at the Technical University of Delft showed, that a weightloss of about 10 mgr. at a pressure of 6 bar and a current of 160A during 15 minutes could be noticed. A weightloss of more than 500 mgr. was noticed at 21 bar. Trials were therefore carried out at TNO on plasma welding. The main results were:
a) No loss of weight was found at 6 bar, 50A for the main arc during 60 minutes. At 100A and a pressure of 4 bar a loss of 1.4 mgr was found for the same time.

b) Starting the pilot arc is more difficult with increasing pressure.
c) When the pilot arc is on, starting the main arc gives no problem up to
the pressure investigated (6 bar).

Fig. 7.   Habitat and diving bell in moon pool of
"Rocky Giant" (special diving vessel)

CONCLUSIONS

It has been proved possible to carry out remote controlled welding in the
dry. An operator can control the welding by observing the image on a monitor. The welds showed no significant defects and are acceptable according
to the API code. Testing the equipment in open water turned out that it can
be handled in a rather rough way. The equipment will be part of a system
consisting of clamping, bevelling, welding and NDT. Because of the permitted
tolerances in joint preparation high demands are set on clamping and mechanical bevelling.

ACKNOWLEDGEMENTS

The work described was part of the project "Neptune", which was co-ordina-
ted by the Dutch Welding Institute. The sponsoring came from a group of industries and the Ministry of Economic Affairs. The testing off-shore was
undertaken by the divers firm Vriens BV and A.C.Z. Marine Engineering in
co-operation with TNO.

Non-destructive Testing Under Water

O. Forli

Det norske Veritas, P.O. Box 300, N-1322 Hovik, Norway

ABSTRACT

A survey is given of the non-destructive testing techniques most commonly applied to welded constructions under water. These techniques are visual examination, magnetic particle testing and manual ultrasonics. Further, other applicable techniques are mentioned alongside some of the peculiarities met with respect to personnel, preparations and special quality assurance measures to be taken when performing non-destructive testing in the underwater environment. The future trends seen are indicated and examples are given of newly developed equipment for more reliable and cost-effective examination, like Det norske Veritas' automated ultrasonic equipment Corroscan.

The survey given corresponds to the content of a document on common practices for underwater non-destructive now being prepared by the IIW, one of the first efforts made towards standardization of non-destructive testing under water. A draft version will be prepared by summer 1983.

KEYWORDS

Non-destructive testing, underwater, magnetic particle testing, visual examination, ultrasonic testing, welds, corrosion.

INTRODUCTION

Non-destructive testing (NDT) is today regularly applied under water. The main application area is for offshore platforms and pipelines, and one of the main driving forces behind taking NDT in use under water was the development of oil and gas fields in the harsh waters of the North Sea.

NDT techniques currently available for underwater use are in most cases direct applications or modifications of techniques used above water. Often the only modifications done are to make the equipment waterproof, to build it into a pressure housing or in an other way adapt it to the underwater environment, and to make it electrically safe for a diver to operate it.

Most of the NDT is still carried out manually, and the final interpretation is left to the diver. Underwater testing often requires many operators due to limited diving time, and it is a problem to find properly qualified divers. The skill of the diving NDT operator is absolutely essential for reliable and efficient testing.

TECHNIQUES AND APPLICATIONS

Visual Examination

The most important underwater NDT method is visual examination. Visual examination is used for gross defect detection and after proper cleaning also for close up examinations of welds etc. On offshore installation the technique is mainly used for detecting:

- mechanical damages and obvious cracks
- coating damages and corrosion
- anode conditions and consumptions

as well as marine growth, scouring and debris.

Visual examination is performed by divers in the water or by remotely operated or other types of underwater vehicles. Still photography, TV cameras and video recording are used as aids.

The quality of visual examination is very dependent on the professional quality of the (diving) inspector, and this is unfortunately an item which in many cases is ignored. Examination often takes place with unqualified divers, not trained and informed on

- what to look for
- where to look
- how to report

which are the 3 basic elements of a visual examination procedure.

Divers should have a certain theoretical background and practical experience to be allowed to perform visual examination, but there does not exist any formal qualification scheme for visual examination of sufficiently high standards today. When using improperly qualified diving inspectors problems may arise in connection with reporting of irrelevant findings and lack of reporting of significant defects.

To minimize these problems, certain quality assurance steps can be taken, it is i.a. considered important to have a professional inspection leader at site in order to obtain as relevant information as possible.

Magnetic Particle Testing

Magnetic particle testing has so far, apart from visual examination, been the dominating underwater NDT-method. The method is suitable for underwater use, in detection of surface breaking cracks, and if correctly used with qualified personnel, the same quality of examination will be obtained as above water. On offshore constructions the main application is in-service`to detect fatigue and other service induced cracks on node and other welds. The method requires the testing areas to be cleaned for marine growth, and this is more timeconsuming than the actual testing.

In most cases fluorescent magnetic particles are used and viewed in ultra-violet light. This testing requires reduced ambient light. Coloured magnetic particles viewed in white light may also be used, but this requires an even more perfectly cleaned test surface and is not as sensitive as fluorescent particles used with ultraviolet light.

Different techniques for magnetization are in use and the choice of technique is mostly dependent on the test object. The most used techniques up to now have been

- current flow (prods)
- coil

Other techniques in use are

- parallel conductors
- electromagnetic yokes
- permanent magnets

Both AC and DC power supplies are in use, but as most of the testing under water is aimed at detecting surface cracks, AC-supply is recommended and is also most frequently in use.

Permanent magnets are not recommended for detection of surface cracks. This is due to uncertainties regarding the obtained magnetic field, especially with complex geometries. They are also time consuming to use and may loose their power with time.

Magnetography gives a permanent record as a paper print-out, but has not been extensively used underwater. Due to some uncertainty with the inter-pretation and lack of experience with the method, acceptance for use has been on the condition that findings could be confirmed by conventional mag-netic particle testing.

Ultrasonic Testing

Ultrasonic testing under water is mainly performed for detection and mapping of corrosion. Weld examination is considered too complicated to perform on a larger scale and is only done for diagnostic purposes and in connection with repairs, tie-ins, etc. This is often done in a habitat.

For thickness measurements digital meters are most commonly in use, the reason being the small size and easy operation, but care should be taken during use. It is easy to make mistakes and it should not be used by personnel not knowing the principles of ultrasound. It is recommended only on smooth and uncorroded surfaces to avoid erroneous readings due to possible paint, water or other layers.

The direct use of an A-scan will give a more reliable reading under all con-ditions. However, the equipment requires exact calibration and qualified operators. If a surface readout is available, this problem can be eli-minated. Internal pitting corrosion can be detected by use of A-scan and angle probes. This gives a positive indication (echo from pit) to the operator, easier to interprete than the shift in a normal probe echo posi-tion.

Corroscan is an automated ultrasonic equipment for internal corrosion mapping of risers and pipelines under water, developed by Det norske Veritas. The equipment consists of an underwater scanner carrying a focused ultrasonic probe interfaced to above water recording and processing equipment giving an on-line wall thickness map of the examined area. Off-line processing contains generation of colour coded wall thickness maps, statistical evaluations and comparison between consecutive examinations to establish corrosion rates.

The equipment has been in use in the North Sea since June 1981, and a large number of risers and pipelines have been successfully inspected in selected areas. Typical accuracies achieved are on pitted areas a wall thickness standard deviation of 0.4 mm with a lateral resolution of 2.5 mm. The scanner can operate on straight pipes and bends down to 3 D with 10 3/4" diameter or larger. A 250 mm section of a 22" diameter pipe can be scanned in about 15 minutes. A skilled diver with no NDT experience can easily be trained to operate the underwater parts of the equipment including mounting of scanner and doing calibration runs.

The Corroscan has been modified to include other applications like corrosion mapping above water and examination of lamellar and other special types of cracking by transmitter/receiver probes and angle probes. Work is in progress to do butt weld examination, and in the future also testing of node welds.

Other Techniques

Crack depth measurements. A tool based on potential drop measurements used in connection with magnetic particle testing, has been developed by Det norske Veritas. The measurements are useful for evaluation of cracks and as input for fracture mechanics evaluations.

Radiography. It is not common to use radiography directly in water, but gamma radiography is used for testing of hyperbaric welds. If applied directly in water it may be necessary to remove water from the beam path by using waterfilled cones, balloons or similar remedies.

Eddy currents. Equipment is now available on the market which is claimed to be able to replace magnetic particle testing for detection of surface cracks in welds. The potential and efficiency of this equipment have, however, not yet been fully established.

PREPARATIONS

Cleaning

For application of most underwater NDT methods cleaning to remove marine growth is necessary. Dependent on type of growth (soft marine growth, or hard deposits) different techniques are applied. The most common ones are water jetting, grinding, wire brushing, abrasive water jetting and the use of needle guns.

In order to avoid masking of surface defects care must be taken with respect to choice of cleaning method. For magnetic particle testing wire brushing and water jetting to bare metal are acceptable. Grinding should only be done in case of extremely hard deposits and cleaning by needle guns avoided.

Location Reference Marking

On large underwater constructions a good location reference system is extremely important to pinpoint spots for repair or repeated in-service inspection. Aids must be available in order to navigate to and identify structural members, nodes, etc. Marking of a structural part should be used locally to identify defects found, areas examined for corrosion with respect to repeated testing of the same area to measure corrosion rate etc.

PERSONNEL

Working Conditions

Doing NDT under water differs from doing the same job above water and it might be difficult to do an examination to the same quality standards as above water. This is due to the following:

- Environmental and working conditions
- Personnel qualification
- Limited on-site surveillance possibilities

A diving NDT operator is faced with a number of difficulties not met by an above water operator like handling diving gear, decompression, limited diving times and threats of diving sickness, marine growth, limited mobility and under special circumstances poor visability, strong currents and splash zone waves. All these items take the time and attention from the actual NDT work. It is, further, difficult to combine the profession of a skilled diver and a skilled NDT operator, and such people are scarce. In addition time is limited for training to establish and maintain competence.

Principles

In principle diving NDT operators and inspectors should be as qualified as their above water counterparts to ensure examination quality. In order to compensate for the difficult working environment special quality assurance measures should be taken (see below).

Qualification Schemes

The only existing certification scheme operating today is a part of the British CSWIP scheme. CSWIP does, however, not have the same requirements on NDT competance for underwater operators as above water.

Det norske Veritas is now issuing a Technical Note on Qualification of Underwater Inspection Personnel. The principle expressed in this technical note is that testing personnel should be as qualified as personnel doing a similar job above water, preferably based on existing certification schemes. It allows, however, also for authorization of earlier non-certified personnel according to a special scheme.

QUALITY ASSURANCE MEASURES

To compensate for the harsh working environment met in underwater NDT, limited NDT operator training and experience, and limitations in direct, on-site surveillance possibilities, the following measures may be taken to ensure examination quality:

- Verification tests

    Verification tests are useful to check adequate functioning of an
    examination system (personnel, equipment, etc.) and often required
    by companies engaging inspection firms, governmental bodies, cer-
    tifying authorities and others prior to an inspection job. Veri-
    fication tests are more frequently used for underwater NDT than
    above water. More comprehensive verification tests may be requi-
    red for novel methods and equipment.

- Special surveillance by inspectors above water using aids as underwater
    TV cameras, direct communication lines to underwater NDT operators
    and/or examination recording equipment above water.

- Stringent requirements to reporting and documentation of examination
    performance and results by e.g. still photography or video recor-
    ding.

- Audit by re-examination by others.

- Drilling of NDT operators for special types of examination.

    STANDARDIZATION

Principles

NDT methods used under water are in principle not very different from
methods applied above water. The same physical properties apply. There is
for instance no practical difference in electromagnetic properties between
air and water, which could influence magnetic particle or eddy current testing. Reference should therefore be made to existing standards and other
recommendations on basic requirements. In order to cope with the special
operational aspects and the minor physical differences, like the water providing couplant during ultrasonic testing or that magnetic particles move
completely immersed in water, it is, however, advantageous to prepare additional guidelines for performing underwater NDT.

State-of-the-Art

IIW work. The IIW has by establishing a special working group (Working
Group 2 of Commission V) on underwater NDT drawn its attention to this special NDT application area. The working group was established 1977, one of
the main reasons being the number of welded structures used in offshore oil
and gas exploration and the need for some sort of guidelines for performing
underwater NDT as a supplement to IIW and other recommendations.

A draft version of a document on Common Practices for Underwater NDT will be
ready by summer 1983. This document addresses requirements to personnel and
procedures and reviews the different techniques and application areas.

Other work. Governmental bodies and certifying authorities have their requirements to underwater NDT. Det norske Veritas is now issuing Technical
Notes with more detailed requirements to procedures, personnel and equipment.

    FUTURE TRENDS

The development of new NDT techniques for underwater use is determined by:

a. Cost-effectiveness of inspection and maintenance
b. Quality of inspection
c. Installations in deeper waters

Offshore inspection and maintenance costs are in waters like the North Sea tremendous and savings related to reduced operation and diving time will have a great impact on these costs. The quality of underwater NDT is today very dependent on having qualified diving inspectors, and improvements are seen in taking critical examination result evaluation away from the diver.

To cope with the above needs, a development of underwater NDT is seen on the following:

a. Automation and remote control of equipment
b. Continuous monitoring
c. Advanced ultrasonic testing
d. Optical and acoustic imaging systems
e. Procedures for optimizing NDT efforts

Extensive research on the above is in progess and may in the future change the underwater NDT scene.

This research includes i.a. automated ultrasonic equipment for weld examination, the use of acoustical holography and ultrasonic time-of-flight. Further, acoustic emission and vibration analysis are addressed as continuous structural integrity monitoring tools. To improve on the quality of visual examination, especially in obscure waters like off the coast of Northern Alaska, imaging systems are being developed, based on sophisticated acoustic or optical principles.

Further, more precise information on examination reliability must be established. This is necessary as the reliability of under water NDT often is questioned and as this forms the basis to optimize inspection efforts.

REFERENCES

The above survey has been prepared from a number of sources: Det norske Veritas in-house experience, communication with a number of official bodies, offshore operators and inspection companies, WG2 of Commission V of IIW as well as a number of papers on offshore inspection and underwater NDT. For further information a few selected references are listed below:

Farley, J.M., and Dikstra, B.J. Review of the Basic Capabilities and Limitations of NDT Techniques Used for Inspection of Offshore Structures. EUR 333, European Petroleum Conference, London, 25 - 28 October 1982.
Bosselaar, H. The State of the Art of Underwater Inspection. EUR 334, European Petroleum Conference, London, 25 - 28 October 1982.
Materials Evaluation (April 1983 Special Issue on Underwater NDT) (in preparation 28.02.1983).
Auchterlounie, A.J. Magnetic Particle Inspection Underwater - the True Potential. Journal of the Society for Underwater Technology, Volume 8 Number 3, Autumn 1982.
Høgmoen, K., and Veierland, S. Corroscan - Automated Ultrasonic Equipment for Corrosion Inspection. International Conference on Pipeline Inspection, Edmonton, Canada, June 13 - 16, 1983.
Haugen, R., and Rangnes, E. A New Instrumentation for Fatigue Crack Depth

Measurement with AC Potential Drop Technique. Second European Conference on Non-Destructive Testing, Wien, September 14 - 16, 1981.

Moncaster, M.B. Underwater Inspection of Welds - An Assessment of Some Techniques and Their Reliability. Journal of the Society for Underwater Technology, Volume 8 Number 3, Autumn 1982.

Sletten, R., Mjelde, K., Fjeld, S., and Lotsberg, I. Optimization of Criteria for Design, Construction and In-service Inspection of Offshore Structures Based on Resource Allocation Techniques. EUR 332, European Petroleum Conference, London 25 - 28 October 1982.

Requirements for the Certification of Diver Inspectors Engaged on Visual Inspection, Ultrasonic Thickness and Lamination Testing, Magnetic Particle Inspection and Cathodic Protection Monitoring under Water. Document No. CSWIP-DIV-7-82.

Youshaw, R., and Dyer, C. Underwater Non-Destructive Testing of Ship Hull Welds. SSC-293 (1979), Ship Structure Committee, US Coast Guard.

Approval of Underwater Inspection Procedures, Certification of NDE Equipment for Underwater Application, Qualification of Underwater Inspection Personnel. Draft Technical Notes Fixed Offshore Installations, Det norske Veritas, 01.12.82.

Information on Common Practices for Underwater Non-Destructive Testing. IIW-Doc. (in preparation).

Requirements to Underwater Welding and Weld Inspection – Qualification of Procedures and Personnel

M. Tystad, G. H. Eide and N. Eikas

Det norske Veritas, Hovik, Norway

ABSTRACT

During the last 15 years dry hyperbaric welding has developed to a recognized and proven joining method of particular importance in connection with installation, maintenance and repair of fixed offshore installations.

As underwater welding is carried out remote from the surface support vessel, supervision of welding performance and inspection of the field joint is often difficult. Further, significant operational restraint is imposed on the welders during performance of the field work. Due to these factors, it becomes of great importance prior to start of field welding to ensure that the work is carried out by qualified welders using procedures qualified under realistic conditions, and that all field equipment is functioning as intended.

This paper is dealing with VERITAS' recommendations regarding requirements to hyperbaric welding and related qualification schemes.

KEYWORDS

Dry hyperbaric welding, underwater welding qualification schemes, underwater weld inspection, requirements to underwater welding.

INTRODUCTION

During the last 15 years underwater welding has developed to a recognized and proven joining method, and has become increasingly important particularly in connection with fixed offshore installations as these installations may require joining under water both in the construction phase and as part of the maintenance and repair work.

A distinction should be made between wet and dry underwater welding. Application of wet welding directly on main structural members and pipeline systems is normally not accepted due to the related high risk of weld cracking.

W–H

Dry underwater welding may be carried out under atmospheric conditions or at a higher pressure (hyperbaric welding). The application of atmospheric underwater welding has so far been limited, partly due to safety aspects and practical operational problems. The metallurgical aspects of atmospheric underwater welding are the same as those for atmospheric welding above water, and will not be further discussed here.

The majority of underwater welds performed on structural members and pipeline systems today, is carried out by dry hyperbaric welding. This is also assumed to be the situation in years to come. The present paper is consequently concentrated on this mode of joining.

Further the paper is mainly reflecting VERITAS' practice and experience in this field, and is directed towards the practical application of underwater welding. This practice and experience is also the basis for VERITAS' requirements to underwater welding as specified in the rules for fixed offshore structures (DnV, 1977) and submarine pipeline systems (DnV, 1981).

DRY HYPERBARIC FIELD WELDING - STATUS

To date extensive field experience in hyperbaric welding is limited to welding at waterdepths down to approximately 150 m, and C-Mn microalloyed/low alloyed steel grades up to and including yield strength level of 450 MPa for pipelines and 360 MPa for structures. So far, only manual welding has been applied.

During the last 10 years, VERITAS has been involved in approximately 100 dry hyperbaric field welding operations on different installations in the North Sea, Indian and Australian waters, at depths down to about 150 m. Some of these operations have been carried out by inert gas shielded flux cored wire welding, but the dominating process is manual metal arc with or without TIG welding of root pass(es).

The quality of these welds have been comparable to that obtained under atmospheric conditions.

Today dry hyperbaric welding is about to be extended to significantly deeper waters. This trend is promoted by several North Sea deep water projects like the transportation of gas from Statfjord to Norway in a pipeline system which will be crossing the Norwegian trench at 300 m waterdepth, and development of deep water oil and gas fields like Troll and Tromsøflaket, located at 300-350 m.

At these waterdepths, welding conditions and welder performance are expected to become increasingly complex and special precautions will have to be taken to ensure operational safety and high quality welding.

REQUIREMENTS TO HYPERBARIC WELDING

As a main principle, an underwater weld should not reduce the reliability of the structural part or pipeline section it belongs to. Consequently, the weld quality requirements would normally be the same as for welding above water.

From a metallurgical point of view, the main differences between welding in atmospheric air and hyperbaric welding are related to the habitat environmental conditions like pressure, gas composition, temperature and humidity. Welding at increased pressure results in an arc constriction which may introduce arc stability problems especially at greater depths. However, by selection of special power sources and/or addition of special ionizing elements to consumables, manual metal arc and TIG welding has been performed with stable arcs at waterdepths down to 320 m. (Knagenhjelm, 1980) and (Corriat, 1979)

The increased pressure also introduces changes in the chemical composition of welds performed with manual metal arc welding. These changes (normally increase in carbon and oxygen, decrease in manganese and silicon) have to be considered when consumables are to be selected.

The problem area of most concern in manual metal arc hyperbaric welding is, however the risk of hydrogen cracking. Normally the main source of hydrogen is the electrode cover. Especially at high pressure and high relative humidity, habitat moisture is rapidly absorbed by this cover (Grong, Ø., Gjermundsen, K., Mathisen, U., 1981) and Knagenhjelm, H.O., Gjermundsen, K., Kvaale, P., Gibson, D, 1982). To minimize the risk of hydrogen cracking, the following recommendations are given:

Initially reduce the amount of hydrogen entering the weld by

ɔ selection of a low hydrogen process (for SMAW, selection of extra low hydrogen electrodes)
ɔ careful handling and transfer of consumables to avoid contamination and minimize moisture pick up
ɔ use of preheat to ensure dry metal surfaces prior to welding

Assist the hydrogen that has entered the weld, to escape by promoting hydrogen diffusion through

ɔ preheating
ɔ postheating (under special critical conditions)

Except for the use of postheat, above mentioned precautions are recommended for all waterdepths, and become increasingly important as depth increases. Postheating is prudent as an extra safety measure for manual metal arc procedures below approximately 200 m waterdepth.

RECOMMENDED ROUTINES FOR ACCEPTANCE OF UNDERWATER WELDING

The fact that underwater welding is carried out remote from the surface vessel makes supervision and inspection difficult. Welding performance can only be observed via TV screens and monitoring of welding parameters. Verbal communication with the welders is often difficult as the characteristic of the welder's voice is changed in the pressurized helium atmosphere. In many cases it is further difficult to carry out a detailed and reliable final inspection of the field joint. Due to these factors, and also taking into consideration the operational restraint imposed on the welders, it is of great importance, prior to start of field welding, to ensure that the work is carried out by adequately qualified welders using qualified procedures, and that all field equipment is functioning as intended. To achieve this, below listed qualification scheme is recommended.

Welding Procedure Specification

A detailed welding procedure specification is to be established. In addition to what is normally required for surface welding, this specification is to give information of the waterdepth, and environmental conditions inside the habitat (e.g. humidity, pressure, gas composition temperature).

A procedure for storage and handling of welding consumables on the support vessel and in the welding habitat as well as sealing and transfer routines to the welding habitat should also be specified.

Welding Procedure Qualification

The specified welding procedure is to be qualified by performing a test under conditions representative for those prevailing under the actual field work. Due to the effects of pressure outlined above, the procedures should be qualified at the maximum pressure for the actual field work. The humidity level and temperature should also be as close to that in the field operation as possible.

The qualification test may be performed in a laboratory hyperbaric facility, at an offshore testing facility or at the site of the actual welding. Normally the preferred location is a laboratory facility. Throughout the duration of welding procedure and welder qualification, the welders are normally living inside a hyperbaric chamber system.

The qualification test welds are to be inspected and tested as specified for the structural section it belongs to, and meet the same requirements.

Normally the qualification programme consists of minimum one complete joint for manual welding, and minimum three joints for mechanized welding systems. This qualification programme may have to be extended in situations where welding will occur under conditions where previous experience is limited, e.g. welding at water-depths significantly beyond those where sufficient field experience is today available, and/or use of welding processes and procedures where limited field experience exists. As an example such an extended programme should include establishment of optimal welding parameters and bevel preparation/fit up and related acceptable tolerance limits. Further, additional documentation on certain critical aspects may be asked for, e.g. reasonable safety against hydrogen cracking when manual metal arc welding is to be applied in deep water.

Qualification of Underwater Welders

Underwater welders should be qualified under actual or simulated conditions for the work in question. Only welders that have passed a relevant welding test above water should be permitted to qualify for welding underwater. Prior to the test, the welders should be given sufficient training to get familiar with the influence of the hyperbaric condition on welding.

To assure maximum flexibility in a field operation all welders should be qualified for all relevant welding positions, and for welding of the complete weld (root, fillers and cap pass).

The approval of welders is normally based on radiographic and visual examination, and mechanical testing (nick break and bend testing).

Normally renewal of certificates for underwater welders is required if welding has been interrupted for more than 6 months. The criteria for type of field welding necessary to keep certificates valid, would have to be evaluated in each separate case.

Confirmation Test Weld

A confirmation test weld may be required made on location prior to staring the field welding to ensure that all systems are functioning as intended, and that sound welds can be performed under site conditions. Prior to starting production welding the test weld is to be accepted both visually and by means of radiographic examination, or ultrasonic plus magnetic particle testing.

When the same welding habitat, equipment and welding procedure are used for consecutive welds under comparable conditions, further confirmation test welds would normally not be asked for.

Examination of the Field Weld/Qualification of Equipment, Procedures and Personnel

Prior to examination of welds in the habitat, related equipment, procedures and personnel should be adequately qualified.

Normally radiographic examination is specified for non destructive testing of pipeline welds, and the actual procedure is then qualified on the surface as radiography is not pressure sensitive. In special situations, e.g. when by accident, water has entered the pipe and radiographic examination is not possible, ultrasonic examination has been applied.

For a potential application of MIG underwater welding (or other processes with significant risks of lack of fusion), ultrasonic examination should be specified.

For jacket leg or tubular joint repair, ultrasonic and/or magnetic particle examination are most often applied. The actual choice of inspection method will depend on geometry and location.

The performance of manual ultrasonic inspection is relatively complicated and procedures and personnel should be qualified on the actual type of joint(s). A slave-screen on the surface showing the ultrasonic signal in addition to monitoring of the actual movement of the transducer may be necessary. However, the trend is to develop remote controlled mechanized UT-equipment allowing better evaluation and documentation of the weld. This type of equipment will most likely be necessary for inspection of hyperbaric welds made at greater depths.

Regarding magnetic particle inspection, special attention should be paid to the choice of technique and equipment. The personnel should be qualified according to recognized standards and should be trained in using the actual test procedure.

REFERENCES

Corriat, G. (1979). Hyperbaric welding in the repair of offshore pipelines and structures. Second International Conference on pipe welding, London.
DnV (1977, reprint with corrections 1981). Rules for the design, construction and inspection of offshore structure.
DnV (1981). Rules for submarine pipeline systems.
Grong, Ø., Gjermundsen, K., Mathisen, U. (1981). Metallurgical aspects in hyperbaric welding. Sveiseteknikk.
Knagenhjelm, H.O. (1980). Hyperbaric welding at 320 msw. Development of adequate welding procedures. Underwater Technology Conference, Bergen.
Knagenhjelm, H.O., Gjermundsen, K., Kvaale, P., Gibson, D. (1982). Hyperbaric TIG welding to 500 m simulated depth. Underwater Technology - 82, Bergen.

User's Requirements and Control of Activities

D. B. J. Thomas

Central Engineering Department, BP International Ltd., London, UK

ABSTRACT

A number of specialist contractors now offer a subsea welding service to
meet a variety of oil industry repair, modification and construction
needs. The technology involved is still in its evolutionary phase, with
few establish rules. Weld quality relies mainly on the skill of the
diver/welder. Described are typical user precautions taken to provide the
best opportunity to acheive welds of acceptable quality, based on the
optimisation of available processes, procedures, pre-planning and overall
supervision. Reference is made to the present status of underwater
welding, associated research studies and its probable future advancement.

KEYWORDS

Welding process status; limited application rules; potential problems;
research studies; industry requirements; future advances.

INTRODUCTION

Reluctant to operate platforms, pipelines and other marine installations
without proven capabilities to undertake the repair of inadvertent damage,
and to cater for design modifications, the oil industry has encouraged the
development of reliable subsea welding systems and procedures. Over the
last decade, subsea welding has advanced to the stage where it is now
often the preferred in-situ, permanent repair method. Its use for new
constructions is less frequent.

Today, a number of specialist offshore contractors offer an underwater
service to meet a variety of user needs down to most water depths of
immediate interest. With the merits and limitations of the various
systems still under debate, and without the benefit of long established
national application standards or codes of practice to provide guidance,
potential users are often forced to submit each individual contractor's
welding proposals to detailed study prior to acceptance. Often a
contractor's proposal will require laboratory studies, or costly full-
scale offshore performance appraisal trials. In addition, whenever a

213

significant advance is required for any proposed novel application the user must be prepared to fund research and development.

Of assistance in 1977 was the introduction by Det Norske Veritas of rules which allowed the use of underwater 'wet' welding for minor repairs to structures, with hyperbaric techniques being required for important repairs and construction. In 1981 the same authority issued 'Rules for Submarine Pipeline Systems', which specified detailed requirements for underwater welded 'tie-ins' within chambers. In the same year the American Welding Society developed a draft recommendation, defining welding and inspection procedures for four categories of weld quality requirements, ranging from critical to non-load bearing attachments. Currently, the British Standards Institute is about to circulate for industry comments a revised version of BS 4515, which provides recommendations for the hyperbaric welding of girth welds in submarine pipelines.

Such formal recognition should engender user confidence in the reliability of subsea welding under controlled conditions. No longer can it be regarded as merely a means for providing a temporary or 'last resort' type of connection.

WELDING PROCESSES

Although friction, explosive, plasma, exothermic and other processes have been under continuous review, to date practical user experience has been limited to the following:-

(a)   Shielded metal-arc welding (SMAW) - using waterproofed electrodes ('in-the-wet'),

(b)   SMAW, semi-automatic gas metal-arc welding (GMAW), gas tungsten-arc welding (GTAW) with filler or flux-cored arc welding (FCAW) - in a dry chamber at a pressure corresponding to the working depth, commonly referred to as hyperbaric welding.

(c)   SMAW, GMAW or GTAW - in a dry chamber at a pressure of one atmosphere, when at working depth.

Mainly due to arc stability problems, and the adverse of effects of the rapid quench experience by the weld area, combined with hydrogen derived from the dissociation of water under the arc, users tend to avoid welding 'in-the-wet' for any critical construction. One atmosphere systems are unpopular due to their engineering complexities. Sufficient data have now been generated to convince both users and certifying authorities that under favourable conditions, manual and semi-automatic hyperbaric welding can result in 'all-position' joints of equivalent quality to those produced at a pressure of 1 bar on land.

USER CONDISIDERATIONS

To acheive optimum results it is obvious that clients must ensure that the contractor pays attention to at least all aspects of welding concern which apply to normal land practice. Into account has to be taken the avoidance of hydrogen cracking in heat-affected zones (HAZ), weld metal solidification cracking, fast fracture as a result of inadequate toughness, lamellar tearing, fatigue and stress-corrosion cracking, as well as possible galvanic corrosion where dissimilar weld metals are used. Exploratory corrosion tests carried out on behalf of BP (TWI, 1975) involving GMAW/Inconel 625 filler welds in Lloyds EH2 plate, indicated that the HAZ had a slightly cathodic potential to the parent plate. Thus no galvanic corrosion should be concentrated in this area. Current measurements on the connected samples in simulated sea water gave a corrosion rate due to galvanic action, if uniformly distributed, of only about 0.1mm/year on the parent plate.

With most underwater welding being applied by manual means, often in hostile situations, weld quality depends significantly upon the skill and concentration of the diver/welder. Usually, his difficulties increase with increasing depth. Variabilities in weld quality must be expected, unless proven non-marginal welding procedures are employed. Any element of experimentation at job location must be precluded, where welds of high integrity are demanded. The best opportunity for success is based on adequate pre-planning, qualified welding procedures, trained diver/welders, good record keeping, well defined weld inspection techniques and overall strict supervision.

No hyperbaric weld repair should be considered for any damaged component before a detailed inspection report is available. Any one or a combination of close visual, video-scan, still photography and magnetic particle inspection methods will normally reveal the extent and nature of any injurious defect. The use of plastic replicas are especially advantageous in providing a permanent record of minor denting, cracking and other features for conclusions to be drawn by specialists away from the job site. Another distinct benefit is that an excavated defect area replica can be used to define precise remedial weld procedures.

In certain cases, an engineering critical appraisal may allow superficial damage to be acceptable as it stands. Even surface cracking in non-critical regions may merely require removal by grinding or monitoring during subsequent service. Should repairs be necessary, it is usual to design and build a special chamber around a full scale model of the structural members concerned. This approach avoids welder/diver access problems, ensures that all tools and equipment are to hand, and facilitates familiarisation with habitat installation and job configuration. Welding inspection and control procedures should be developed and agreed before the chamber is lowered and assembled at the work location. To avoid the high cost of qualifying welding procedures offshore, it is customary to establish basic welding procedures in small scale laboratory pressure vessels, followed by their qualification at the contractor's manned hyperbaric facility. The same facility is used for the training and approval of welders, as well as for the evaluation of non-destructive examination techniques.

Ideally, the design of underwater repairs or modifications should be simple and not require precision installation. Good 'in-built' tolerances to 'fit-up' will avoid time consuming activities. The contractor must be fully conversant with the proposed work, and ensure that immediate 'back-up' is available for all failed equipment. Any tool required to expedite the work should be considered, including special welding fixtures and jigs.

Throughout repairs the supervising engineer must be in the diving station or control module to monitor welding operations effectively. For most hyperbaric welding, voltage, amperage and shielding gas flows are controlled by an experienced technician in constant communication with the welder by CCTV and audio. The Welding Institute (TWI) has recently developed a solid state device for the measurement, display of any deviation, and recording from preferred values of the main variables of current, voltage and wire feed rate. Such continuous monitoring could provide a valuable quality assurance tool. A combination of a proven welding procedure, appropriate NDE technique, and the continuous recording of any welding conditions should provide users with a reasonable proof of weld integrity. The introduction of this device should alleviate the distinct disadvantage of the welding engineer or inspector having, limited or no direct access to the work area, and the diver/welder, except by CCTV and aural means.

Where the proposed weld joint configuration, material composition and thickness would normally entail post-weld heat-treatment (PWHT) investigators are attempting to develop welding procedures which minimise the need for this requirement. Emphasise on tempering techniques, reduction of stress intensification zones, assurance that all joint ligaments possess adequate toughness and stringent NDE procedures are producing promising results. In all probability, it will be demonstrated that the fast fracture resistance of 'as welded' condition repairs will be dominated more by the deposited weld metal properties than its HAZ. Considerable metallographic, residual stress, hardness and fracture toughness studies will be required before such special welding procedures designed to avoid PWHT are widely accepted.

MATERIALS

North sea users tend to require underwater welding procedures to cover structural and linepipe steels with a strength range of 270 to 690 N/mm^2. Typically, structural tubulars are supplied in accordance with the requirements of BS 4360 Gr. 43, 50D or modified 50D, with pipe in control-rolled, or quenched and tempered variants of API 5LX Gr. 60 or 65. For structural repairs, it is usual to aim for a HAZ hardness that is not in excess of about 300 HV10, where there is a risk of stress-corrosion cracking. In sea water environments, opinions exist that hardnesses of 450 HV10 are a potential problem. Where hydrogen levels may be higher as a result of cathodic protection the hardness limit should be well below this level. Hardness requirements for submarine pipeline repair welds vary for each particular duty with the experience that material softer than Rc 22 (248 HV10) has behaved satisfactorily in sour service. Should doubts arise regarding the justification of the acceptance of higher hardness levels, the options are appraisal by stress-corrosion cracking tests or PWHT by electric resistance elements. With the modern available line pipe steels and the wall thickness normally involved, PWHT is seldom necessary.

A feature which should be taken into account by the user is the hyperbaric cold-cracking behaviour of each combination of material chemical composition, thickness, welding process and water depth concerned. An example of such a study, undertaken by the Cranfield Insitute of Technology on behalf of BP (CIT, 1982), established that a 100°C preheat was sufficient to produce crack free welds on Tekken test samples in 25 mm wall thickness plate at a pressure equivalent to 200 m depth.

Steel :   CE (11W)   = 0.45   Pcm = 0.22
Welding Process : FCAW   Thermal input : 1.81 kJ/mm
Welding Position : Flat

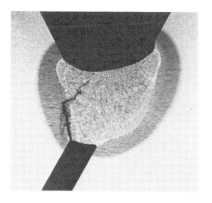   

Fig : 1   No preheat - hydrogen crack   Fig : 2 100°C preheat - no crack

For both tests, the maximum hardnesses for weld metal and HAZ respectively
was 268 and 345 HV5. Based on the limited laboratory pressure chamber
results obtained, it was tentatively suggested that plate cooling in the
argon atmosphere concerned was the same as under normal atmosphere
conditions. Thus, if the only pressure dependent factor influencing
cracking was the deposited hydrogen level, it is possible that experiments
at 1 bar pressure could simulate the influence of ambient pressure on cold
cracking even at hyperbaric pressure. The main conclusion was that
preheat had a pronounced effect on cold cracking. It should be remembered
that the Tekken test has a higher restraint than is usually encountered in
practice, and does not include the beneficial tempering effects of
subsequent weld passes.

UNDERWATER WELDING - PRESENT STATUS

Qualified (usually to BS Std. 4515 or API.1104 and supplementary user
requirements) pipe welding procedures exist for SMAW and GTAW processes
down to 300 m depths, with GTAW welds being superior to all other process
welds, even in the shallower depths. In 1981, Sub-Sea Offshore Ltd. U.K.
demonstrated that FCAW was satisfactory at 180 m as a standby, contingency
repair method for the BP Magnus 24in O.D. X 19 mm. w.t. API 5L Gr.X65 oil
line (Thomas, 1981). The same paper describes a FCAW pipeline repair
carried out earlier in 1976 at a depth of 70 m.

Whilst various investigators justifiably claim benefits from the use of
alternative welding processes, it is generally agreed that the replacement
of the welder/diver with mechanised welding would enable joints of more
consistent quality to be produced. As the depth increases, decompression
times are increased, and the capabilities of the diver in terms of
dexterity, concentration and productivity are reduced. Even in the
shallower depths, downtime due to adverse weather and tide conditions can
result in highly inefficient working. A few diving contractors are

developing mechanised orbital welding systems, with a limited degree of manned intervention. The welding processes likely to be an early offer will be solid filler GTAW or GMAW with filler addition. Despite its impressive manually applied operational record to date, the FCAW process is unlikely to be favoured. This is due to its complicated chemistry, which increases with depth, combined with deslagging, fume and spatter disadvantages. A further difficulty is the provision of flux-cored consumables capable of being wound on to small diameter "take-off" spools.

Due to the lack of comparative data, an objective choice between the GTAW and GMAW processes for hyperbaric orbital welding is not readily apparent.

Keen to ensure that underwater welding capabilities existed to meet its anticipated future requirements, BP commissioned TWI (TWI,1975) to study the effect of pressure on GMAW arc characteristics in a laboratory chamber at a range of pressures equivalent down to a depth of 130 m (=14 bar). High speed cine photography was used to examine metal transfer mode with electrode positive polarity (DCEP), and electrode negative polarity (DCEN) in various shielding gases, as well as with pulsed GMAW (DCEP and DCEN). With conventional DCEP welding in argon or helium based gases, the arc became progressively constricted as pressure increased. At 14 bar a strong, single spot cathode jet was formed which repelled the molten droplet, causing spatter and heavy fume. The GMAW/DCEN arc became progressively more stable as pressure was increased and weld bead shape improved at 14 bar pressure with low fume levels.

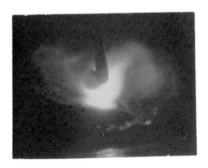 

Fig 3
Argon based DCEP, 5 bar
pressure, repulsive transfer
high fume, constricted arc

Fig 4
Argon based DCEN, 14 bar
pressure, smaller molten
drop size, less fume

218

In 1981 BP placed a contract with CIT, in colaboration with TWI, for a comprehensive review of automatic welding for underwater applications (CIT 1981). There followed a commission to assess the hyperbaric orbital welding capability of the GMAW and GTAW processes (CIT, 1982). Successful laboratory trials were carried out with both processes at pressures equivalent to depths greater than 300 m. An indication of the weld quality achieved on the BS 4360 Gr. 50D steel samples concerned at 20 bar pressure is shown below.

Fig 5.
GMAW section, overhead position

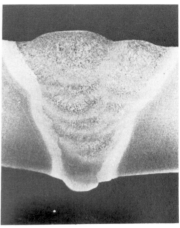

Fig. 6
GTAW section, flat position

During testing considerable data were gained for the optimisation of welding parameters, joint designs, weave patterns, interpass temperature and other factors to achieve acceptable weld and HAZ hardnesses. For GMAW fill passes at 20 bar pressure weld metal deposition rate was 1 to 2 kg/hr. and for GTAW it was 1.5 kg/hr. The use of narrower gap joint preparations for the GTAW process would have increased its productivity without sacrificing quality. To avoid inter-run defects with GMAW, fill run weave patterns were important.

CONCLUDING REMARKS

In simulated hyperbaric conditions welding has been successfully carried out at around 500 m. Opinions have been expressed that the realistic physiological limit for a diver is some 500 m in the forseeable future. Thus, any welding system devised for greater depths is either likely to incorporate a 'one-atmosphere' environment around the weld or a water-proofed welding head controlled by an ADS unit ('Jim' or 'Wasp' as developed by Oceaneering Int.). The concept of a subsea truly automated orbital machine is beyond the reach of present day technology. However, the notion of a pre-programmed system with minimum direct saturation diver/welder is a realistic target, as is a system capable of operation by remote control. Fortunately, with a few notable exceptions, user interest is likely to be concentrated to waters located on the Continental Shelves, where the average depth is some 90 m, and seldom exceeds 300 m. Beyond this depth welding problems will be resolved only when the market forces demand solutions.

Most users have neglected 'in-the-wet' welding methods. With the undoubted attraction of low cost and simplicity, the industry should re-examine their potential, if only for work of secondary importance, and in shallow waters. The processes meriting further studies would appear to be SMAW, GMAW and friction welding.

Providing collaboration continues between main clients, contractors, classification societies, industrial, university and government laboratories, many advances can be anticipated in this challenging technology both in operational performance and cost effectiveness.

ACKNOWLEDGEMENT

Permission to publish this paper has been given by The British Petroleum Company p.l.c.

REFERENCES

TWI Unpublished Report (1975). The Welding Institute, U.K. for BP.
CIT Unpublished Report (1982). Cranfield Institute of Technology, U.K. for BP.
Thomas, D.B.J. and K.R. Doble (1981). Steel requirements for submarine piplines. Steel for Linepipe and Pipeline Fittings Conf., The Metals Soc., London, paper No. 3.
TWI Unpublished Report (1975). The Welding Institute, U.K. for BP.
CIT Unpublished Reports (1981/82). Cranfield Institue of Technology in collaboration with The Welding Institute, U.K. for BP.

340/41G

An Overview of "Specifications for Underwater Welding" AWS D3.6

C. E. Grubbs*, M. F. Bluem** and P. T. Delaune***

*D & W Underwater Welding Services, Inc., USA
**Exxon Company, USA
***Delaune and Associates, USA

ABSTRACT

Extensive use of underwater welding resulted in the need for specifications
that would allow users of underwater welding to conveniently specify and
obtain underwater welding of a predictable performance level. This paper
provides an overview of how American Welding Society's "Specifications for
Underwater Welding" were developed and acquaints the reader with the contents
of the Specification.

Five basic underwater welding methods are described, including One Atmosphere,
Dry Hyperbaric and Wet. Rather than a sharp distinction between each weld
method, intermediate degrees of weldment and welder protection from the water
are in use. Metal transfer characteristics, solidification behavior, mechanical properties, etc. can vary with pressure and some methods of underwater
welding may produce properties that differ from what is usual with welds
made above water. Because of this, the specification defines four weld types.
Each weld type has a specific set of quality requirements which shall be verified during welding procedure qualification. The specification provides
information required for qualifying welding procedures and welders for each
of the four weld Types plus inspection and nondestructive examination of
weldments.

KEYWORDS

Underwater Welding Specifications, American Welding Society, Subsea Welding,
Hyperbaric Welding, Wet Welding, Welding Specifications, Underwater Inspection.

INTRODUCTION

The American Welding Society's "Specification for Underwater Welding" (D3.6)
is a milestone in the progress of underwater construction and repair technology. The specification was developed in response to the needs of the offshore industry and is designed to allow a user of underwater welding to conveniently specify and obtain underwater welding of a predictable performance
level.

The AWS D3.6 technical committee that developed the specification was formed in 1974 as a sub-committee of the AWS D3 committee on "Welding in Marine Construction" in the Spring of 1974. The committee members were selected so that about one third represented potential users of underwater welding and about one third were involved in providing underwater welding services. The balance was made up of representatives of regulatory agencies and others with a high degree of technical and practical knowledge of welding. The initial membership of fourteen was gradually increased and ultimately twenty six members, representing twenty four U.S. and foreign companies, contributed valuable time and effort to the development of the specification. In addition, input was received from U.S. and overseas experts who were not members of the committee. During the nine years of development of the specification, innumerable meetings of the committee and work groups were held and members participated in various underwater welding technical conferences in the U.S. and abroad.

From the beginning, the goal of the committee was to identify essential variables associated with the various modes of underwater welding and develop specifications for qualifying welding procedures and welders as well as specification for inspection and nondestructive examination of the welds made at the underwater work site. The anticipated result is that the specification will provide substantial improvements in the manner in which future underwater construction and repair projects are contracted for and executed.

The committee elected to use AWS D1.1 "Structural Welding Code ⌐ Steel" as a basis for the format of the new specification. Scope of the specification progressively developed as the committee considered the following:
1. For what purpose will underwater welding be used.
2. What methods of underwater welding will be most commonly used. (Dry Hyperbaric, Wet, etc.)
3. Capabilities and limitations of the various methods.
4. Essential variables pertinent to the various methods.
5. Welds made by different methods produce different types of welds with different performance or quality levels. (The specification therefore deals with four types of welds: A,B,C,and O).
6. Acceptance standards for the four types of welds.
7. Quality assurance, including underwater inspection and nondestructive examination, on site confirmation weld, etc.
8. Safe practices for underwater welding and associated tasks.
9. What does industry need to conveniently contract for underwater welding with consistently predictable results.

In developing the specification, the committee defined five basic methods of underwater welding as follows:
1. Welding in a pressure vessel in which the pressure is reduced to approximately one (1) atmosphere, independent of depth (one atmosphere welding).
2. Welding at ambiet pressure in a large chamber from which water has been displaced in an atmoshpere such that the welder/diver does not work in diving equipment (habitat welding).
3. Welding at ambiet pressure in a simple openbottomed dry chamber which accomodates ⌐ as a minimum ⌐ the head and shoulders of the welder/diver in full or modified diving equipment (dry chamber welding).
4. Welding at ambient pressure in a small transparent enclosure with the welder/diver outside in the water (dry spot welding).
5. Welding at ambient pressure with the welder/diver in the water

222

without any physical barrier between the water and the welding arc (wet welding).

Rather than a sharp distinction between each method, intermediate degrees of weldment and welder protection from the water are in use. Metal transfer characteristics, solidification behavior, apperance, mechanical properties, etc. can vary with pressure and each method of underwater welding may produce properties that differ from what is usual with welds made above water. Because of this the specification defines four weld types.

Each weld type has a specific set of quality requirements which shall be verified during welding procedure qualification. (The specification does not address the selection of which type weld meets the requirements of a particular application. The selection of the type of welds to be used is left up to the customer).

Type O welds meet the requirements of a governing "in-air" code, specification or other mandatory document as well as additional requirements defined in the "Specifications for Underwater Welding".

Type A defines a stand-alone set of requirements for underwater welds which are also intended to be suitable for comparable applications and design stresses as their above water counterparts by virtue of having comparable properties and testing requirements specified.

Type B welds are defined by an intermediate set of mechanical and examination requirements. They are intended for less critical application where reduced ductility, increased porosity, etc., can be tolerated. The suitability of Type B welds for a particular application should be evaluated on a fitness for purpose basis.

Type C welds satify lesser requirements than the other types and are intended for applications where the load bearing function is not a primary consideration. The selection of Type C welds shall include a determination that their use will not impair the integrity of the primary structure by creating fracture initiation sites.

As illustrated by Appendix A, (Contents - "Specification for Underwater Welding") the specification provides information required for qualifying welding procedures and welders for each of the four weld Types plus inspection and nondestructive examination of weldments. See Appendix A for other information provided in the specification. The subcommittee on "Specifications for Underwater Welding" will undertake to modify and revise the specification to keep current with technologial changes and general input from users of the specification. Suggestions for improvements or questions on interpretation are welcome and should be addressed to American Welding Society, P.O. Box 351040, Miami, FL 33135, U.S.A.

APPENDIX "A"

SPECIFICATIONS FOR UNDERWATER WELDING, AWS D3.6

CONTENTS

224

Mechanized Underwater Welding Using Austenitic Consumables

A. A. Ignatushenko, A. V. Denisenko and Yu. V. Djachenko

E. O. Paton Welding Institute of the Academy of Sciences of the Ukrainian SSR, USSR

ABSTRACT

One of the possible alternatives of improving the quality of welded joints of steels, difficult-to-weld in underwater conditions ($C_e \geq 0.40$), is considered. The results of preliminary experiments on mechanized underwater flux-cored wire welding, providing the austenitic structure of the weld metal, are given. The measurements of diffusive-mobile hydrogen, contained in austenitic weld metal, are made and compared with hydrogen saturation of ferritic welds. The corrosion resistance of underwater welded joints is investigated. It has been concluded that the application of the mechanized underwater welding using austenitic welding consumables is promising enough.

KEYWORDS

Wet underwater welding. mechanization, flux-cored wire, hydrogen, austenite, cold cracks, corrosion resistance.

Underwater welding is widely used in construction and repair of various structures. Moreover, the role and the importance of underwater welding considerably increase due to the widening of the scope of operations on the continental shelf. The perspectives of future underwater welding application are quite attractive, e.g. the construction of off-shore floating platforms, towns and airports, the laying out of underwater service lines, the installation of large capacity gas and oil storage tanks and others. The gained experience of the application of "wet" and "dry" methods of underwater welding allows to estimate their advantages and drawbacks.

Wet underwater welding provides an exceptionally high mobility, allows to operate in difficult-to-reach places of complex three-dimensional metal structures, does not require high expenditures and is performed using a simple equipment. The wet underwater welding drawbacks are caused,as was noted by Tsai and Masubuchi (1977), by intensive cooling of welds, resulting in the formation of quenching structures, and the high content of diffusion-mobile hydrogen in weld metal, i.e. such factors that cause the occurence of cold underbead cracks, the decrease of toughness and impact resistance in the metal deposited under water in certain conditions. Low carbon and low-alloy steels, their carbon equivalent being < 0.40, are underwater welded well enough using electrodes with low-carbon steel core. Recently, in the practice of underwater-engineering jobs the higher and high strength steels, corresponding to X-60, X-65, X-70 classes according to American standard API-1104 found their application. It is difficult to obtain sound joints of these steels in welding with electrodes, providing the deposited metal ferritic structure. At the same time, according to data of Grubbs and Seth (1977), the application of electrodes with stainless steel or nickel alloy core allows to obtain welds free of underbead cracks, even of steels with 0.696 carbon equivalent, thus ensuring high mechanical properties of welded joints. The results of work, made by Helburn (1979), prove the efficiency of the application of electrodes with nickel alloy core. The studies, carried out by Sadowski (1980), showed that the application of electrodes with stainless steel or nickel-base alloy core can prevent the cold underbead crack formation in "black" steel welding. The improvement of the weld quality in "wet" manual arc welding can be achieved by the development of the low-carbon (C=0.03% max.) cored electrodes, by using such electrodes, which provide the deposited metal with the increased "tolerance" to hydrogen, i.e. the electrodes with nickel alloy or stainless steel core for welding steels with high carbon equivalent, by applying the precipitation-hardening steels for the underwater metal structures. In any case, the use of the electrodes providing the austenitic structure of the deposited metal allows to eliminate cold cracking in welding steels, having carbon equivalent of ≥ 0.40. However. the wet underwater manual arc welding method has some drawbacks, i.e. the low efficiency, difficulty of making the welds in different spatial positions, the coating lamination, especially when operating at a large depth, the incomplete coating burning, the arc burning instability the formation of a chemically heterogeneous interlayer in the weld zone in welding with austenitic steel-cored electrodes.

The elimination of the drawbacks peculiar to the manual arc stick electrode welding and the widening of the technological possibilities of mechanized underwater welding became feasible by using the new welding consumable, such as flux-cored wire, specially developed for welding the higher strength steels and providing the austenitic weld structure.

The technology of its manufacture does not differ from that usually employed in the flux-cored wire manufacture. The chosen sheath and core composition is such that it provides the austenitic structure of the deposited metal. The forming slag layer, covering the weld, is of a basic type.

The said flux-cored wire was used to weld butt joints of steels of 17ГIC, X-60, X-65 and other grades, difficult to weld under water by the wet method. The mechanical properties obtained in welding X-60 - X65 steels are presented in Table.

TABLE Mechanical properties of the welded joints made by using the austenitic type flux-cored wire

Mechanical properties	Yield strength σ_y, MPa	Ultimate strength σ_u, MPa	Elongation δ, %	Reduction in area ψ,%	Impact toughness at +20°C a_H, kgm weld metal	HAZ
Weld metal	346	623	38-40	55.6	18.7	18.7
Base metal Steel X-65	449-450	597-604	33.3-38.0	55.8	20.0	-

Note: The Table shows the average values of three mea- surements.

The Table shows that the values of ultimate strength and of yield strength are in relationship characteristic of austeni- tic metal, the values of elongation and reduction in area are on a rather high level, impact toughness, particularly in the HAZ, which is seemed to be the weak point in welding some X-60 - X-65 type steels, is also high.

The susceptibility to cold cracking in the HAZ due to hydrogen was determined in welding of restrained V-grooved test speci- mens. In one-pass X-60 and 17ГIC steel plate welding, the hyd- rogen cracks were not revealed.

Fig.1 shows the distribution of hardness in the cross-section of the X-65 steel plate butt joint, made by using the wire pro- viding the austenitic weld metal structure. It should be noted, that the distribution of hardness in the joints welded by using the austenitic and ferritic type wires is similar; the maximum values of hardness in the HAZ are approximately the same.

Fig.2 shows the character of the hardness changes while pas- sing through the line of fusion of beads, deposited on the 08Г2МФБ steel plate by using the flux-cored wires, providing the austenitic (curve 1) and ferritic (curve 2) structures of the deposited metal.

The sound base metal-weld metal fusion is provided in austeni- tic type flux-cored wire welding of the higher strength steels. The interlayer, i.e. decarburized diffusion zone, characteris- tic of austenitic steel-cored electrode welding, is,as a rule, absent. The weld structure is a purely austenitic one with etched polygonization boundaries. Austenite grains in the over-

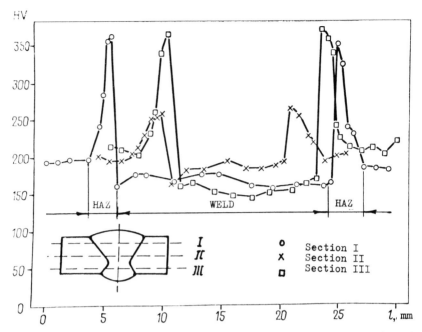

Fig.1. The distribution of hardness in a butt joint made of
steel X-65 by using the austenitic type flux-cored
wire.

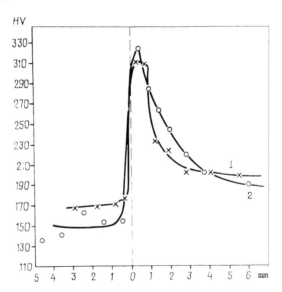

Fig.2. The character of hardness changing in the HAZ of deposits having the austenitic-1 and the ferritic-2 structures.
(base metal - steel 08 Γ 2МФБ   )

heating zone are surrounded by a ferritic network, the maximum grain being No. 3-4 (0.125 - 0.111mm), at the weld root - No.6. Separate regions of martensitic structure, their hardness being 400 H_μ , are formed in the fusion zone of some specimens. The typical fusion line is of the form shown in Fig.3.

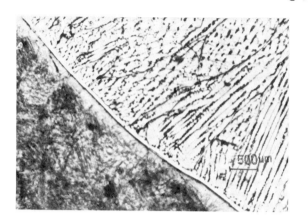

Fig.3. The typical fusion line in austenitic
flux-cored wire welding of the higher
strength steels.

At present there is no doubt that hydrogen together with structural factors contributes to cold crack formation. The fact, that in the considered case with comparatively high values of hardness the cold cracks are not detected must be accounted for different behaviours of hydrogen in austenitic and ferritic welds (structures). Actually, under other conditions being equal, there are no cold underbead cracks in the welds having the austenitic structure, while in the ferritic structure welds they are observed. (Fig.4,5)

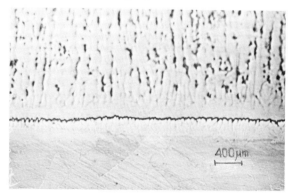

Fig.4. The fusion line in austenitic type flux-cored wire welding of steel 17ГIC (No cracks in the HAZ.)

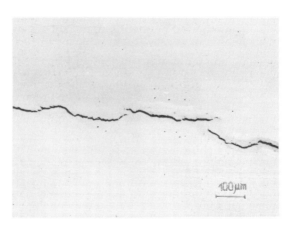

Fig.5. The fusion line in ferritic type flux-
cored wire welding of steel 17Г IC
(cracks in the HAZ are parallel to the
fusion line).

To prove the said assumption, which had been formerly used for
the explanation of cracks absence in the HAZ of austenitic
welds, made under water by using stick electrodes, Koibushi
and Yokota (1981), the comparative measurements of the content
of diffusion-mobile and residual hydrogen were made in the
welds of the austenitic and ferritic structures and the dependencies of these contents on the main welding parameters were
determined (Fig.6,7).

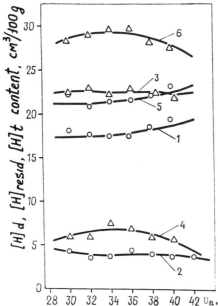

Fig.6. The dependen-
ce of the total hyd-
rogen content on the
arc voltage:
in ferritic type wire
welding; in austenitic
type wire welding
$1,2,3 - H_{diff}$,
H_{resid}, H_{total} in
ferritic wire welding
$4,5,6$ - the same in
austenitic wire wel-
ding.

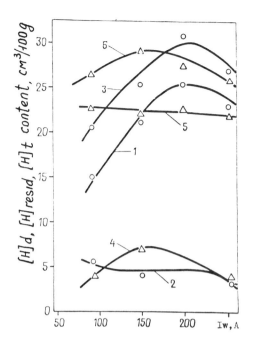

Fig.7. The dependence of the total hydrogen content on the current value: in ferritic type wire welding; in austenitic type wire welding.

1,2,3- $H_{diff.}$, $H_{resid.}$, H_{total} in ferritic type wire welding; 4,5,6 - $H_{diff.}$, $H_{resid.}$, H_{total} in austenitic type wire welding.

The welding conditions were chosen within the ranges practically used for welds in various spatial positions. The content of diffusion mobile hydrogen was determined by the chromatographic method, using the installation and the procedure developed at the E.O.Paton Electric Welding Institute by Pokhodnya and Paltsevich (1980). The content of residual hydrogen was determined by the method of reduction melting in the stream of high purity nitrogen using the analyser RH-2. For the ferritic structure welds it amounts to 4-5 cm^3 per 100 g, while for the austenitic structure welds it being 23 cm^3 per 100 g and does not depend on the condition variation. According to the results as a whole it may be noted, that with the contents of hydrogen in the welds of austenitic and ferritic structures being approximately equal, the quantity of diffusion mobile hydrogen in the former is considerably lower. Hydrogen in the welds of austenitic structure is in the bound state, its diffusion movements are to a great degree blocked. It may be assumed that there occur no increases of the diffusion mobile hydrogen concentrations, dangerous from the cold cracking point of view, in the HAZ higher hardness regions. This, probably,

233

accounts for the underbead cracks being absent in the HAZ of the underwater welds of the austenitic structure with the relatively high hardnesses of structural components.

Generally,the underwater welding with welding consumables, pro viding the austenitic structure of the deposited metal, includes the joining the "black" steel with a high-alloy weld. When submerging such welded joint into sea water, at the boundary of heterogeneity the stress potential occurs and the accelerated corrosive wear is observed. To verify the technical acceptability of the proposed version of joining the higher strength steels with the high-alloy weld, they were corrosive medium resistance tested. Corrosion tests were performed according to the procedure of the E.O.Paton Electric Welding Institute with regard to the requirements of the industry standard (1969). It is established that the joints of the "black" higher strength steels, made with the austenitic welds are subjected, as it was expected, to accelerated corrosion wear within the HAZ. At the same time, the high-alloy weld is absolutely insensible to the corrosive medium influence, but the wear in the HAZ is 1.4-1.5 times greater than that of the base metal plate (see profilogram a., Fig.8).   For comparison, Fig.9 shows the profilogram of the specimen welded by the flux-cored wire, providing the ferritic weld. It is possible to eliminate the increased corrosion wear in the HAZ by using the protective adhesive-base coatings, the underwater application technology of which was tested and used in industry.The protective coating of the said type allowed to considerably decrease the cor rosion wear of both the ferritic and austenitic welded joints and completely eliminate the wear in the HAZ of the welded joint, the weld of which has the austenitic structure (Fig.8, 9, dotted lines).Adhesives were reported by Shanaev (1980).

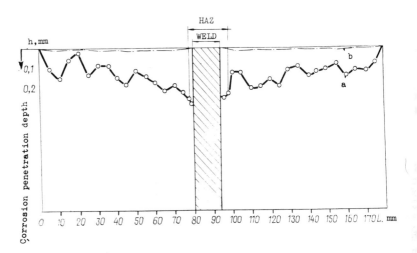

Fig.8.   The profilogram of the specimen welded by using the austenitic type flux-cored wire: a) without protective coating; b)with protective coating.

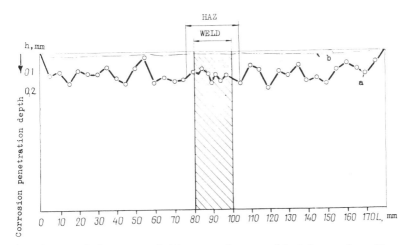

Fig.9. The profilogram of the specimen welded by using the ferritic type flux-cored wire: a)without protective coating; b)with protective coating.

CONCLUSIONS

1. The flux-cored wire has been developed for mechanized underwater welding of higher and high-strength steels, the pipe steels in particular, this wire providing the austenitic structure of the deposited metal and the high mechanical properties of weld of steels 17ГIC, X-60, X-65 and others, difficult to weld under water by the wet method. The high mechanical properties are ensured due to the decrease of diffusion-mobile hydrogen content in the deposited metal.
2. The application of austenitic welding consumables in wet welding permits to widen the technological possibilities of the latter in improving the quality of welds of steels with the high value ($C_e \geqslant 0.40$) of carbon equivalent.
3. The corrosion properties of the welded joints made with the austenitic welds can be considerably improved due to the protective adhesive coating application.

REFERENCES

Tsai, Chong-Liang and Masubushi, Koichi Interpretive Report on Underwater Welding. Welding Research Council Bulletin No.224, February (1977).
Grubbs C.E. and Seth O.W. Underwater wet welding with manual arc electrodes. "Underwater welding for Offshore Installations" Abington, (1977) pp.17-34.
Helburn, Steve Underwater welders repair drilling rigs. "Welding Design and Fabrication" July,(1979), pp.21-27
Sadovski E.P. Underwater Wet Welding Mild Steel with Nickel Base and Stainless Steel Electrodes. Paper presented at the 61st AWS Annual Meeting held in Los Angeles, California, during April 14-18, 1980.

Koibushi Masao and Yokota Takeo Underwater Wet Welding with
Ni, Fe-Ni and Stainless Steel Electrodes",
J.Jap.Weld Soc, 50,5, pp.489-495.
Pokhodnja I.K., Paltsevich A.P. Chromatographic method for
determining the quantity of diffusion hydrogen in welds.
J. "Avtomaticheskaya svarka",(1980) No.1, pp.37-39.
"The weld and welded joint metal of hull steels. Technical
requirements and methods for determining the properties.
Industry standard OH9-173-69.
Shanaev Zh.I. Underwater technical works using adhesives.
J. "Stroitelstvo truboprovodov", (1980), No.1, pp.29-32.

SECTION III

PHYSICAL, METALLURGICAL AND MECHANICAL PROBLEMS / PROBLEMES PHYSIQUES, METALLURGIQUES ET MECANIQUES

Effect of Pressure and Process Parameters on Weld Joint Properties of St E 36 and St E 47 Dry FCA-MAG Welds

H. Hoffmeister*, G. Huismann* and H. G. Schafstall**

*Hochschule der Bundeswehr, Hamburg, Federal Republic of Germany
**GKSS Forschungszentrum, Geesthacht, Federal Republic of Germany

ABSTRACT

Dry hyperbaric MAG welding of St E 36 and St E 47 at (Ar + 1 bar air) pressures up to 30 bars and constant mass flow rate of Ar+5% O_2 shielding gas shows the following results:
1. Increasing pressure causes excessive short circuiting and weld spatter at increasing voltages, constant welding speeds and wire feed rates.
2. Increased nitrogen pick up of the weld metal with increasing pressure and voltage takes place.
3. Increasing amounts of plate material elements are introduced into the weld metal at increased pressure.
4. CVN toughnesses are deteriorated at increased pressures due to nitrogen pick up which yet may be balanced by microalloying elements.
5. The conditions of laminar flow of the shielding gas appear not to provide sufficient shielding capacity for the arc and weld pool.

KEYWORDS

Dry hyperbaric MAG welding, effect of pressure, process parameters, mechanical properties, shielding gas flow.

INTRODUCTION

In hyperbaric welding of microalloyed medium strength steels the achievement of specified weld joint quality requires to consider the effects of pressure on the arc welding process as well as of the welding conditions on the microstructure of multilayer welds. While pressure dependent conditions for stable arc welding processes and their effect on weld geometry and weld metal chemistry are fairly well understood (Schafstall, H.G. and R. Schaefer, 1979) the influence of habitat pressurized gas composition on weld joint quality despite some work (Yamada and Ozaki, 1979) seems not so well established and recognized. This applies also to residual amounts of air in the habitat which may cause weld metal nitrogen pick up and, consequently, decrease of toughness. In the present work, therefore, the effect of a

239

partial air pressure of 1 bar and argon pressures up to 29 bars on process parameters of FCA-MAG welding of St E 36 and St E 47 butt welds, weld metal composition and multilayer weld metal toughness is investigated.

EXPERIMENTAL PROCEDURE

The experiments were carried out in a 10 m^3 pressure vessel of the GKSS research center, Geesthacht, at total pressures of 1, 2, 5, 11, 21, 29 and 30 bars, leaving 1 bar of air atmosphere within and pressurizing by the addition. A CP power source with 60 V maximum open circuit voltage and 400 A maximum current was used. St E 36 and St E 47 330x150x15 mm plate samples (Table 1) were prepared by a V-groove 60° opening angle, with up to 3 mm root face at increasing pressure.
Ceramic backing strips were attached. Basic flux cored seamless filler wire "Fluxofil 31", 1,2 mm diameter was used at a constant wire feed rate of 12 m min^{-1}. The shielding gas was argon + 5 % O_2 which, for the purpose of approximating laminar flow conditions according to (Allum and others, 1980) was used at constant flow rate of 15 Nl min^{-1}. Though, at constant shielding gas mass flow rate, increasing chamber pressures would possibly deteriorate the arc shielding effect, laminar flow is not only in accordance with usual practice at 1 bar but was also chosen for a defined experimental basis. Depending on pressure, between 3 and 4 layers containing up to 8 beads were deposited. The time-temperature history was recorded by thermocouples positioned in the root pass and melting back to the fusion line. The travelling speed was between 26 and 31 cm min^{-1}. Before starting, appropriate voltages for a stable process were selected by a test weld. In this procedure, the limits of the stable process at increasing pressure were determined by excessive short circuiting at low and arc burn back to contact tube at too high voltages. The metal droplet transfer was observed and controlled by a camera and continuous oscillograph - measurements of voltages and currents.

RESULTS AND DISCUSSION

Process and Weld Metal Chemistry

Figure 1 reveals the well known (Kobayashi and others, 1975) increase of welding voltage with pressure from 26-29 V at 1 bar up to about 41 V at 30 bars.

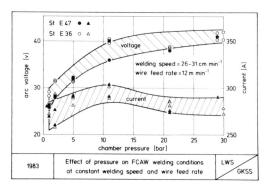

Fig. 1.

The corresponding welding currents increase from a range between 260 and 290 A at 1 bar to peak values between 305 and 290 A at 11 bars but decrease to 270-290 A at a pressure of 30 bars. The metal droplet transfer at low pressures showed only few short circuiting and small droplets, which, at increasing voltage, could be maintained up to 11 bars. At increasing pressure, however, the decrease of the current of the constricted arc column was accompanied by an increase in short circuiting frequency and droplet size, Fig. 2. In addition, increasing spattering occurred. This seems to be in contrast to earlier results of Stalker and Salter (1975) who used lower wire feed rates and massive wires in GMA welding and found a transition from globular short circuiting transfer to spray transfer above 8 bars. On the other hand, Yoshinori and others (1974) results are in accordance with the present authors in that increasing pressure enforces more short circuiting. As a consequence of voltages and currents, Fig. 3 shows

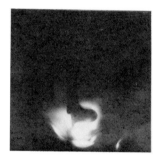

Fig. 2: Droplets at 5 bar

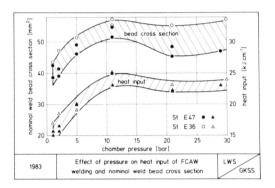

Fig. 3.

increasing heat inputs from about 16 kJ cm^{-1} at 1 bar up to peak values between 23 and 25 kJ cm^{-1} at 11 bars but slightly decreasing to about 23 kJ cm^{-1} at 29-30 bars. The effect of pressure and welding conditions on deposition of wire metal and fusion of base plate material is reflected by the upper scatter band in Fig. 3 of nominal cross sections per bead. These were calculated from the total weld metal cross section and the corresponding number of beads. They also reveal peak values at 11 bars. Together with less nominal heat input as well as loss of spatter droplets and their heat content at higher pressures slightly smaller bead sections are resulting. Moreover, as in TIG-welding (Allum and others, 1980) at higher pressures the flow of the shielding gas is likely to be disturbed thus additional heat losses from the arc may occur. Looking at the weld metal analysises, Fig. 4 and 5 and of the base plate materials and wire, Table 1, increasing pressure and corresponding variations of welding parameters are resulting in an increase of Nb, V, Ni and C. For Ni, Nb and V this is to be explained by base material fusion due to the deeper digging arc in the weld pool, while, for C, additional pick up from the flux may have contributed, as, for the SMA process, was found by Christensen and Gjermundsen (1976). A slight decrease of Al in both weld metals should be mainly attributed to its reaction with oxygen from the shielding gas during droplet transfer and at the increasingly stirred weld pool surface. While the sources of Al are both the base plates and the flux cored wire, its contents do fall below the plate materials' values at increasing pressure. However, at the same time, the contents of Si and Mn are not significantly affected giving average values of 1,46 % Mn and

0,37 % Si in the St E 36 respectively 1,64 % Mn and 0,51 % Si in the St E 47
weld metal. Nitrogen is increased by pressure, the contents of the St E 36
weld metals revealing peak values of 0,049 % N at 11 bars while both weld
metals acquire about 0,025 % N at 29 and 30 bars. Keeping in mind a constant
partial nitrogen pressure of about 0,8 bar, the nitrogen, pick up at higher
pressures should be

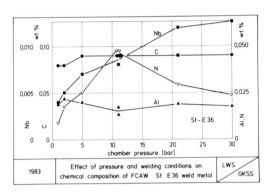

Fig. 4.

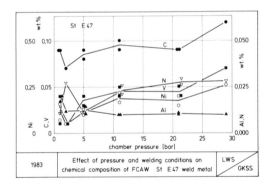

Fig. 5.

explained by decreasing arc shielding capacity of the constant mass gas flow
which was chosen deliberately for the sake of laminar conditions. As, with
increasing pressure, more nitrogen from the chamber atmosphere may have access
to the arc droplet transfer, thicker droplets however with smaller specific
surfaces develop thus possibly slowing down the rate of nitrogen transfer to
the weld metal
As already in Figs. 1, 3, 4, 5 also the variations of the cooling times $t_{8/5}$
of the root passes, indicate a basic change of process characteristics at
pressures above 11 bars. Starting with $t_{8/5}$ values between 10 and 17 s at
1 bar, peak values between 20 and 30 s are reached at 11 bar, again followed
by an obvious drop down to the 1 bar level when reaching 21 and 30 bars. It
seems, that the disturbed shielding gas flow as well as the change over to a
more uncontrolled droplet transfer above 11 bar are at least two main reasons
for a lower process heat efficiency.

Due to multilayer technique, no obvious effect of the weld root cooling times on hardness of the root weld metal (220-255 HV) nor on

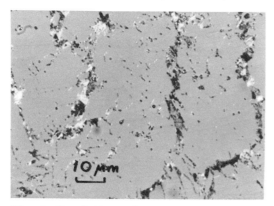

Fig. 6: Precipitates in St E 47 root weld metal at 21 bar

the adjacent HAZ (260-290 HV) can be seen. According to Fig. 6 root weld metals reveal precipitations predominantly at grain boundaries of primary grains. From etching behaviour under potentiostatic conditions with NaOH, Nb-, V-, or Ti-carbonitrides seem to prevail. The microalloying elements in the weld metal precipitate as a result of weld metal chemistry and weld root thermal weld cycles which, for 1 and 11 bar, are represented in Fig. 7. Following the cooling down to about 200 °C at the end of the first root pass, decreasing peak temperatures between 1000 and 600 °C depending on pressure and process parameters are periodically gained in the course of the entire welding procedure, which lasts about for 20 min. According to Strassburger, (1976), these weld reheat treatments satisfy the conditions for at least partial precipitation of Nb-, V-, Ti-carbonitrides. Due to the relatively high nitrogen content in the weld metal, the comparatively short times of weld cycle heat treatment and small contents of Al the visible particles may mainly represent nitrides.

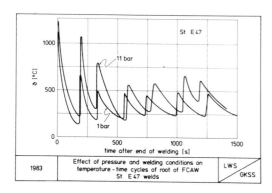

Fig. 7.

Resulting from the multilayer technique and the weld thermal cycles, most of the weld metal is grain refined while only a small proportion remains visibly unaffected in the as welded state, the microstructure consisting of grain boundary and acicular ferrite together with some aligned carbide-ferrite aggregates. Yet even those sections of the weld metal which do not reveal precipitations may undergo submicroscopic changes at lower temperatures than the ones responsible for visible precipitations. Thus, the CVN toughness of the weld metals, Fig. 8 is determined

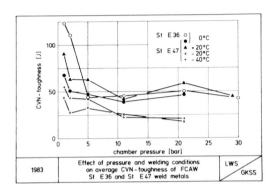

Fig. 8.

mostly by the grain refined and weld-heat treated microstructure. At testing temperatures of -40 to +20 °C the CVN toughnesses are dropping sharply with increasing pressure, starting, as an example for the St E 36 with a 0 °C-value of 125 J at 1 bar and falling below 50 J at 15 bar but keeping this level up to 30 bars. Thus, for the prevailing experimental conditions the first increase of pressure of a 1 bar air + Ar atmosphere leads to severe deterioration of weld metal toughness, the fracture surfaces being characterized by grain boundary- and dimple- as well as cleavage fracture. At first sight this may be attributed only to the effect of increasing nitrogen pick up, Fig. 9.

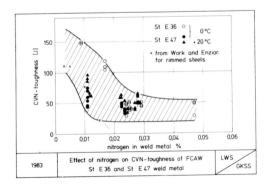

Fig. 9.

The slope of CVN toughness with nitrogen between 0.009 and 0.02 % is in accordance with earlier Izod-results of rimmed steels by Work and Enzian (1945) the latter occupying the lower limit of the scatter band.
Yet taking into consideration the presence of nitride forming elements in the weld metal the formal "unbalance factor":

$$B = 14 \cdot \%Al/27 + 14 \cdot \%Ti/48 + 14 \cdot \%V/51 + 14 \cdot \%Nb/93 - N$$

after Fehervari and Rittinger (1972) can be used to characterize the effect on toughness of surplus nitrogen. The factor "B" is based on the assumption of all Nb, V, Ti, Al being combined with nitrogen within the corresponding nitrides. For the St E 36 weld metal, peak toughness values, Fig. 10, are achieved at about B = +0.005 indicating no surplus nitrogen.

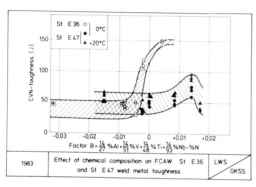

Fig. 10.

In an attempt to quantify the amount of precipitates according to Fig. 6 by the point counting method, no significant variations with pressure could be detected, the precipitated agglomerates amounting to between 5 and 9 % by area. For the St E 47 weld metal peak toughnesses are gained at about B = +0.015. According to Tsubai and Terashima, (1982) higher B-values than for the maximum toughnesses are indicating secondary hardening by surplus of microalloying elements which, for multipass welds, is also indicated by Dolby (1981). In assessing the toughness of St E 47 weld metal it should be pointed out hat the applied filler wire was undermatching with respect to the strength of the plate material.

CONCLUSIONS

The following conclusions can be drawn:
1. Hyperbaric MAG welding in an atmosphere containing nitrogen may cause nitrogen pick up of the weld metal due to increased access of the habitat atmosphere to the arc droplet transfer with increasing pressure.
2. The nitrogen pick up may be of the order which once was typical for ancient rimmed steels. Severe embrittlement of the weld metal, therefore may take place.
3. In hyperbaric welding of microalloyed steels the microalloying elements may provide a certain balance for nitrogen in the weld metal. Yet the amount and character of the precipitates as well as their effect on mechanical properties will be substantially out of control.
4. Considerations to avoid nitrogen pick up, preferably by improving process and gas flow characteristics in hyperbaric welding may take into account

that laminar shielding gas flow conditions are not providing arc and weld pool protection from habitat atmosphere.

REFERENCES

Allum, C.J., B.E. Pinfold and J.H. Nixon (1980), Weld. Res. Supp. p. 199s-207s
Christensen, N. and K. Gjermundsen (1976) IIW-doc. 212-384-76
Dolby, R.E. (1981). IIW-doc. IX-1213-81
Fehervari, A. and J. Rittinger (1972). IIW-doc. XII-B-1o9-72
Kobayashi, T., R. Kigucki and M. Honda (1975). Sec. Intern. Symp. Japan Weld. Soc. paper 2-2-(4)
Schafstall, H.-G. and R. Schaefer (1979). Schw. u. Schn. 31, 9, p. 374/81
Stalker, A.W. and G.R. Salter (1975). The Welding Institute, RR-SMT-R 7505
Straßburger, C. (1976). Verlag Stahleisen GmbH, Düsseldorf
Tsuboi, J. and H. Terashima (1982). IIW-doc. IX-1246-82
Work and Enzian (1945). Trans. AIME 162, p. 723
Yamada, S. and H. Ozaki (1979). IIW-doc. CREAU 17-79
Yoshinori, J., Y. Noboyuki and K. Akira (1974). The Sumitomo Research No. 12

TABLE 1 Chemical Composition of plates and deposited Weld Metal Wt.-%

	C	Si	Mn	P	S	Ni	V	Mo	Al	Ti	Nb
St E 36	.13	.23	1.39	.021	.010	.02	.06	.01	.043	.003	.026
St E 47	.15	.53	1.44	.022	.016	.54	.16	.02	.012	.002	.003
Weld metal at 1 bar	.06	.42	1.40	.013	.012	.01	.01	.01	.01	.009	.004

ACKNOWLEDGEMENT

The experiments were carried out by cand. Ing. N. Born and cand. Ing. W. Müller as part of their work for the degree of Dipl.-Ing.

Investigation of Magnetism and its Elimination in Submarine Pipelines

D. N. Waller and J. P. Gaudin*

Comex Services, Marseille, France
(*formerly Institut de Soudure, Paris)

ABSTRACT

The occurence of magnetism in ferromagnetic materials and its effects upon arc stability have been frequently reported. This paper describes the construction of an arc blow simulator and defines a test procedure and gives results for measuring the intensities and directions of magnetic fields that occur in the single V preparation for circumferential hyperbaric pipeline welding. The same simulator was used to investigate methods of annuling magnetism to enable equipment to be constructed for actual hyperbaric production welding.

KEYWORDS

Magnetic deflection / Arc blow / Arc welding / Pipelines / Hyperbaric welding / Demagnetisation

INTRODUCTION

Although considerable reported work exists on the subject of arc blow, little information exists on the particular practical problems encountered during hyperbaric circumferential welding of pipelines.

Following field welding problems, it was decided to construct a simulator to reproduce and examine as many magnetic conditions as possible and also to assist with the design and testing of demagnetising equipment for the hyperbaric chamber.

Typical site welding problems previously encountered included :
- Inability to make tack welds on a vertical riser pipe
- Inability to establish an arc for root welds for mid line horizontal axis tie-ins
- Area of excessive penetration coincident with previous reports of magnetic flux in the same root area

Observations were also noted that :
- Only segments of the pipe circumference were affected
- Strengths of flux within the root gap were sufficiently strong to allow electrodes to be magnetically suspended
- The worst magnetism was experienced when the pipeline axis was close to the magnetic north-south axis of the earth's field
- Degaussing techniques using DC coils wrapped around the pipe and connected to a welding power source output only partially removed the magnetic arc blow to a condition of excessive arc deflection
- Large and variable magnetic fields were experienced with a riser connection thought to be associated with platform cathodic protection system as anodes were in chamber vicinity
- Pup pieces cut on surface mechanically had remanant magnetism on the bevel face whereas flame cut and hand ground bevels did not display any magnetism even when cut from the same pipe diameter

Logistically the effect on hyperbaric welding is dramatic if magnetism prevents welding. Diver/Welder work programmes are delayed whilst demagnetising current represents variables outside of the reach of the welders workstation, extra welder effort, communications to surface and time is needed to reduce arc blow. Extra dives have to be made because of the nett low welding productivity which represents direct additional barge time cost and increases the risk of not completing hyperbaric welds within a "weather-window" time period.

Factors contributing to arc blow for hyperbaric welds in ferromagnetic (C-Mn) steels include :

Remanent magnetism

- Pipe manufacturing methods :
  . pipe forming
  . longitudinal welding
  . handling pipe materials with electromagnets
  . magnetic particle inspection

- Hyperbaric worksite :
  . pipe cutting methods
  . magnetic particle inspection of weld preparation

Other induced magnetism

- Earths field same as pipe line direction
- Platform cathodic protection sacrificial currents
- Magnetic fields associated with preheat elements
- Asymmetrical weld earth paths with respect to weld arc axis

Other factors

- Carbon manganese pipeline steels with refined grain structures are relatively magnetically hard having large coercivities with the capability of producing and maintaining high flux densities across a gap (the weld preparation)

248

- The chamber atmosphere (usually helium-oxygen mixture) is not considered to be an additional factor as the permeability of the gas is the same as for air or vacuum. An apparant advantage for combatting magnetism of hyperbaric welding is the increased arc stiffness as the arc constricts with increased chamber pressure. However, this should be offset against the disadvantages of arc welding under pressure and in a helium atmosphere where the combined effects of a restricted arc and the faster freezing rate of the helium gas reduces the weldpool size significantly. This makes welder technique difficult, trying to compensate and maintain a deflecting arc on a smaller "target" (weldpool) than normally achievable on surface.

ARC BLOW SIMULATOR

A literature survey has revealed large reported differences in measured values of magnetic intensity on preweld preparations, providing no absolute reference levels. Also most previous research has been carried out on dimensionnally reduced test pieces with empirical apparatus such as simulating the arc by mercury deflections. In view of the specific practical objectives, a test bench was constructed to hold a horizontally supported full size pipe specimen i.e. 24 inches diameter and 2,4 meters long which is similar to the exposed pipe length within a hyperbaric chamber. The test bench was designed to induce two principal types of magnetic field, longitudinal (transverse to the weld direction) and tangential (parallel to the weld direction) with fields strengths adjustable up to 600 gauss. The test rig was also capable of supporting a full size external (Dearman) alignment clamp preheating equipment, demagnetising coils and permitting normal welding access (fig. 1). If necessary, the entire test bench can be assembled in the Comex pressure simulation laboratory (the Hydrosphere) at Marseilles.

A Hall effectprobe was adapted to measure accurately and separately, tangential, perpendicular and longitudinal magnetic fluxes at a chamfer and at 4 heights within the V-groove weld prep (root, 5mm from root of V mid position and top of V). The digital display/calibration unit of flux density and polarity was marinised to 150 meters.

TESTING AND RESULTS

Using the Hall probe and with a non magnetic positionning aid, a plot was made of a typical naturally remanent magnetic field which occured in a bevel from 2 separate pieces of pipe. The field was examined transversly across the bevel at the pipe's 12 o'c position and plots were made of the components of magnetic field for the pipe's longitudinal, tangential and radial (vertical) components. The results are shown in fig.2. A series of tests was made measuring intensities of the magnetic field around the entire circumference of the weld preparation before and after any artificially induced magnetic fields. These measurements have shown extensive magnetic heterogeneity

around the pipe before and after the influence of linear
external longitudinal applied fields (from coils wrapped
precisely concentrically around the pipe). Fig.3 shows
this heterogeneity, demonstrating the varied magnetic
history of the pipe and particularly in the region of the
longitudinal pipe seams and also at the top of the
chamfer . Measurements were made of magnetism 5mm a-
bove the pipe interior surface which is the mid region
of the arc length for the critical root weld pass and
where arc blow normally occurs. It was determined during
welding trials that the limit of arc stability for MMA
welding with 2,5mm electrodes was at 70 gauss longitudinal field (NS poles on either side of chamfer) at 1 atmosphere air pressure. This value would be probably reduced further for the hyperbaric arc/plasma.

Several measurements were taken of the induced magnetism
on each bevel separately and then were brought together
in the ready to weld presentation. No relationship was
found between separate and coupled magnetic field strengths ; typical measurements are 2,4 gauss (tube 1),
24,4 gauss (tube 2) and on combination for V preparation
(3mm gap) 150 gauss (measured within root gap). It is
impossible to predict therefore the likely preweld magnetism by measuring weld preparations at the time of machining the individual bevels.

The influence of the position of the earth connection
was investigated using a copper stick "electrode" tou-
ching into the weld groove. The complexity of the earth
path alternatives that existed within the pipe, particularly with the pipe,particularly with the pipe alignment
clamp in position prevented accurate analysis. It was
decided to standardise for all testing on 2 circumferential tensionned earth belts symetrically placed about the
weld groove.

The main test programme to determine methods of annuling
magnetism used 2 coils placed on either side of the bevel. Tests were made with the current flowing in the
same direction through the demagnetising coils. This
produces a longitudinal field and creates NS poles on the
two chamfer faces. By applying this demagnetising field
in the opposite direction to that (SN) of the remanent
and standing field, the magnetic flux at a particular
point in the chamfer is annuled and arc welding becomes
possible. To counteract the heterogeneity round the pipe
and possible reversals of polarity, the field strength of
the demagnetising coils plus their polarity have to be
adjusted. For cases where the remanent magnetism on the
bevels was of the same polarity, this method did not
give a stable arc welding condition.

Further tests were carried out with the currents flowing
in opposite directions through the two demagnetising
coils to effectively produce two equal polarities of
equal intensities (e.g. : two South poles on the two
faces of a non magnetic bevel). Welding with such equal
and opposite fields was found to be very stable, the arc

preferring to remain parallel to the perpendicular magnetic lines of force. This technique was then applied to
various "remanent" magnetic field types in the bevel i.e
with bevel faces having NS, SN, NN (different intensities)
and SS (different intensities). In all cases the most
stable arc was found when the opposite demagnetising
fields were adjusted to induce the same nett polarity
and intensity on each bevel face and then to augment
equally the ampere turns in each coil to give vertical
magnetic lines of force into the weld preparation (fig 4).
A practical benefit of this approach is that the demagnetising coils can be some distance (behind clamps and preheating pads) from the weld and still produce the vertical flux lines needed for arc stability.

DISCUSSION AND APPLICATION OF METHOD

The various demagnetising methods which exist cannot all
be successfully applied for the hyperbaric pipe weld
situation. Progressive and diminishing cycling through
the BH hysterisis curve is only successful for eliminating true remanent magnetism within the pipe. As other,
more dominant, magnetic induction factors are also in
permanent operation, such as the earth's fields, this
approach is only partially successful. The same is true
for moving an AC coil along the pipe. Other suggestions
of shunting the magnetism through iron filings within the
weld V prep. or shunting with massive iron "bridges"
are not practically possible. Also precluded are the
workshop options to reverse electrode polarity or use
AC instead of DC as they represent essential variables
and would have to be separately qualified as a procedure.

The principales described have, since the experimental
work, been successfully applied in two offshore situations where magnetism was experienced. One of these was
a 16 inch. diameter pipeline in the North Sea and the
other a 40 inch. diameter pipeline in the Indian Ocean.
Although both fields were in different hemispheres, the
pipelines were both oriented very close to the earth's
N-S magnetic axis.

Fig. 1   Test pipe in test bench during a weld trial.

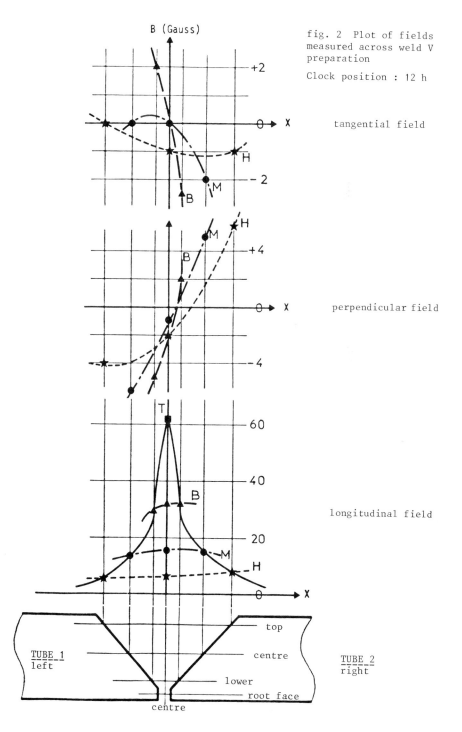

fig. 2 Plot of fields measured across weld V preparation

Clock position : 12 h

tangential field

perpendicular field

longitudinal field

B (Gauss)

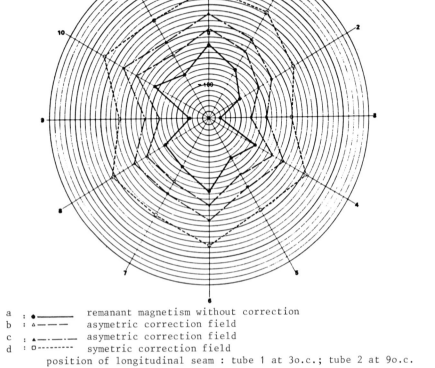

a : •————— remanant magnetism without correction
b : △——— asymetric correction field
c : ▲—·—·— asymetric correction field
d : □········· symetric correction field
　　position of longitudinal seam : tube 1 at 3o.c.; tube 2 at 9o.c.

fig. 3 Plot of magnetic field in longitudinal direction

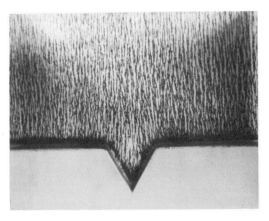

fig. 4 Iron filing demonstra
　　　　tion of perpendicular
　　　　magnetic field giving
　　　　maximum arc stability

A Study on the Metallurgical Properties of Steel Welds with Underwater Gravity Welding

S. Ando and T. Asahina

College of Industrial Technology, Nihon University, Narashino-shi, Chiba, Japan

ABSTRACT

The main object of this study is to clarify the effects of water pressure on the metallurgical properties of steel welds such as blowhole occurrence, crack sensitivity, formation of diffusible hydrogen and oxide inclusions etc. with underwater gravity welding. Underwater welding were carried out at water pressure 0.03 to 10 kgf/cm^2 (0.003 to 1 MPa) in the experimental tank and some facts were made clear.

KEYWORDS

Underwater welding; gravity welding; water pressure; high tensile strength steel; coverd electrode; crack; blowhole; diffusible hydrogen; non-metallic inclusion.

INTRODUCTION

Many researches (1) have been carried out on the study of underwater welding with covered electrodes.
Authors have engaged in the study of underwater welding with gravity welding and a part of their investigation was introduced by Dr. Hamasaki (2).
However, about the influences of water pressure on the metallurgical properties of welds with underwater gravity welding, further investigations have been required.
This study aimes to clarify the relation between water pressure and the metallurgical properties of welds with gravity welding and further contribute to developments of base metals and electrodes suitable for underwater welding.

MATERIALS AND EXPERIMENTAL METHODS

Table 1 and 2 show the properties of base metals and electrodes. Base metals are mild steel (JISG3106,SM41) and low carbon 50 kgf/mm^2 class high tensile strength steel (3) (L.C.50HT). D4327, 24, 40, 26 are iron powder type and iron oxide, high titanium oxide, ilmenite and low hydrogen type respectively (JISZ3211, D4324 is trial electrode) and D308, 309, 310 are austenitic electrodes (JISZ3221).

TABLE 1 Chemical Compositions and Mechanical Properties of Base Metals

| Materials | Thickness (mm) | Chemical composition (%) | | | | | | | | Mechanical properties | | |
		C	Si	Mn	P	S	V	Ti	Ceq.※	Y.S. (kgf/mm²)	T.S. (kgf/mm²)	E. (%)
SM 41	25	0.15	0.25	0.77	0.020	0.011	—	—	0.29	28	45	30
L.C. 50HT	25	0.04	0.30	1.27	0.016	0.006	0.03	0.01	0.26	38	56	24

※ Ceq. = C + 1/6 Mn + 1/24 Si (%)

TABLE 2 Chemical Compositions and Mechanical Properties of Deposited Metals

| Electrode type | Electrode diameter (mm) | Chemical composition (%) | | | | | | | Mechanical properties | | |
		C	Si	Mn	P	S	Ni	Cr	Y.S. (kgf/mm²)	T.S. (kgf/mm²)	E. (%)
D 4327	5	0.08	0.26	0.63	0.016	0.012	—	—	41.8	47.8	30.6
D 4324	5	0.06	0.86	0.80	0.030	0.013	—	—	49.4	54.8	30.0
D 4340	5	0.08	0.11	0.39	0.015	0.012	—	—	42.3	48.0	32.0
D 4326	5	0.09	0.63	0.90	0.016	0.006	—	—	43.6	48.8	32.8
D 308	5	0.05	0.33	1.55	0.015	0.011	10.36	20.31	—	60.5	43.5
D 309	5	0.06	0.33	1.81	0.015	0.012	13.23	24.18	—	61.8	36.8
D 310	5	0.12	0.34	1.70	0.023	0.008	21.37	27.04	—	62.1	43.2

In experimental water tank (Max. water pressure 10kgf/cm², correspond to 100m water depth, ϕ 1500 x 2700 mm), mainly single bead welding on base metals were done with gravity welding (Fig. 1) and several researches were carried out on the metallurgical properties of welds. Welding conditions are as follows: Iron powder type electrodes (except D4326) and austenitic electrodes –A C, welding current 280 A, arc voltage 40 V (austenitic electrodes 30 V), beading ratio 1.0 and electrode angle 45°. D4326–D C R P, 290 A, 30V, 1.4, 80° respectively.

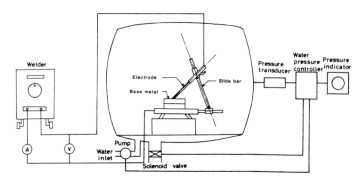

Fig. 1. Water pressure tank and underwater gravity welding apparatus.

EXPERIMENTAL RESULTS

Blowholes in Weld Metals

Fig. 2 shows the relation between water pressure and blowhole occurrence tendencies (specific gravity) of weld metals. Blowholes of each weld metal increase as increasing water pressure and clear differences are recognized between each electrode and D4324 shows low blowholes. According to their experiment by partial dry method with gravity welding, D4327 shows low blowholes to water pressure 4kgf/cm².

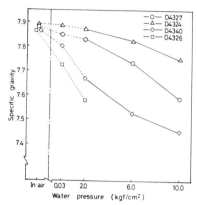

Fig. 2. Relation between water pressure and specific gravity of weld metals.(Base metal:SM41)

Cracks in Welds

Fig. 3 shows relation between crack sensitivity of welds and holding time(4) underwater pressure and shallow water depth . In this experiment, low carbon 50 kgf/mm² high tensile steel was used as base metal. In each case, heat affected zone cracks are not recognized (Max. Hv about 250). In case of D4327, weld metal crack is not recognized, but weld metal microfissures occurred immediately after welding in D4324. In D308, weld metal micro cracks occurred immediately and growing tendencies of cracks are recognized as increasing holding time. About ten minutes later after welding, micro cracks in bond are often recognized in D310 weld metal. In welding of SM41 with D4340 and D4326 , HAZ cracks (Max. Hv about 400) and weld metal cracks occurred immediately.

In Y-Groove Cracking Test(JISZ3158) with low carbon 50 kgf/mm² high tensile steel and D4327 electrode, HAZ crack did not occurred after two days, but delayed cracks due to hydrogen occurred in the root of weld metals during two weeks. Hot cracks in bead surfaces occurred immediately in D310.

In Implant Test (5) with SM41 and D4327, fracture did not occur in applied stress 30 kgf/mm² and holding seven days and with D310, fracture did not occur in 10 kgf/mm² and seven days. Fracture in high applied stress occurred from underbead cracks with D4327 and micro cracks in bond with D310.

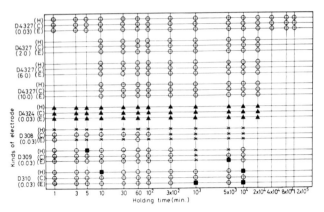

Fig. 3. Time-cracking diagram of weld metal cracks.

Notes ○ No crack   ■ Weld metal fine crack near bond

▲ Weld metal fine crack   × Weld metal crack

(H) Head   (C) Center   (E) End

Figures in ( ) show water pressure (kgf/cm²)

Base metal L C 50H T

Diffusible Hydrogen Contents of Weld Metals

Fig. 4 shows relation between water pressure and diffusible hydrogen contents of deposited metals by glycerin replacement method (correspond to JISZ3113) with six kinds of iron powder type electrodes.

Diffusible hydrogen shows most high value at 0.03 kgf/cm² and then decreases as increasing water pressure.

D4326 and D308 show remarkably high value of diffusible hydrogen as compared with in air and a small diffusible hydrogen appeared in D310.

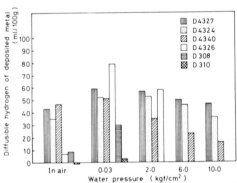

Fig. 4. Relation between water pressure and diffusible hydrogen of deposited metals. (Base metal : SM41)

In immersion test of iron powder type electrode coatings, water absorption ratio of coatings have tendencies to increase as increasing water pressure and immersion time and D4324 shows low influences of above two factors and low water absorption ratio.

Chemical Compositions and Non-Metallic Inclusions of Weld Metals

Fig. 5 and 6 show relation between water pressure and chemical compositions and non-metallic inclusions by I_2-CH_3OH method in weld metals with four electrodes. In case of D4327 and D4340, FeO increases remarkably from about 2 kgf/cm^2 and increases of FeO are low in D4324 and D4326.

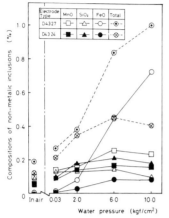

Fig. 5. Relation between water pressure and compositions of non-metalic inclusions of weld metals.
(Base metal: L C 50HT)

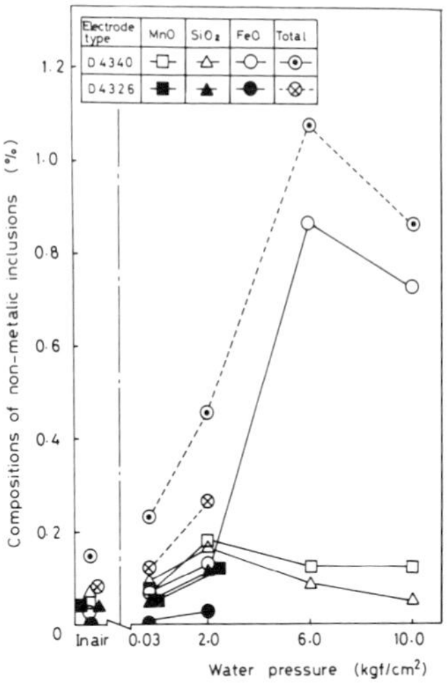

Fig. 6. Relation between water pressure and compositions of non-metalic inclusions of weld metals.
(Base metal: SM 41)

Relation between apparent Si and Mn contents and true Si and Mn contents by deduction of Si and Mn contents as SiO_2 and MnO from apparent Si and Mn contents are shown in Köber u. Oelsen diagram (Fig. 7) (6,7).
Oxidation of fusion metal are promoted as increasing water pressure and true Si and Mn almost are lost under high water pressure as D4340.
D4324 maintain relatively high true Si and Mn under high water pressure, so blowhole occurrences are low.

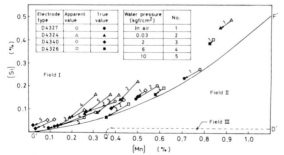

Fig. 7. Relation between water pressure and apparent and true Mn,Si contents in weld metals.
(Base metal: D4327,D4324 ; L·C·50HT
D4340,D4326 ; SM 41)

Fig. 8 shows relation between water pressure and purity degree of weld metals (correspond to JISG0555). Purity degree of weld metals become higher as increasing water pressure and are in order of D4340, D4324, D4327, D4326 and D4326 shows remarkably low purity degree. Many inclusions of about 30 μ size exist in D4327 and many of about 5 μ size in D4324 weld metal.

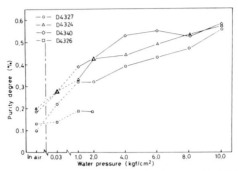

Fig. 8. Relation between water pressure and purity degree of weld metals.
(Base metal: SM41)

In case of austenitic electrodes, due to oxidation of fusion metal and dilution by penetration, D308 weld metal become martensitic structures and H_2 evolution from weld metal becomes large, so diffusible hydrogen shows high value. In D310, some parts of bond are diluted by base metal and become martensitic structures and evolve H_2 gas. Then micro cracks in bond occur.

Gas Compositions of Blowholes in Weld Metals and Bubbles in Arc Atomospheres

Table 3 shows relation between water pressure and gas compositions of blowhole in weld metals and bubbles in arc atomospheres by gas chromatography. Gas compositions of blowholes in each weld metal show more than 99% H_2, and effects of water pressure are not recognized. Principal gas compositions of bubbles are also H_2 but under in blowholes and CO gas exists more than blowholes. These phenomena explain occurrence of CO reaction during welding.

TABLE 3 Relation Between Water Pressure and
Gas Compositions of Blowholes and Babbles

Electrode type	Water pressure (kgf/cm^2)	Blowhole gas (%)			Bubble gas (%)		
		H_2	CO	CH_4	H_2	CO	CH_4
D4327	2	99.9	—	0.1	92.4	7.6	0.02
	6	99.7	0.2	0.1	89.3	10.7	0.05
	10	99.3	0.6	0.1	93.0	7.0	0.09
D4324	2	99.4	0.2	0.4	85.4	14.6	—
	6	99.3	0.2	0.5	87.7	12.3	—
	10	99.4	—	0.6	86.3	13.7	0.04
D4340	2	99.9	—	0.1	77.7	22.2	0.05
	6	99.7	0.2	0.1	79.4	20.5	0.06
	10	99.3	0.1	0.6	83.5	16.4	0.07
D4326	2	—	—	—	76.8	23.6	—

SUMMARY

1) Low carbon 50 kgf/mm^2 high tensile steel shows very low crack sensitivity of HAZ with underwater gravity welding, so this steel are thought to be suitable for underwater welding.

2) Iron powder, iron oxide type electrode shows low crack sensitivity of weld metal compared with other electrodes and blowhole occurrence tendency is relatively low to water pressure about 2 kgf/cm^2.

3) For formation of cracks and blowholes in steel welds underwater pressure, a series of metallurgical phenomena as oxidation, deoxidation, gas absorption and evolution during welding become important factors.

4) Developments of electrodes suitable for underwater welding and improvements of welding apparatus will contribute to increase of reliability of underwater gravity welding.

ACKNOWLEDGEMENT

This research was supported by the National Sience Foundation of Ministry of Education and the College of Industrial Technology, Nihon University. Low carbon 50 kgf/mm^2 high tensile steel and trial iron powder high titanium oxide type electrode were supplied through the courtesy of Nippon Steel Corporation and Nippon Steel Welding Products & Engineering Co., Ltd. respectively. Also the students of their laboratory engaged earnestry with this research. The authors would like to express their gratitude to whom concerned.

REFERENCES

1) Dadian, M. (1976). Review of literature on the weldability underwater of steels. Welding in the World, Vol.14, No.3/4, 80-99.

2) Hamasaki, M. (1976). Development in Underwater Welding in Japan. Seminar on underwater welding, London, Welding Institute.

3) Onoe, Y., S. Sekino. Y. Shiomi and M. Sato (1974). A 50 kg/mm^2 High Tensile Strength Steel with Good Weldability. The Iron and Steel Institute of Japan, Vol.60, No.8, 1144-1154.

4) Kobayashi, T. and I. Aoshima (1971). Toe and Underbead Cracking in Low-Alloy Steels. Transactions of the Japan Welding Society, Vol.2, No.1, 70-76.

5) Granjon, H. (1969). The 'implant' method for studying the weldability of high strength steels. Metal Construction and British Welding Journal, Vol.1, No.11, 509-515.

6) Körber, F. und W. Oelsen (1933). Die Grundlagen der Desoxydation mit Mangan und Silizium. Mitteilungen aus den Kaiser-Wilhelm-Institute für Eisenforshung-Bd, XV. lieferung 21.

7) Sekiguchi, H. (1964). Theory and Proposal on Steel Fusion Welding and Their Applications. Nikkan Kogyo Shinbun Ltd.

Prediction of Cooling Rate and Hardness of Base Metal in the Underwater Welding by Local Cavity Process

A. Matsunawa*, K. Nishiguchi** and I. Okamoto*

*Welding Research Institute, Osaka University, 11-1 Mihogaoka, Ibaraki, Osaka 567, Japan
**Faculty of Engineering, Osaka University, 2-1 Yamadaoka, Suita, Osaka 565, Japan

ABSTRACT

The paper describes the analytical solution of temperature distribution during the underwater welding by local cavity method and its application to prediction of the cooling rate and hardness of base metal. Theoretical calculations are compared with experimental data, and the guiding line of selecting optimum welding parameters is discussed.

KEYWORDS

Heat conduction; Temperature distribution; Thermal cycling; Cooling rate; Hardness; Underwater welding; Moving source; Moving boundary; Prediction.

INTRODUCTION

In the underwater welding by local cavity methods (Nishiguchi, 1975, 1977; Hamasaki, 1973, 1975), the arc stability, bead formation and metallurgical problems have been greatly improved compared with those in wet processes. The cooling rate of base metal, which gives eminent effects on hydrogen diffusion and hardness, is controllable in principle in the local cavity process by selecting cavity size. The relation between the quenching speed and the size of dry zone is, however, much dependent on various welding variables, but the problem has not been well analysed yet. The aims of this paper are to establish a theoretical method to estimate the thermal field of workpiece in a local cavity process and to apply it to the hardness prediction of the underwater welded steel plate.

PRINCIPLE OF UNDERWATER WELDING BY FLUID STABILIZED
LOCAL CAVITY PROCESS

The principle of process which Nishiguchi and others (1975) developed is illustrated in Fig. 1. A welding torch is surrounded by a divergent ring water jet of high speed which entrains the inside gas and water by the flow and they are smoothly exhauseted to the outer region. The gas is fed either sin-

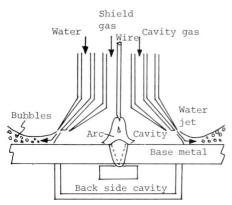

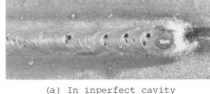

(a) In inperfect cavity

(b) In perfect cavity

Fig. 1. Underwater welding process in   Fig. 2. Bead appearance of MIG
fluid stabilized local cavity   underwater welding

gly or multiply into the torch, and then a dry zone is steadily formed in the
whole region surrounded by the water jet. The water jet acts as a shutt-off-
wall to the exterior water due to its high momentum. The obtained cavity
is stable and durable to the external and internal perturbations and also
to the torch travel. The stability of arc and bead formation are excellent
as seen in Fig. 2, if the cavity is perfectly formed.

Feasibility studies of this process have been extensively conducted by Tamura
(1976), Nishio(1977) and Shinada(1978), and they developed fully automatic
MIG welding systems. They showed that the metallurgical and mechanical properties of welded joints were satisfactory if the welding variables were adequately selected. However, so far as the welding is performed inside a small
cavity, the base metal receives more or less the effects of environmental
water. Among all, the hardness and hydrogen absorption are the most significant problems in this process.

ANALYSIS OF TEMPERATURE FIELD DURING UNDERWATER WELDING

In order to evaluate the temperature distribution during underwater welding,
the authors (Matsunawa and others, 1980) developed the simplified analytical
models of heat conduction for local cavity process. It was assumed that the
semi-spherical or cylindrical cooling boundary of constant temperature ($T = T_0$)
moved together with the moving point or linear heat source of constant
intensity and speed, as illustrated in Fig. 3. The basic equation of heat
conduction was analytically solved by the same method with Rosenthal's (1946)
under the boundary conditions of;

$$T = T_0 \quad \text{at} \quad r = R, \text{ and}$$

$$-2\pi r^2 K(\partial T/\partial r) \to q \quad \text{as} \quad r \to 0 \quad \text{(for three-dimensional heat flow), or}$$

$$-2\pi r K(\partial T/\partial r) \to q/h \quad \text{as} \quad r \to 0 \quad \text{(for two-dimensional heat flow).}$$

The objective solutions of temperature distribution inside the cavity are;

Three-dimensional heat flow:

$$T - T_0 = \frac{q}{2\pi K} \frac{1}{r} exp \ [-\lambda vr (\ 1 \ + cos \ \alpha \ cos \ \beta] \ \frac{1 - exp[-2\lambda v(R-r)]}{1 - exp[-2\lambda vR]} \quad (1)$$

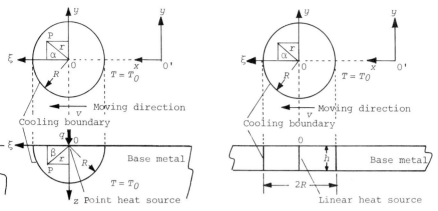

Fig. 3. Heat conduction modelling with equiradial cooling boundary around the moving heat source

TABLE 1   Nomenclature Used

q : heat input per unit time	R : radius of cooling boundary
h : thickness of plate	R_0 : radius of water jet nozzle
v : speed of source	r : radial distance from heat source
t : time	
x, y, z : distances in fixed coordinate	n : dimensionless heat input ($= \lambda vq/[2\ K(T_f - T_0)]$)
ξ, y, z : distances in moving coordinate ($\xi = x - vt$)	θ : normalized temperature ($= (T - T_0)/(T_f - T_0)$)
T : temperature	Λ : dimensionless radius of cooling boundary ($= \lambda vR$)
T_0 : cooling boundary temperature	
T_f : reference temperature (fusion temperature in this paper)	ρ : normalized radius ($= r/R$)
K : heat conductivity	H : dimensionless plate thickness ($= \lambda vh$)
$1/(2\lambda)$: thermal diffusivity	

or in dimensionless expression by Christensen (1965),

$$\frac{\theta}{n} = \frac{1}{\rho\Lambda}\, exp\,[-\rho\Lambda(1 + cos\ \alpha\ cos\ \beta)]\frac{1 - exp[-2\Lambda(1-\rho)]}{1 - exp[-2\Lambda]} \qquad (2)$$

Two-dimensional heat flow:

$$T - T_0 = \frac{q}{2\pi hK}\, exp[-\rho\Lambda cos\ \alpha\,]\cdot K_0(\rho\Lambda)\cdot[1 - \frac{K_0(\Lambda)\cdot I_0(\rho\Lambda)}{I_0(\Lambda)\cdot K_0(\rho\Lambda)}] \qquad (3)$$

$$\frac{\theta H}{n} = \frac{1}{\Lambda}\, exp\,[-\rho\Lambda cos\ \alpha]\cdot K_0(\rho\Lambda)\cdot[1 - \frac{K_0(\Lambda)\cdot I_0(\rho\Lambda)}{I_0(\Lambda)\cdot K_0(\rho\Lambda)}] \qquad (4)$$

where, K_0 and I_0 : Modified Bessel functions of 0 order.

These equations coincide completely with the Rosenthal's or Christensen's solutions in case of semi-infinite plate, where the radius of cooling boundary (R or Λ) becomes infinity. Figure 4 shows an example of calculated temperature distribution in case of linear heat source.

The temporal change of temperature at a certain point (x, y, z) in the fixed coordinate is related with the temperature gradient at the corresponding

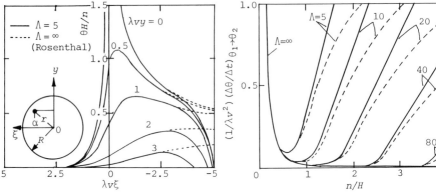

Fig. 4. Temperature distribution inside the region bounded by circular cooling boundary (Two-dimensional heat flow).

Fig. 5. Effect of heat input on average cooling rate from θ_1 to θ_2 along moving line and fusion boundary.

point (ξ, y, x) in the moving coordinate.

$$\frac{\partial T(x, y, z, t)}{\partial t} = \frac{\partial T(\xi, y, z)}{\partial \xi} \frac{\partial \xi}{\partial t} = -v \frac{\partial T(\xi, y, z)}{\partial \xi} \tag{5}$$

The above relation can be expressed in dimensionless form as;

$$\frac{1}{\lambda v^2} \frac{\partial \theta}{\partial t} = -\frac{\partial \theta}{\partial (\lambda v \xi)}$$

Here, the factor $(1/\lambda v^2)$ that has the unit of time is a parameter for non dimensional transformation. Thus, the average cooling rate at the particula temperature range $\Delta \theta = \theta_1 - \theta_2$ is defined as follow.

$$\frac{1}{\lambda v^2} \left(\frac{\Delta \theta}{\Delta t} \right)_{\theta_1 \to \theta_2} = -\frac{\theta_1 - \theta_2}{\lambda v (\xi_1 - \xi_2)} \tag{6}$$

Notations ξ_1 and ξ_2 in the above equation are the distances in ξ-direction fro the source where θ adopts θ_1 and θ_2 respectively. It is, therefore, possibl to calculate the average cooling rate of point and linear hea sources, combining the equations (6) with (2) or (4). Figure 5 shows th calculated results of average cooling rate vs. heat input in case of two-di mensional heat flow, in which the cooling rate along the moving axis an fusion boundary (bond line) are presented by solid and broken lines. In th calculation, θ_1 and θ_2 were chosen 0.533 (=800°C/1,500°C) and 0.333 (=500°C 1,500 °C) respectively, considering the hardness prediction of steel. Whe the heat input is increased under the condition of constant radius of coolin boundary (Λ), the dimensionless cooling rate reduces along that of Rosentha curve, but it rises remarkably in the range above a certain critical hea input.

MEASUREMENT OF THERMAL CYCLE DURING UNDERWATER WELDING
AND COMPARISON WITH THEORETICAL CALCULATION

Since the effect of cooling boundary on the temperature distribution is mor eminent in case of two-dimensional heat flow, the underwater welding of thi plate by Argon TIG arcs were demonstrated, forming the cavities of the sam size on both sides of the plate. In this experiment were employed two kin

TABLE 2 Chemical Composition of Steel Used

C	Si	Mn	P	S	Ni	Cr	Mo	V	B
0.091	0.22	0.72	0.013	0.015	0.02	0.03	0.003	0.005	0.0003

of water jet nozzles having the radius of R_Q =50 mm and R_Q =34 mm, and TIG arcs of 150 to 450 A were used. Welding speed was adequately selected so as to get the full penetration bead in order to realize the two-dimensional heat flow as far as possible. The chemical composition of mild steel (6 mm in thickness) is tabulated in TABLE 2.

In Fig. 6 are compared the differences between the measured and calculated thermal cycles under the low and high heat input conditions. In the calculation, physical constants were adopted as; K = 33.6 J/s·m·°C (=0.08 cal/scmC) $1/(2\lambda)$ =1.00x10^{-5} m^2/s (=0.100 cm^2/s), T_Q=0°C and T_f=1,500 °C. The thermal efficiency of arc was 55% which was the average value (55$\pm$6%)of whole experimental data. The actual size of circular cavity was measured by direct observation of dry zone formed on a transparent plate, and the equivalent radii were R =61 mm for R_Q =50 mm nozzle and R =37 mm for R_Q =34 mm one. The calculation and measurement under the low heat input condition coincides each other fairly well. In the high heat input, on the contrary, the calculated temperature shows a similar profile to the measured one, but the quenching in the vicinity of cooling boundary delays somewhat from the calculation and the curve tails gently in the range below 150 °C, which suggests that the heat transfer is actually not high as assumed in the theory.

Figure 7 shows the corelation between the experimental and theoretical cooling rate from 800 °C (θ_1=0.533) to 500 °C (θ_2=0.333). Here, it should be noted that the dimensionless cooling rate defined in eq. (6) shows smaller value when the heat input is increased proportionally with the increase of welding speed, since the definition includes the square term of moving speed in the denominator. As seen in Fig. 7, the half of plots stay within the error band of $\pm$10%, but there are considerable deviation under the conditions of low heat input (circled by A) and very high heat input (B).

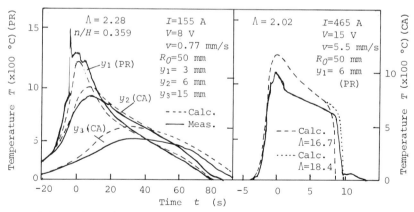

(a) Low heat input (Slow speed)   (b) High heat input (Fast speed)

Fig. 6. Comparison between measured and calculated thermal cycles (y_1, y_2 and y_3 are the distances from center line of bead where thermocouples (PR and CA) were installed.).

267

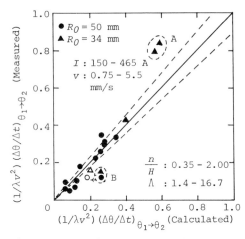
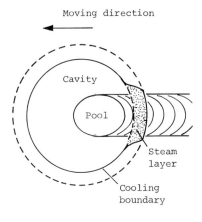

Fig. 7. Comparison between calculated and measured average cooling rate from 800 °C to .500 °C.

Fig. 8. Worse heat transfer by formation of steam layer in high heat input.

A main cause of underestimation of the calculated cooling rate in the low heat input (circle A in Fig. 7) is primarily based on the fact that thermal properties such as K and 2λ have been assumed constant. It was verified that the numerical calculation considering temperature dependent constants showed better accuracy than that by analytical method.

On the other hand, overestimation of calculated cooling rate in higher heat input (circle B in Fig. 7) is presumably due to the worse heat transfer at the position of water impingement by the formation of steam layer on the surface, as schematically illustrated in Fig. 8. If the effective radius of cooling boundary is taken to be $R = 67$ mm instead of the original size of 61 mm, the calculated temperature profile at the rear part of cavity is improved greatly as shown by dotted curve in Fig. 6(b), and thus the calculated cooling rate shows better accuracy as indicated by the open marks in Fig. 7. However, the quantitative relations between the effective radius of cooling boundary and amount of heat input have not been obtained yet.

PREDICTION OF HARDNESS AND GUIDE-LINE OF SELECTING WLEDING VARIABLES

It is well known that the hardness of steel has close corelation with the amount of martensitic phase in microstructures. By the regression analyses of seventy kinds of steel, Arata and coworkers (1978, 1979) have proposed the hardness prediction equations which contain the cooling time from 800 °C to 500 °C and the chemical composition terms (Carbon equivalent) of steel. Basic formulas and their calculated results for the steel used (TABLE 2) are shown in TABLE 3 and Fig. 9.

It is, therefore, possible to estimate the hardness of plate welded under the water, using the previous equations (1) - (4) and (6). In Fig. 10 are compared the calculated and measured hardness at the boundary of fusion zone and HAZ (bond part). It is obvious that the hardness estimation is quite satisfactory within the accuracy of $\pm 10\%$.

TABLE 3   Hardness Prediction

Formulas   (Arata,1978)

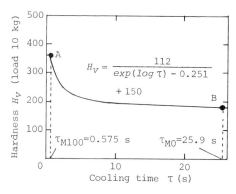

At point A (100% Martensite):
$H_V(\tau_{M100}) = 835[C] + 287$
$log\,\tau_{M100} = 2.55([C] + [Mn]/6.3$
$+ [Si]/3.6) - 0.92$

At point B (0% Martensite):
$H_V(\tau_{M0}) = 273([C] + [Mn]/13$
$+ [Si]/9.7) + 133$
$log\,\tau_{M0} = -0.37([C] - [Mn]/1.1$
$- [Si]/0.44) + 1.02$

Hardness at arbitrary time:
$\tau > \tau_{M100}: H_V = b/(exp(log\,\tau)+a) + 150$
$\tau < \tau_{M100}: H_V = 835[C] + 287$

In figure: $H_V = \dfrac{112}{exp(log\,\tau) - 0.251} + 150$

$\tau_{M100} = 0.575$ s $\tau_{M0} = 25.9$ s

Fig. 9.   Calculated hardness of
steel used.

.s described above, a mathematical approach to evaluate the thermal fields
.nd hardness of base metal is rather effective as the first order approxima-
.ion.   This means that the optimum welding variables can be determined by the
.alculation.   For example, if one desires to get an underwater welded joint
.aving the same hardness with that obtained in air, one must select the heat
.nput and cavity size in accordance with the guide-line diagram.   An example
.s shown in Fig. 11 for the two-dimensional heat flow, in which the critical
.urve shows that the average cooling rate defined in eq. (6) is 10% higher
.han that calculated from Rosenthal's solution.

.o far described was the case of a circular cavity.   However, the rapid cool-
.ng occurs only at the rear part of cavity, and its front half does not give
.nfluence on temperature distribution.   Therefore, the cavity size becomes
.nnecessarily large if the cooling rate must be regulated under the high heat
.nput condition, which eventually restricts the feasibility of process.   This
.ifficulty may be reduced greatly by adopting a rectangular cavity elongated
.ehind the heat source.   Figure 12 shows the calculated temperature distribu-

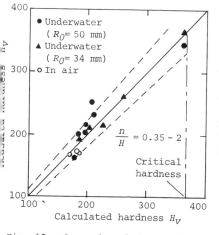

Fig. 10.   Comparison between the
measured and calculated
hardness.

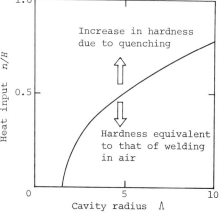

Fig. 11.   Guide-line diagram for
selection of heat input
and cavity radius.

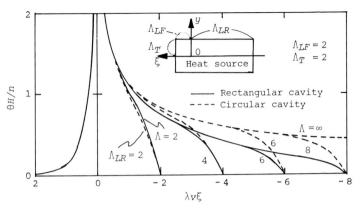

Fig. 12. Temperature distribution inside rectangular cavity

tions in the rectangular cavities that have very narrow width. In the figur
is also shown the calculated results for circular cavity of the same radiu
with the rear half length of rectangular cavity. The rapid cooling near th
rear end can be greatly improved in a rectangular cavity in spite of muc
smaller cavity area. In conclusion, the described method will be also usefu
for controlling the cooling rate and hardness at their desired values.

REFERENCES

Arata, Y. and others (1978). New concept of weldability. Proc. 2nd Internt.
    Colloq. Electron Beam Welding and Melting, Avignon, France, 169-176.
Arata, Y. and others (1979). Weldability concept on hardness prediction.
    Trans. JWRI, Osaka Univ., 8, 1, 43-52.
Christensen, N. and others (1965). Distribution of temperatures in arc
    welding. British Welding J., 54, 12, 54-75.
Hamasaki, M. and Sakakibara, J. (1973). Underwater welding of mild steel
    with water curtain type CO_2 arc welding method. J. Japan Welding Soc.
    (JWS), 42, 9, 897-906. (in Japanese).
Hamasaki, M., Sakakibara, J. and Watanabe, M. (1975). Underwater butt weldin
    of mild steel with water curtain type CO_2 arc welding method. Trans.
    JWRI, Osaka Univ., 6, 2, 3-9.
Matsunawa, A., Takemata, H. and Okamoto, I. (1980). Analysis of temperature
    field with equiradial cooling boundary around moving heat sources.
    Trans. JWRI, Osaka Univ., 9, 1, 11-18.
Nishiguchi, K., Matsunawa, A. and others (1975). Development of underwater
    welding with local cavity formation method. Proc. 2nd Internt. Sympo.
    JWS., Osaka, Japan, Paper No. 2-2-(6), 315-320.
Nishiguchi, K., Matsunawa, A., Tamura, M. and Wada, H. (1977). Underwater
    welding in a fluid stabilized local cavity. IIW, Doc. XII-B-214-77.
Nishio, Y. and others (1977). Development of underwater welding system. IIW,
    Doc. XII-B-213-77.
Rosenthal, D. (1946). The theory of moving sources of heat and its applica-
    tion to metal treatments. Trans. ASME, November, 849-866.
Shinada, K. and others (1978). Development of automatic underwater welding
    system. IIW, Doc. XII-B-237-78.
Tamura, M., Nishio, Y. and Wada, H. (1976). Development of automatic under-
    water welding with local cavity formation method. Offshore Technology
    Conf., Dallas, Texas, U.S.A., Paper No. OTC 2646.

Cold Cracking Susceptibility of Welds Obtained by Wet Underwater Welding

S. H. v.d. Brink* and G. W. Boltje**

*Metaalinstituut TNO, P.O. Box 541, 7300 AM Apeldoorn, The Netherlands
**Department of Metallurgy, Delft University of Technology, Delft, The Netherlands

ABSTRACT

Cold cracking and detorioration of mechanical properties can be serious problems in wet underwater welding. This paper describes the influence of hydrogen of MMA welds in Fe 410, Fe 510 and API-X60.
Rutile ferritic and austenitic electrodes were used. Implant tests were carried out with an apparatus suitable for underwater welding in the North Sea at a depth of 10 m. The cracking susceptibility has been determined for both single bead and multipass welds. It is generally assumed that the diffusible hydrogen in the weld metal and heat affected zone is generated from the surrounding water. Implant testing underwater shows, however, that a substantial part of the hydrogen emerges from the coating of the electrodes. Acceptable cracking properties can be obtained if the electrodes are dried prior to welding. Though the hydrogen problem is rather small when austenitic welds are made, the crack susceptibility of the weld metal causes a low implant strength.

KEYWORDS

Wet underwater welding, manual metal arc welding, rutile electrodes, austenitic electrodes, implant testing, implant strength, hydrogen content, Fe 410, Fe 510, API-X60.

INTRODUCTION

A number of objections of both executive and metallurgical nature, connected with wet underwater welding with coated electrodes are the reason that this technique is sometimes considered less suitable for reliable welded joints. The metallurgical difficulties mainly relate to brittle joints that are susceptible to cracking. In order to obtain quantative data, which also enable to establish small differences in crack susceptibility in a reproducible way, implant tests have been carried out in the laboratory as well as in deep salt waters. In either case both ferritic and austenitic-ferrictic weld metals were used as well as different types of steel. In addition to implant rupture strength and hydrogen contents, mechanical characteristics have been determined.

EXPERIMENTS

Implant measurements

For the experiments an implant machine has been used based on a triple tensile machine of the lever type. With this machine three implant specimens
(with helical notch) can be tested independently and simultaneously with a
constant load during 24 hours (1).
The tests were carried out in the laboratory (salt water, depth 50 cm) and
in salt water (North Sea, depth 10 m).
By means of the staircase method (2), it was established at which tensile
stress there is a 50% chance that a notched test bar ruptures as a result of
hydrogen cracking (implant rupture strength, σ_{50}). In Fig. 1 an example is
given. About 15 test specimens are needed for a reliable result.

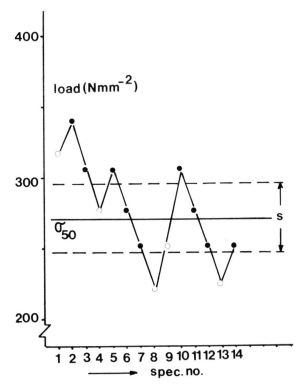

Fig. 1. Determination of σ_{50} with the staircase method
(. = rupture, o = no rupture)

Steel and welding consumables

The chemical composition of the steel used is given in Table 1 and its mechanical characteristics in Table 2.
The electrodes are indicated in the text with code a, b or c. Electrode a is
a rutile electrode (type AWS - E6030). Electrode b has an oxidizing coating

(type ASW - 6020). Electrode c is an austenitic-ferritic one (29 Cr-12 Ni). The diameter of the electrodes is 5 mm. Electrode a and b have been welded with straight polarity, electrode c with reverse polarity. Welding parameters have been stated wherever necessary.

TABLE 1   The chemical composition of the examined steel types

Material	Chemical composition									
	C	Mn	Si	P	S	Cu	Ni	Cr	Mo	C_{eq} [*]
Fe 410 ship's plate	0.17	0.84	0.14	<0.01	0.020	0.03	–	0.01	–	0.31
Fe 510 construction steel	0.18	1.29	0.62	0.03	0.022	0.01	–	0.02	–	0.40
API X 60 pipe material	0.19	1.35	0.57	0.04	0.008	0.01	0.03	0.06	0.04	0.44

[*]
$$C_{eq} = C + \frac{Mn}{6} + \frac{Cr + Mo + V}{5} + \frac{Ni + Cu}{15}$$

TABLE 2   The mechanical characteristics of the examined steel types

Material	yield-strength σ_v (N/mm^2)	tensile-strength σ_B (N/mm^2)	elongation δ(%)	contraction Ψ(%)	notched [*]	
					tensile-strength σ_B (N/mm^2)	notched factor k
Fe 410	256	406	38.6	72.6	380	1.07
Fe 510	352	510	34.5	72.5	452	1.13
API X 60	–	553	25.9	79.8	502	1.10

[*] Testbar with the same diameter and circular notch with the same geometry as the implant specimen.

Hydrogen determination

From both the electrodes used at sea and in the laboratory the quantity of diffusible hydrogen (3) as well as the residual hydrogen have been determined in duplicate.

RESULTS AND DISCUSSION

In Table 3 are stated the implant rupture strength, the scatter as well as the hydrogen content for the experiments at sea, together with the findings from the previous laboratory tests.

TABLE 3   Results for hydrogen determination and implant tests for various electrodes and steeltypes (electrode a with rutile coating, b with oxidizing coating, c gives austenitic-ferritic weld metal)

combination electrode - steel	welding place	heat input MJ/m	hydrogen content ml/100 gr weld metal, 0°C and 1 atm		Vickers hardness (100 gr) in H.A.Z.	Implant strength σ_{50} (N/mm^2)	scatter $\hat{s}$ (N/mm^2)
			diffusible	total			
a Fe 510	sea	2.8	35.5	38.2	550	272	53
a Fe 510	lab.	2.4	57.9	68		275	28
a Fe 510 (2 layers)	sea	2.3/2.4	33.9	36.7	400	628	9
b Fe 410	sea	3.3	37.6	40.9	550	553	173
b Fe 410	lab.	2.4	24.1	65.1		>515	5
b Fe 510	sea	3.3			475	338	64
b Fe 510	lab.	2.4				262	24
b API-X 60	sea	3.4			510	381	47
c Fe 510	sea	2.7	20.4	29.1	525	284	39

The implant strength of Fe 510 with electrode a often used in practice, is very low: well below the yield strength. This strength can be drastically improved by welding in two layers, with the second one annealing the first one; the practical disadvantage being that the second layer must be applied within a few minutes after the first one. With electrode b a higher implant rupture strength is the result in a comparable steel owing to the lower hydrogen content. Increasing carbon equivalent otherwise causes the implant strength to drop.

Finally, electrode c (austenitic-ferritic) gives a relatively low strength as a result of a microstructure susceptible to cracking (martensite with high nickel content) with a high hardness on the boundary of the weld. The conformity between the strength results at sea and those in the laboratory is striking. The hydrogen measurements, however, show large differences which cannot be explained at the moment.

Furthermore, the importance of drying the electrode prior to welding should be noted. It appeared that as a result of the long waiting periods at sea the electrodes got damp. For example it was then found for electrode b and steel Fe 410 that the implant strength was approx. 125 N/mm^{-2} lower than for well dried electrodes.

It is generally accepted that the hydrogen which penetrates the weld metal and the heat affected zone is directly coming from the surrounding water. The present results, however, of the implant tests clearly show that the major part of the hydrogen comes from the electrode coating.

Further to this, experiments were carried out with electrode b with welding in Fe 510 (25 mm thickness, V-joint, opening angle 75°). A macro section is shown in Fig. 2.

At 0°C the notch toughness of the final layers proved to be 47 J on an average. The hardness in the heat affected zone was 320 H$_V$10, in the weld metal 240 H$_V$10.

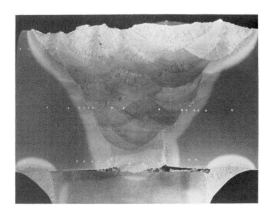

Fig. 2. Macro picture of a weld with electrode b
(Fe 510, thickness 25 mm) welded under water

CONCLUSIONS

1. With the implant test it is possible to obtain quantative data in a reproducible way on the cold cracking behaviour at wet underwater welding.
2. From the electrodes examined those used in practice produce the worst results with respect to cold cracking.
3. Although the hydrogen problem is reduced if austenitic or austenitic-ferritic electrodes are used, the final result is still a low implant strength. Namely, the cracking susceptibility of the microstructure has become larger.
4. The best perspective is offered by an electrode of the oxidizing type; without purposely aiming at improvements the cold cracking behaviour is already acceptable.
5. Also in the case of wet underwater welding it is important to keep the electrodes dry as long as possible.
6. Laboratory experiments with wet underwater welding give a correct picture of cold cracking under practical conditions i.e. welding in approx. 10 m deep water in the North Sea.

ACKNOWLEDGEMENT

This work is mainly carried out in the context of the Neptunus project under auspices of the Dutch Welding Institute.

REFERENCES

1. Report "Neptunus project" 1981, Dutch Welding Institute.
2. Christensen, N., (1976). IIW-doc. IX-983-76.
3. ISO-standard 3690-1977.

Moisture Absorption of Basic Electrodes Under Pressure up to 33 bar

P. E. Kvaale, K. Gjermundsen and N. Christensen

SINTEF, N-7034 Trondheim-NTH, Norway

ABSTRACT

The kinetics of moistening under pressures up to 33 bar are closely similar to those reported under normal pressure. A model is proposed for estimating the water content on the basis of initial water content, vapour pressure in the habitat and time of exposure. Tentative times of premissible exposure are suggested.

KEYWORDS

Hyperbaric MMA welding; habitat humidity; exposure time.

INTRODUCTION

Much work has been reported on moisture absorption of electrodes under normal atmospheric pressure. To the authors' knowledge no data are available for the conditions expected in operations at a depth of 300 to 350m. The present paper provides experimental data under such conditions for an electrode of AWS type E 8018-C1.

It is evident from published data that different makes of electrodes within the same AWS class cover a wide range of absorption rates. Most probably such differences depend on the composition, which is seldom reported. Information of this kind has been obtained with electrodes formulated to represent an average composition of E 7016 and E 7018 types commonly used in 1973. The results, although not obtained under hyperbaric conditions, have been included for the purpose of comparison.

MATERIALS AND EXPERIMENTAL PROCEDURE

A commercial type E 8018-C1 electrode of 2.5mm core and 4.5mm coating dia. has been employed. The same make has been used in previous studies of hydrogen, oxygen and carbon absoprtion in hyperbaric welding (Christensen and Gjermundsen, 1976, 1977; Grong, 1979; Knagenhjelm and colleagues, 1982). The mass ratio of coating to wire is 0.62.

Electrodes pre-baked to an initial moisture level of 0.10% H_2O were pressurised in dry helium and introduced through a gate into the test chamber where they were exposed to a slow stream of helium of constant humidity at $28^{\circ}C$. The amount of absorbed moisture was determined after each experiment by combustion analysis.

The E 7016 and E 7018 "prototype electrodes" for measurements of moisture absorption at atmospheric pressure were formulated and prepared by Magnusson (1973). Details of composition are given in Table 1. Moisture absorption was monitored by weighing in a chamber of constant humidity.

TABLE 1   Specifications for "1973 Prototype" Electrodes

Electrode	E 7016	E 7018
Iron powder; vol.wt. 2.9 kg/l	370 g	–
Iron powder; vol.wt. 3.7 kg/l	–	600 g
Ferromanganese	30 g	20 g
Ferrosilicon	75 g	50 g
Rutile	20 g	20 g
Limestone	280 g	185 g
Fluorspar	220 g	120 g
Cekol or Na/K alginate	5 g	5 g
Potassium silicate; ratio 3.3; 1000cP	200 g	190 g
Die dia. for core wire 3.25mm	5.65mm	6.80mm
4.00mm	6.80mm	8.05mm
5.00mm	8.15mm	10.00mm
Final baking at $370^{\circ}C$; final moisture $< 0.3\%H_2O$		

RESULTS

Moisture Absorption under Hyperbaric Conditions

The increase of water content is shown in Table 2, and in graphical presentation in Fig. 1. Included in Fig. 1 are also data from a manned test summarised by Knagenhjelm and colleagues, where the relative humidity fluctuated considerably during the test, with an average estimated to roughly 80%.

278

TABLE 2 Water Absorption under Hyperbaric Conditions (wt.%)

Pressure and humidity	Time of exposure, minutes									
	0.2	1	2	3	5	10	15	30	60	120
16 bar - 95%	0.00	0.09	–	0.20	0.28	–	–	–	–	–
16 bar - 70%	0.05	0.07	–	0.11	0.14	0.19	0.25	0.30	0.50	–
31 bar - 95%	0.12	0.15	0.19	0.25	0.28	0.40	0.46	0.75	1.04	1.73
31 bar - 70%	0.02	0.04	–	0.06	0.10	0.15	0.22	0.29	0.41	–
31 bar - 50%	0.01	–	–	0.03	0.05	0.10	0.10	0.15	–	–

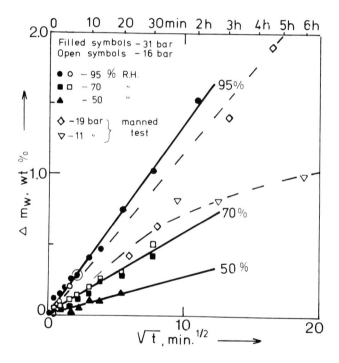

Fig. 1. Observed Moisture Absorption Δm_w under Pressure

Data at Atmospheric Pressure

Results obtained with the "1973 prototype" electrodes (Christensen, 1974; 1976) are shown in Figs. 2 and 3. For the sake of easy presentation the experimental points have not been included in the low range of Fig. 2 (34% R.H.)

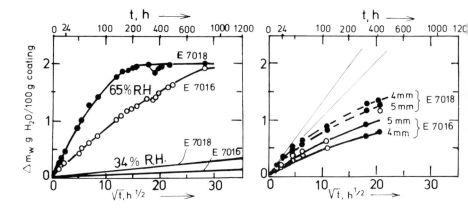

Fig. 2. Moistening at 25°C and Relative Humidities 34 and 65%.

Fig. 3. Moistening at 15°C and 65% Relative Humidity.

DISCUSSION

Inspection of Fig. 1 reveals that the rate-controlling step in moisture absorption does not depend on the external pressure. It is strongly dependent, however, on the partial pressure of water. Similar observations have been reported from measurements made at atmospheric pressure, e.g.; by Evans (1977).

Comparison of Fig. 1 with the early stages of moistening from Figs. 2 and 3 also indicates similarities. The slope of the lower curve of Fig. 1, appertaining to 28°C/50% R_{r}H. or about 14 Torr, is 0.028% per min$^{\frac{1}{2}}$, as compared to 0.022% per min$^{\frac{1}{2}}$ from Fig. 2 for E 7018 at 25°C/65% R.H. (about 15 Torr). In the latter case saturation has been attained at 2.15% H_2O after about 400 hours. At 34% R.H. saturation is approached after 1200 hours; for E 7018 the saturation value is estimated to about 0.5% H_2O. Indications of approach to saturation are also seen from Fig. 1 for data obtained at 11 bar in a manned test; however, in this case a constant relative humidity could not be maintained.

It is seen from Figs. 2 and 3 that both the rate of absorption and the amounts absorbed by electrode E 7018 at 15°C/65% R.H. are much greater than at 25°C/34% R.H. The vapour pressures are nearly identical, however: 8.3 Torr at 15°C and 8.1 at 25°C. It may be inferred that the observed over-all rate is probably not controlled by transfer of water through a gas film at the surface, so much more as the absorption rate then would be expected in dependent of time.

Assuming diffusion of water into the coating as a rate-controlling step, as has also been done by Chew (1976), the course of the process can be predicted from published general solutions. For simple geometries the fractional saturation $F = \Delta m / \Delta m_s$ is given as a function of the dimensionless parameter $u = \sqrt{Dt}/L$, where D is the diffusivity and L a characteristic linear dimension of the geometry under consideration. Taking as a rough approximation diffusion into a coating of thickness L equivalent to the first stages of diffusion into a wide plate of thickness 2L, the initial slope dF/du is 1.128 (Darken and Gurry, 1953). The diffusivity is thus given by the water absorption Δm_s at saturation and the initial slope $d\Delta m/d(\sqrt{t})$:

$$D = (L/1.128\Delta m_s)^2 (d\Delta m/d\sqrt{t})^2$$

For diffusion of water into the E 7018 coating at $25^\circ C/65\%$ R.H. this gives $D = 4 \times 10^{-6} mm^2/s$. At $25^\circ C/34\%$ R.H. and at $15^\circ C/65\%$ R.H. the saturation values have not been attained; a very rough estimate gives $D \approx 2 \times 10^{-6} mm^2/s$. It may be concluded that the diffusivity is not strongly temperature-dependent.

Application of the data of Fig. 1 in terms of diffusivities cannot be made until saturation values have been established for the chosen combinations of temperature and relative humidity. It is of course possible to apply the measurements directly, with the limitation that predictions cannot be made for combinations not tested.

Taking as a tentative limit a permissible amount of 0.2% water and assuming an initial content of 0.1%, the following exposure times should not be exceeded:

> At 95% R.H.: 30 seconds
> At 70% R.H.:   3 minutes
> At 50% R.H.: 15 minutes.

An exposure time of 30 seconds is clearly unacceptable in practical operation. It must be emphasized that the choice of 0.2% total water as an upper limit is very conservative, because a substantial fraction of the absorbed moisture will not penetrate far beyond the surface of the electrode, and will therefore be more easily lost on resistance heating than water remaining from the manufacturing process.

For a safe procedure of handling electrodes under hyperbaric conditions, it will be necessary to examine this effect under increased pressure.

ACKNOWLEDGEMENT

The authors gratefully acknowledge permission from Messrs. Norsk Hydro a.s. to publish the present data.

REFERENCES

Chew, B. (1976). Moisture loss and gain by some basic flux covered
    electrodes. Weld J. 56, 127s-134s.
Christensen, N. (1974; 1976). Reports to the Royal Norwegian Council for
    Scientific and Industrial Research. NTNF-B0520.4728
Christensen, N. and K. Gjermundsen (1976, 1977). Effects of pressure on
    weld metal chemistry. IIW Doc. II-212-384-76; supplement II-212-395-77.
Darken, L.S. and R.W. Gurry (1953). In R.F. Mehl (Ed.), Physical Chemistry
    of Metals, Mc Graw-Hill, London.
Evans, G.M. (1977). Moisture absorption characteristics of Tenacito 65R
    electrodes. Oerlikon Schweissmitt. 35 No. 79, 4-8.
Grong, Ø. (1979). Metallurgical Engineer's Thesis. The Norwegian Institute
    of Technology.

Knagenhjelm, H.O., K. Gjermundsen, P. Kvaale and D. Gibson (1982).
    Hyperbaric TIG welding to 500m simulated depth. In Underwater
    Technology Conference, Bergen 1982.
Magnusson, E.J. (1973). Personal communication from Messrs. ESAB,
    Gothenburg.

Sampling and Analysis of Welding Fumes and Gases Produced Under Experimental Hyperbaric Conditions

D. A. Worrall* and D. E. Gibson**

*Materials Department, The Welding Institute, Cambridge, UK
**Stolt-Nielsens Rederi A/S, Oslo, Norway

ABSTRACT

Welding fumes and gases produced during the tungsten inert gas (TIG) welding and manual metal arc (MMA) welding of carbon manganese steels have been sampled during a manned hyperbaric welding experiment. Concentrations of oxides of nitrogen, carbon monoxide, carbon dioxide, argon and nitrogen were found to be below the control limits established for the working chamber. Concentrations of particulates and ozone frequently exceeded their control limits, in particular during TIG welding. In this experiment, control of the pollutants arising during the MMA process was more effective than that during TIG welding, since the TIG arc is much more sensitive to the disturbance caused by local exhaust ventilation, and fume capture close to its source was therefore more difficult.

KEYWORDS

Hyperbaric environment; MMA welding; GTA welding; fumes; ozone; ventilation equipment; protection.

INTRODUCTION

The Deep Ex II experimental diving research project, which was undertaken using the hyperbaric test facilities of the Norwegian Underwater Technology Centre, Bergen, Norway, included manned hyperbaric welding tests. Various aspects of the semi-automatic tungsten inert gas (TIG) welding of carbon manganese steels were studied to a simulated depth of 500m. The moisture pick up of basic coated manual metal arc (MMA) electrodes and the resultant influence on weld metal hydrogen content were also evaluated, to a maximum depth of 300m. The parent materials used were a structural steel conforming to the German Standard ST-52-3N, and a pipeline steel in accordance with API 5LX65.

In order to ensure the safety of the diver-welders, it was essential to monitor and control the pollutants produced during welding. The methods of monitoring and sampling procedures used were designed to give, as far as possible, a rapid feedback of results which would provide not only a basis

for controlling the safety aspects of the experiment, but also data for a
general study of hyperbaric welding pollutants. Due to the inevitable
retrospective nature of pollutant concentration results, the divers were
equipped during welding sessions with masks, which were supplied with
breathing gas from an external source. The results were therefore utilised
primarily to assess the environment in terms of safety before the masks were
removed at the end of each session.

POLLUTANT SPECIES

Particulate welding fume contains a range of elements which arise from the
consumable, its coating (if any) and the work-piece. Hyperbaric atmospheres,
in which the partial pressure of oxygen is maintained within the range
0.3-0.4 bar, do not support the combustion of many materials which burn
readily in air at sea level, and it is probable that during hyperbaric
welding, particulates are formed containing metals in a lower oxidation state
than is the case in air. Therefore, control limits derived in respect of
metal oxide fume may not necessarily be applicable to hyperbaric
environments.

There is a range of pollutant gases which may be evolved during welding in
air, and the following species were considered to be relevant to hyperbaric
welding by the Deep Ex sponsors. Nitric oxide and nitrogen dioxide (oxides
of nitrogen) are formed by thermal oxidation of nitrogen, which would enter
the welding chamber via the transfer lock. Ozone is produced by the action
of ultra violet (uv) radiation on oxygen, while the decomposition of
carbonates in MMA electrode coatings leads to emissions of carbon monoxide
and carbon dioxide. Argon, which was used as shielding gas in some TIG
trials, and nitrogen were also of significance due to their influence on
decompression schedules.

The general policy adopted by the sponsors was that exposure to contaminants
should be kept as low as was reasonably practicable. Concentration limits
were specified for each of the above species and breathing masks were
removed only when none of these limits were exceeded. The limits used are
shown in Table 1, and were based on the Threshold Limit Values for schedules
involving a working week of 40 hours as documented by ACGIH (1980), although
some modifications were made in consideration of the high pressure.

SAMPLING AND ANALYSIS

This experiment provided the opportunity to assess the suitability of some
analytical methods and equipment for use under hyperbaric conditions.
Facilities were made available for the analysis of each pollutant by two
methods, in order to provide cover in the event of failure of non-proven
techniques.

The most reliable method of measuring particulate concentrations is
gravimetry; small pumps are used to draw gas samples through pre-weighed
filters which are subsequently re-weighed. Rotameters cannot be used to
measure gas flowrates under hyperbaric conditions, since they are influenced
by the density of the ambient atmosphere. The sampled volume was therefore
measured by means of direct displacement gas meters, the responses of which
are independent of gas density. A respirable aerosol continuous monitor
operating on the light scattering principle was tried as the second method
for measuring particulates. The sampling points are illustrated in Fig. 1.

The methods of monitoring pollutant gas concentrations are given in Table 2.
Two separate gas chromatographs were used, the first for continuous

monitoring of carbon monoxide and the second for argon, nitrogen and carbon dioxide. On dive day 15, pollutant gas concentrations were measured sequentially at points A, B, C and a point 20cm from the arc; on subsequent days, all measurements were made at point A.

ENVIRONMENTAL CONTROL

The hyperbaric welding chamber (and the other chambers in the complex) were connected to external environmental control units (ECUs), the function of which is to remove moisture, absorb carbon dioxide and control the gas temperature. Carbon dioxide is absorbed by soda lime which, being alkaline, also absorbs to some extent the acid gases nitric oxide and nitrogen dioxide. Carbon monoxide was removed by means of a platinised alumina catalyst in a unit which was added to the working chamber ECU system. No specific measures were taken for ozone removal, since ozone is unstable and may break down by interaction with any of a number of substances either in the gas circulating system or inside the chamber. Argon and nitrogen were introduced to the working chamber in only small quantities and, although the facility to flush the chamber with uncontaminated gas was available, in practice these gases did not reach their joint concentration limit.

A small fume extractor unit was installed in the welding chamber for the removal of particulates. Due to limitations of time and available space, the flow capacity of the system was limited to about $1m^3$/minute, and manually positioned local extraction was utilised in order to achieve the best possible efficiency. For MMA welding, this took the form of a small nozzle of area $70cm^2$, which was maintained by a welder's assistant in a position approximately 5cm from the arc, without detrimental effects on the welding process. The TIG process is more susceptible to disturbance by draughts, and a larger hood ($500cm^2$) was built for positioning directly above the TIG torch. The shielding gas proved to be extremely sensitive to disturbance when the hood was moved closer than about 30cm, so that fume removal was less efficient than during MMA welding.

The extracted gas was filtered through high efficiency particulate filters of nominal pore size 0.3 micron, and the filtered gas outlet was connected in line with the ECU system (also rated at $1m^3$/min) during welding.

RESULTS

The results of concentration measurements for each species were used to calculate a time weighted average (TWA) concentration by means of the equation

$$\text{TWA concentration} = \frac{C_1 T_1 + C_2 T_2 + \ldots C_n T_n}{T}$$

where C_1, C_2, $\ldots C_n$ are individual concentrations, T_1, T_2, $\ldots T_n$ are periods of time in which individual results were obtained and T is the sum of the sampling periods. For particulates, the values C_1, C_2, $\ldots C_n$ each relate to a mean value calculated from results obtained simultaneously at the sampling points A, B and C (Fig. 1). Good agreement was in any case observed between the results obtained at the three points. For pollutant gases, readings were taken individually and relate to point A for samples taken after dive day 15 as described above.

The duration of the working sessions varied on a day to day basis, and the results are expressed as time weighted averages for the total time during which samples were taken.

Particulates

Gravimetry proved to be a successful technique in terms of providing a series of fume concentration results. The results were influenced by the duration of arcing periods within sampling periods, and therefore Table 3 contains an arcing duty cycle column (arcing duty cycle = the ratio of arcing time to sampling period duration, expressed as a percentage) in addition to TWA and peak concentrations of particulates.

The TWA values frequently exceeded $5mg/m^3$ by a margin which was in general greater during TIG welding than during MMA welding. The lowest TWA concentrations occurred in the sessions when the arcing duty cycle was at relatively low values. At depths of 300m and 100m, MMA welding tests immediately followed the TIG trials, and concentrations measured during the former process were enhanced by residual TIG fume. At 200m, there was an overnight break between TIG and MMA welding, and the results on day 29 reflect most accurately the degree of control achieved during MMA tests.

Concentration results were not provided by the respirable aerosol continuous monitor, which failed soon after the chamber was pressurised.

Pollutant Gases

For each of the pollutant gases monitored, the first method listed in Table 2 successfully provided a series of results during each working session; a few results were obtained by the second method, but these were used more to assess the validity of the techniques than to generate additional data. Problems were encountered in the application of chemical methods to ozone and nitrogen dioxide. These methods each require the use of a bubbler apparatus containing chemical reagents which must either enter and leave the working chamber via the transfer lock (with the risk of reagent losses during rapid pressure changes) or be operated outside the chamber (with inevitable losses of the monitored species e.g. by adsorption in the sample lines).

Ozone concentration results, measured using gas detector tubes, are given in Table 4. Time weighted average concentrations exceeded the control limit of 0.1 microbar in each TIG session; in MMA sessions, little or no ozone was detected. Where combined results are given for TIG/MMA work periods, nearly all the ozone was generated during the TIG process. Ozone production was greater during bead on plate and capping runs than during welding in grooves, and the maximum concentration recorded was coincident with the use of helium, as opposed to argon, shielding gas.

Concentrations of oxides of nitrogen, carbon monoxide, carbon dioxide, argon and nitrogen did not exceed the concentration limits at any time; the highest recorded TWA and peak concentrations of these substances are given in Table 5, the results being obtained by the first monitoring method (Table 2). It is notable that carbon monoxide was detected during TIG welding; the highest TWA concentration was 4.6 microbar, compared with 24 microbar during MMA welding.

DISCUSSION

Assessment of Sampling Techniques

For particulates, the gravimetric method which is in widespread use at sea level was modified in respect of the variations in the density of the chamber atmosphere, which occurred continuously as a result of the decompression schedule. Satisfactory time averaged results were obtained, and good agreement was observed between the concentrations at the sampling points A, B and C; therefore, the fume was distributed uniformly throughout the chamber. The failure of the respirable aerosol continuous monitor is attributable to the difficulties of operating complex electronic devices in pressurised heliox environments.

For gaseous contaminants, gas detector tubes provided rapid, reliable results at each pressure, but this method is not ideal for assessing time averaged concentrations, since quantities of tubes are consumed and the availability of a diver to carry out sampling is essential. Long term detector tubes may provide TWA results with only small demands on labour, but the fluctuations which occur, for example, in ozone concentrations, would not be recorded. Chemical methods are not recommended for hyperbaric applications due to the difficulties described above, and instrumental techniques are possible alternatives for analysing ozone and oxides of nitrogen.

Decompression of samples prior to analysis has little influence on methods utilising the mass spectrometer or gas chromatograph, since the sample is introduced into the instrument in a fixed quantity and it is possible to achieve a great deal of sensitivity. These methods are suitable for the determination of small quantities of stable gases (carbon monoxide, argon and nitrogen). The infra red analyser was also used successfully at atmospheric pressure, with a suitable in-line pressure reducer, to monitor hyperbaric carbon dioxide, which is found at relatively high concentrations.

Pollutant Concentrations

Particulate concentrations were higher during TIG welding than during MMA welding; this reflects the difficulties encountered in TIG fume capture rather than differences in fume emission rates from the two processes. Control of particulates is essential for either process, since the closed environment does not allow fume dispersal into a large volume, as might be anticipated in the case of factories and workshops. In a chamber of volume $50m^3$ (as in this case), the generation of only 250mg of fume will produce a concentration of $5mg/m^3$ in the absence of control measures; such a quantity may be emitted during the consumption of a single MMA electrode or only a few minutes of TIG welding.

During the TIG sessions, increased ozone production during bead on plate and capping runs arose as a result of the larger solid angle being irradiated from the arc. Higher concentrations when using helium compared with argon shielding gas are caused by the higher welding voltage associated with helium; the resultant greater arc length leads to a rise in uv radiation intensity. The provision of a means of ozone control is desirable during manned hyperbaric TIG welding. This could take the form of an activated charcoal filter stage within the gas recirculating system. Alternatively, uv radiation emissions could be reduced by the incorporation of a mechanical shroud in the welding torch. This approach would impair the welder's view of the arc and is therefore more applicable to operations controlled remotely e.g. using a fibre optic and/or television system.

The other pollutant gases were controlled adequately by the measures taken during this experiment, although attention is drawn to the need for continuous monitoring of carbon monoxide and carbon dioxide, to ensure that the platinised alumina catalyst and soda lime absorbing medium can be replaced when a decrease in efficiency is observed. The mechanism for the generation of carbon monoxide during TIG welding was not identified unambiguously, but the decomposition of carbon dioxide thermally in the arc region or under the influence of uv radiation, and the oxidation of TIG filler wire lubricant, are potential contributors of this gas.

Ventilation Control

The effectiveness of exhaust ventilation depends on the flowrate and the position and geometry of the inlet. The degree of control achieved over contaminants generated during welding was much greater during MMA than during TIG tests. In the latter case, persistent problems occurred due to the generation of particulates and ozone, and when the exhaust inlet was moved close to the arc, the shielding gas was disturbed. The most promising alternative is to increase the exhaust flowrate and it is suggested that a high flowrate recirculating system with an efficient filtration system be used in future studies. The selection of argon shielding gas rather than helium would minimise arc disturbance due to the higher density (and inertia) of the former, and ozone generation would also be reduced.

Comparison with Subsea Operation

In a real subsea working situation, a higher welding duty cycle is to be expected than was achieved during this study. The internal volume of the welding habitat (typically 25-30m^3) would be lower than that of the working chamber in this experiment (50m^3). One would therefore anticipate increased quantities of pollutants generated during welding, and exposure would be for several hours in each session.

When extraction equipment is provided in subsea habitats, it is normally more powerful than the small unit available for this experiment, and better control of fumes and gases should be achieved when the exhaust inlet is positioned carefully. It is not possible to predict the actual concentrations of pollutants which may arise during the wide range of subsea activities, and in view of the difficulties experienced in respect of contaminant control in this study, when one diver was available for adjusting continuously the inlet position, respiratory protection is recommended as the best means of minimising exposure. Complete protection is in any case necessary in the absence of either concentration measurements indicating that the environment is sufficiently free from pollutants, or control techniques which have proved themselves to be capable of dealing adequately with the fumes and gases generated.

CONCLUSIONS

1. Gravimetric techniques can be applied successfully to the determination of particulate concentrations during hyperbaric welding. In this experiment, attempts to use a respirable aerosol continuous monitor were not successful.

2. Pollutant gas concentrations can be measured at intervals under hyperbaric conditions by using appropriate detector tubes. Continuous monitoring by means of instruments such as the gas chromatograph and the mass spectrometer, in which analysis is conducted remote from the pressurised environment, may be carried out after decompression of the sample. The use of chemical methods is subject to disadvantages which were not overcome during Deep Ex II.

3. Fumes arising during MMA welding can be captured close to the arc by local exhaust techniques. During TIG welding, low flowrate local

exhaust did not provide an adequate degree of control in this study, and shield gas disturbance occurred at exhaust inlet positions less than 30cm from the arc. The use of a higher flowrate system at a greater distance from the arc is proposed for future studies.

4. The quantities of ozone generated during hyperbaric TIG welding are such that specific control measures are required. An activated charcoal filter stage in the gas recirculating system would prevent the accumulation of ozone in the working chamber. A mechanical shroud fitted to the welding torch would reduce the generation rate; during remotely controlled, automated welding, this method could be particularly effective.

5. Carbon monoxide and carbon dioxide can be controlled adequately by platinised alumina and soda lime, the efficiency of which can be maintained by replacement either at times indicated by continuous pollutant gas monitoring, or at pre-defined intervals based on the minimum lifetime.

6. In a real subsea working situation, the lack of facilities and diver time for contaminant monitoring indicate a requirement for complete respiratory protection for diver-welders in the absence of proven techniques for pollutant control.

REFERENCE

ACGIH (1980). Documentation of the Threshold Limit Values. Fourth Edn. 1980. American Conference of Governmental Industrial Hygienists, P O Box 1937, Cincinnati, Ohio 45201.

ACKNOWLEDGEMENTS

Particular acknowledgement is made to Norske Shell and Norsk Hydro for the financial support provided for the above work programme.

TABLE 1 Concentration Limits Adopted for Pollutant Species in the Working Chamber

Species	Partial pressure (microbar)
NO_x*	5
O_3	10^{-1}
CO	50
CO_2	5×10^3
Ar)) N_2)	10^6

Particulate fume controlled to $5mg/m^3$ at all depths.

* NO_x = NO + NO_2. Combined limit based on the more stringent limit for NO_2.

TABLE 2 Methods Used for Monitoring Pollutant Gas Concentrations

Species	First method	Second method
NO_x	Gas detector tubes	Chemical method*
O_3	Gas detector tubes	Chemical method[†]
CO	Gas chromatograph 1	Gas detector tubes
CO_2	Infra red analyser	Gas chromatograph 2
Ar	Mass spectrometer	Gas chromatograph 2
N_2	Mass spectrometer	Gas chromatograph 2

* Oxidation of iodide ions in solution and
  photometric determination of free iodine

[†] Photometry of an azo dye (Saltzman method).
  Determines NO_2 only.

TABLE 3 Particulate Fume Concentrations Measured by Gravimetry

Dive day number	Depth (m)	Welding process	Arcing duty cycle (%)	Total sampling time (min)	Total duration of welding session (min)	Concentration of particulates (mg/m³)	
						Peak[(1)]	TWA[(2)]
14	500	TIG	1.6	129	372	3.0	1.5
15	500	TIG	1.0	239	326	9.5	4.4
22	400	TIG	6.3	355	515	22.2	9.8
25	300	TIG	11.1	316	395	23.8	13.2
25	300	MMA	5.8	72	90	6.4	5.8
28	200	TIG	21.8	135	180	8.3	5.2
29	200	MMA	5.6	60	67	4.1	3.3
31	100	TIG	9.3	192	360	21.3	15.9
31	100	MMA	8.7	51	45	10.1	10.0

Notes

(1) Peak concentrations relate to results obtained at individual sampling points, illustrated in Fig. 1.

(2) TWA concentrations calculated from mean concentrations obtained from the three sampling points.

TABLE 4 Ozone Concentrations Measured using Gas Detector Tubes

Dive day number	Depth (m)	Welding process	Arcing duty cycle (%)	Total sampling time (min)	Total duration of welding session (min)	Ozone concentration (microbar)	
						Highest reading	TWA
15	500	TIG	1.0	80	326	0.8	0.23
22	400	TIG	6.3	184	515	1.8	0.69
25	300	TIG/MMA	10.1	244	485	2.8	0.88
28	200	TIG	21.8	108	180	1.4	0.62
29	200	MMA	5.6	20	67	N.D.	N.D.
31	100	TIG/MMA	9.2	84	405	0.8	0.43

N.D. = not detected

TABLE 5   Highest Recorded Concentrations for Pollutant
          Gases other than Ozone

| Pollutant gas | Highest recorded concentration (microbar) | |
	TWA	Peak
NO_x	<0.5	<0.5
CO	24	48
CO_2	3.5×10^3	3.8×10^3
Ar)) N_2)	5.6×10^5	7.9×10^5

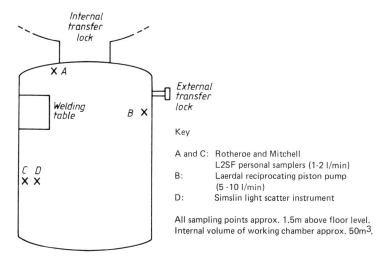

Fig.1. Plan view of working chamber, illustrating sampling points.

Key

A and C: Rotheroe and Mitchell
          L2SF personal samplers (1-2 l/min)
B: Laerdal reciprocating piston pump
          (5 -10 l/min)
D: Simslin light scatter instrument

All sampling points approx. 1.5m above floor level.
Internal volume of working chamber approx. $50m^3$.

.

Gas Shielding Welding – The Reaction of Oxygen in Normal and Hyperbaric Environments

F. R. Coe and J. Moreton

The Welding Institute, Abington, Cambridge, UK

ABSTRACT

Using techniques of chemical thermodynamics, theoretical models can be developed relating weld oxygen and carbon contents to the partial pressure of oxygen or carbon dioxide in the shielding gas during gas shielded welding. These have proved successful in predicting weld oxygen concentrations for known carbon levels in wire and weld with both $Ar-O_2$ and $Ar-CO_2$ mixtures, at normal and elevated pressures. The paper briefly describes the models, and illustrates their successful application using both new experimental results and literature data.

KEYWORDS

GMMA WELDING; MIG WELDING; CO2 WELDING; HYPERBARIC ENVIRONMENT; SHIELDING GASES; OXYGEN; CARBON; MOLTEN POOL; COMPOSITION; THERMODYNAMICS.

INTRODUCTION

In the gas shielded welding of steel using solid wires it is readily observed that the total oxygen content of the weld is greater than the sum of contributions from plate and wire. Given adequate shielding to eliminate air, and noting that oxygen is necessary for arc stability, heterogeneous reaction equations should be capable of describing the distribution of oxygen between gas and liquid phases during welding.

It has been shown (COE & MORETON, 1979) that for shielding gases rich in CO_2 the Fe - C - O reaction can be expressed quantitatively by the equation:-

$$\log P_{CO_2} \, [\%O]^2 \, [\%C] = 9175/T - 2.304 \quad\text{------------------------(1)}$$

where P_{CO_2} is the partial pressure of CO_2, T is the reaction temperature ($^\circ$K) and $[\%O]$, $[\%C]$ are the weight % concentrations of oxygen and carbon respectively in the weld metal. In turn, the relation between P_{CO_2} and V (the volume % of CO_2) in the gas mixture is given by:-

$$P_{CO_2} = \frac{2P(100-\alpha)/100}{2 + \frac{\alpha}{100} + 2(\frac{100-V}{V})} \quad\text{----------------------(2)}$$

where P is the total gas pressure and α is the degree of dissociation of CO_2.

In the case of argon-oxygen shielding; the appropriate equation is:-

$$[\%O] = 0.0726 \ P_{O_2}^{\frac{1}{2}}$$ ----------------------------------(3)

where P_{O_2} is the partial pressure of oxygen.

By selecting values of T close to the solidus for steel and converting partial pressures to volume concentrations in the gas shield, it is possible to produce curves predicting the balance between oxygen and carbon in the weld metal and the oxygen content of the shielding gas. The curves shown in Fig.1, refer to the theoretical situation when welding at normal atmospheric pressure.

For argon-oxygen shielding gases the left hand side of Fig.1, is used; the position of the plot depends on the value selected for T but the weld oxygen level is independent of the carbon level. For CO_2 shielding gases the right hand side of the graph applies, and indicates a balance between weld carbon and oxygen which depends on the CO_2 level in the shield. Oxygen in the shield arises from the dissociation of CO_2 and the position of the family of curves depends on the value of α, the degree of dissociation used in the calculation. This, in turn, is influenced by the value used for T.

Hyperbaric welding with higher gas pressures is also described by equation (1) and (3),and the calculation of the effective degree of dissociation of CO_2 at elevated pressures has been examined by Moreton and Boothby, 1981. There, it was shown that the theory for CO_2 containing gases could be tested by comparing calculated values for the parameter $[\%O]^2 \ [\%C]$, from equation (1), with those obtained using analysed weld oxygen and carbon contents.

APPLICATION OF THEORY TO LITERATURE RESULTS

A comparison of test data from the published literature of Widgery (1974), Shackleton and Ruckdeschel (1976) and Stout et al (1970) against the theoretically predicted curves showed an encouraging degree of corres-pondence (Coe and Moreton, 1979). For Ar - O_2 gas mixtures, the results of Shackleton and Stout are illustrated in Fig. 2, and show good agreement with the curve corresponding to a dissociation temperature for O_2 of $1800^{\circ}C$. Fig. 3, shows the theoretical relationship between weld metal carbon and oxygen levels for varying proportions (5 to 100%) of CO_2 in the Ar - CO_2 shield. The results of Shackleton and Ruckdeschel (1976),Widgery (1974), and more recently, Schultz (1981) and Agusa et al (1982) show reasonable agreement with the relevant theoretical curves. However, the need for extremely careful analysis of carbon, and particularly oxygen, in weld metal was highlighted, and it has been shown elsewhere (Stevens and Moreton, 1979) that a number of weld metal samples taken along the length and across the thickness of the weld are essential. Further experimental investigation involving such careful analysis was thought worthwhile to examine the theory in both normal welding and at elevated pressures. These tests, (Coe and Moreton 1979, and Moreton and Boothby 1981), are described below.

EXPERIMENTAL PROCEDURES AND RESULTS

Wire Selection

Four wires, B, C, D and E, with carbon contents ranging from 0.05 to 0.11% were used. Wire B was a commercial, coppered, 1.2mm wire to BS 2901, Part 1,

1970, A18. Wires C & D were experimental, uncoppered, 1.0mm wires. Wire E was a maraging, uncoppered 1.2mm wire. All were degreased before welding.

Welding Procedure

Welds were made in a V-preparation of Si-killed, Al-treated mild steel plate using the buttering technique indicated in Fig.4 (a-c), to minimise dilution. Plate, wire and weld analyses are recorded in Table 1 and welding conditions for the final run in Table 2. Hyperbaric welds were made in the chamber described by Stalker and Salter, 1975, at pressures of 5 and 10 bar so that, for CO_2 mixtures the critical point of CO_2 would not be approached (Moreton and Boothby, 1981).

Sampling Procedure and Chemical Analysis

Welds normally 250mm long, were sectioned for analysis as indicated in Fig. 4 (d-f). The procedures for oxygen and carbon analyses described by Stevens and Moreton, 1979, were used and the results are given in Table 2. The results for carbon in test welds 1-18 are each the average of 3 separate analyses; in tests 19-24 up to 20 separate analyses were made. The oxygen results each represent an average of between 11 and 28 separate analyses.

Normal Pressures

For tests 1 to 9 inclusive with $Ar-CO_2$ gases at normal pressures, results are plotted in Fig.3. In all cases the test results fall within the band representing 2 standard deviation limits for oxygen and carbon analysis applied to the theoretical curve. For tests 10 to 18 inclusive with $Ar-O_2$ shielding gases at normal pressures, results are plotted in Fig.2.

Elevated Pressures

Welds 2, 20 and 22 in Table 2 were made with wire B and $Ar-20\%$ CO_2 shielding gas at 1, 5 and 10 bar (absolute) pressures respectively. Experimental and theoretical values for the parameter $[\%O]^2 [\%C]$ are compared in Table 3. In deriving the theoretical values, deviations from ideal gas behaviour have been allowed for by adopting a value for α of 51% at 5 bar and 39% at 10 bar, respectively for $Ar-20\%$ CO_2 (Moreton and Boothby, 1981). Welds 11, 19 and 21 also used wire B, but with $Ar-2\%$ O_2 shielding gas at 1, 5 and 10 bar pressures. In this case oxygen is not near its critical condition so that the experimental data have been entered into Fig.2, making the following assumptions:-

(a) $Ar-2\%$ O_2 at 5 bar provides an oxygen partial pressure in the shield equivalent to 10% O_2 at 1 bar (absolute).

(b) $Ar-2\%$ O_2 at 10 bar provides an oxygen partial pressure in the shield equivalent to 20% O_2 at 1 bar.

DISCUSSION AND CONCLUSIONS

CO_2 - containing gas shields

Both data from the literature and also current experimental results support the contention that the Fe-C-O reaction (equation (1)) predicts the balance between weld oxygen and carbon content at least over the range 0.03 to 0.11%C.

295

The agreement between experiment and theory is encouraging but not exact and several factors may be responsible for this:-

(a)   Values for 2 standard deviations in weld oxygen and carbon analyses are shown in Fig.3. In practice, these amount to ± 20% at 0.03%C but only ±6% at 0.1%C. Similarly for weld oxygen analyses this may be ±10% to ±50% of the measured level. The need for representative sampling and multiple analyses is thus essential to minimise the uncertainty in the placing of experimental points in Fig.3.

(b)   The theory does not account for contributions of plate oxygen to the final weld metal content as a consequence of dilution. It proposes that oxygen uptake is determined by equation (1) before droplet detachment and the experiments test this by using a buttering technique. However, in routine welding if dilution occurs test data may not fit the predicted curves. On the other hand weld carbon levels are invariably equal to or greater than the corresponding wire carbon levels and this increase must have arisen from CO_2 in the shield.

(c)   The use of equations (1) and (2) requires values for both α and T. The difficulty of choosing these for a high temperature arc situation has been discussed (Coe and Moreton, 1979) so that the choice of position of the theoretical curves used in Fig.3, may be subject to variations in welding conditions.

Argon-oxygen gas shields

Both data from the literature and also current experimental results support the contention that the Fe-O reaction (equation 3) predicts the balance between weld oxygen content and shielding gas composition. The use of equation (3), based on the dissociation of oxygen, requires again the selection of a value for T. A value of $1600^{\circ}C$ was originally selected (Coe and Moreton, 1979), but the results in Fig. 2, suggest that a temperature of $1800^{\circ}C$ places the theoretical curve in a position closer to the experimental data.

Shielding Gas selection when welding at atmospheric pressure

The agreement between theory and experiment suggests that Fig. 1, can indicate shielding gas mixtures which will yield required weld oxygen and carbon levels. The balance between weld oxygen and carbon levels when using CO_2 rich gases is defined by equation (1). The equivalence of a 100% CO_2 shield to one of Ar-17% O_2 (Fig.1), is in excellent agreement with the conclusions of Cordoroy and Wallwork (1980) who, working from an entirely different set of premises, predicted the equivalence to be Ar-20% O_2. Furthermore, Fig. 1, indicates an equivalence between Ar-5% CO_2 and Ar-7% O_2.

Hyperbaric welding

For CO_2-rich gases, equations (1) and (2) allow high pressure systems to be examined through the manipulation of P in equation (2), together with selection of values for α and T. A set of curves, similar to those in Fig.1, but specific to each chosen pressure is obtained. The curves shift to the right (i.e. to higher weld oxygen levels) with increasing pressure but comparison of theory with experiment is more conveniently made (Coe and

Moreton, 1979) via the parameter $[O\%]^2$ $[\%C]$ in equation 1. Table 3 shows that a close comparison is obtained and the results in Table 2 confirm that the effect of increased pressure is to raise the weld oxygen level for a minimal increase in carbon level. This would be expected from the application of Le Chatelier's principle to equation (1). A similar effect would be expected with equation (3) for Ar-O_2 mixtures and the results in Table 2 confirm an increase in weld oxygen level with pressure. However, Fig.2, suggests that the effect of total gas pressure on oxygen partial pressure may have been inexactly estimated when entering the data.

It is concluded that conventional reaction chemistry provides a basis for selecting shielding gas mixtures to maintain control over weld oxygen and carbon contents in both normal and hyperbaric operation.

REFERENCES

Agusa, K., Kosho, M., Nishiyama, N., Kamada, A., and Nakano, Y. Matching ferritic filler MIG welding of 9% Ni steel. Kawasaki Steel Technical Report, (6) Sept. 1982, 11pp.

Coe, F.R. and Moreton, J., The Chemistry of oxygen in gas shielded welding of steel. Welding Institute Report, 9285.01/79/165. 3 July,1979.

Corderoy, D.J.H., and Wallwork, G.R., Gas/weld metal reactions in MIG arc plasma. Weld pool Chemistry and Metallurgy - International Conf. London, April 15 - 17, 1980. Publ. The Welding Inst., Abington, Cambs 1980. Paper 12, pp 147-153.

Moreton, J., and Boothby, P.J., The Chemistry of oxygen in gas shielded welding of steel, Part II. Welding Inst. Report 9285.01/82/270. 3 May 1981.

Schultz, J.P., Influence du gas de protection sur la composition chimique du metal depose en soudage à l'arc electrique sous flux gazeux - procedes MIG-MAG. Journees D'Information, Métallurgie de la Zone Fondue Société Française de Métallurgie - Section Sud-Est. Paper 04, 1981

Shackleton, D.N., and Ruckdeschel, W.E.W., Welding Res. International 1976, 6,11.

Stalker, A.W., and Salter, G.R. A preliminary study of the effects of increased pressure on the welding arc. Welding Inst. Report, 3412/6/74, Dec. 1974 (Also Technology Reports Centre Report RR-SMT-R-7504, April, 1975).

Stevens, S.M., and Moreton, J., Sampling and analysis for oxygen Welding Inst. Res. Bulletin, 20 (5) May 1979, 146-151.

Stout, R.D., Machmeir, P.M., and Quattrone, R., Welding J. Research Suppl. 49 (11) Nov. 1970. 521s-530s.

Widgery, D.J., Deoxidation practice and toughness of mild steel weld metal. Welding Inst. Members' Report, M/78/74, 1974.

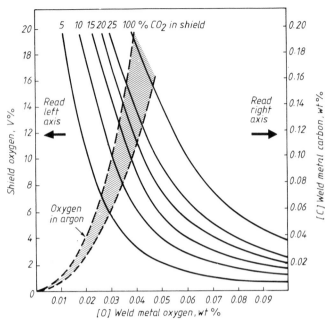

Fig. 1. Comparison of Ar-O$_2$ and Ar-CO$_2$ shielding gases for resultant oxygen content of weld metal.

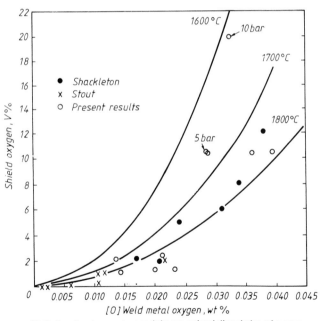

Fig. 2. Reaction temperature and the arc-assisted dissociation of oxygen, showing experimental results.

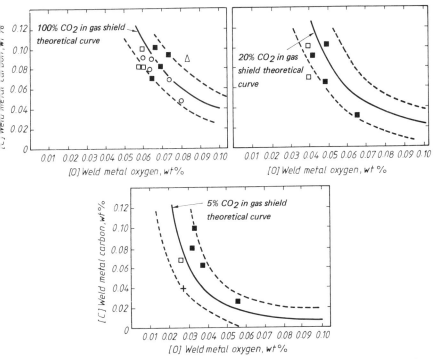

Fig.3. The relationship between weld metal carbon and oxygen concentrations for varying proportions of CO_2 in the Ar-CO_2 gas shield, and agreement of experimental results. Present results ■, Shackleton □, Widgery ○, Schultz △, Agusa +. − − − − Typical 2s.d. limits on O, C results.

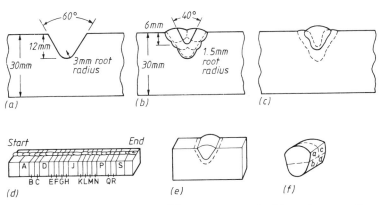

Fig.4. Weld preparation and sectioning: (a) original plate preparation (b) secondary preparation of buttered groove (c) final welds (d) typical marking of plate into sections − A, D, J, P, S for spectrographic analysis; C, F, H, L, N, R for carbon analysis; B, E, G, K, M, Q for oxygen analysis (e) identification of weld area by polishing and etching (f) final division of each section into pieces a, b, c, d, for O/N analysis.

Table 1   Chemical analysis of plates, wires and welds

	Element, wt%													
	C	S	P	Si	Mn	Ni	Cr	Mo	V	Cu	Nb	Ti	Al	Co
Plate	0.17	0.015	0.019	0.24	0.81	< 0.01	<0.01	<0.01	<0.01	0.02	<0.005	<0.01	0.031	<0.01
Wire B	0.08	0.018	0.026	0.78	1.50	0.04	0.02	0.01	<0.01	0.18	<0.005	<0.01	0.007	<0.01
Wire C	0.05	0.010	0.008	0.55	1.14	< 0.01	0.03	<0.01	0.01	<0.01	<0.005	0.21	0.050	<0.01
Wire D	0.11	0.010	<0.005	0.50	1.11	< 0.01	0.04	<0.01	0.02	<0.01	0.006	0.40	0.13	<0.01
Wire E	0.010*	0.010	0.004	0.17	0.59	16.1	–	3.25	–	–	–	0.31	0.17	5.99
Weld 1	0.080* / 0.08	0.018	0.024	0.66	1.23	0.04	0.01	<0.01	<0.01	0.17	<0.005	<0.01	<0.005	<0.01
Weld 2	0.091* / 0.09	0.019	0.024	0.60	1.15	0.03	0.01	<0.01	<0.01	0.16	<0.005	<0.01	0.006	<0.01
Weld 3	0.093* / 0.10	0.017	0.024	0.47	0.99	0.03	0.01	<0.01	<0.01	0.15	<0.005	<0.01	<0.005	<0.01
Weld 4	0.063* / 0.07	0.009	0.007	0.49	1.01	< 0.01	0.01	<0.01	<0.01	<0.01	<0.005	0.11	0.028	<0.01
Weld 5	0.064* / 0.07	0.008	0.007	0.45	0.95	< 0.01	0.01	<0.01	<0.01	<0.01	<0.005	0.08	0.021	<0.01
Weld 6	0.070* / 0.08	0.010	0.007	0.35	0.78	< 0.01	<0.01	<0.01	<0.01	<0.01	<0.005	0.06	0.015	<0.01
Weld 7	0.103* / 0.11	0.009	<0.005	0.47	1.04	< 0.01	0.01	<0.01	0.01	<0.01	<0.005	0.24	0.071	<0.01
Weld 8	0.103* / 0.12	0.009	<0.005	0.44	0.95	< 0.01	0.01	<0.01	<0.01	0.01	<0.005	0.18	0.055	<0.01
Weld 9	0.103* / 0.11	0.009	<0.005	0.32	0.76	< 0.01	0.01	<0.01	<0.01	0.01	<0.005	0.15	0.041	<0.01
Weld 10	0.08	0.020	0.025	0.74	1.27	0.04	0.01	<0.01	<0.01	0.17	<0.005	<0.01	0.007	<0.01
Weld 11	0.08	0.020	0.025	0.74	1.30	0.04	0.01	<0.01	<0.01	0.18	<0.005	<0.01	0.008	<0.01
Weld 12	0.07	0.019	0.025	0.66	1.16	0.04	0.01	<0.01	<0.01	0.17	<0.005	<0.01	0.007	<0.01
Weld 13	0.05	0.010	0.006	0.59	1.09	< 0.01	0.01	<0.01	<0.01	0.08	<0.005	0.13	0.032	<0.01
Weld 14	0.06	0.010	0.007	0.52	1.03	< 0.01	0.01	<0.01	<0.01	<0.01	<0.005	0.10	0.023	<0.01
Weld 15	0.05	0.009	0.007	0.47	0.90	< 0.01	0.01	<0.01	<0.01	<0.01	<0.005	0.11	0.023	<0.01
Weld 16	0.09	0.008	<0.005	0.52	1.08	< 0.01	0.02	<0.01	0.01	<0.01	<0.005	0.36	0.093	<0.01
Weld 17	0.09	0.009	<0.005	0.53	1.09	< 0.01	0.02	<0.01	0.01	<0.01	<0.005	0.34	0.080	<0.01
Weld 18	0.09	0.010	<0.005	0.47	0.92	< 0.01	0.01	<0.01	0.01	<0.01	<0.005	0.27	0.078	<0.01
Weld 19	0.083*	–	–	–	–	–	–	–	–	–	–	–	–	–
Weld 20	0.098*	–	–	–	–	–	–	–	–	–	–	–	–	–
Weld 21	0.068*	–	–	–	–	–	–	–	–	–	–	–	–	–
Weld 22	0.106*	–	–	–	–	–	–	–	–	–	–	–	–	–
Weld 23	0.027*	0.023+	0.004	0.01	0.33	15.1	<0.01	3.1	–	0.04	–	0.013†	0.009†	4.97
Weld 24	0.031*	0.028+	0.006	0.02	0.30	14.9	<0.01	3.1	–	0.04	–	0.002†	0.010†	5.20

NOTES: Most analysis results were obtained on the emission spectrograph; those marked * by Leco C analysis, those marked + by X-ray fluorescence analysis, and those marked † were diluted and remelted twice during X-ray fluorescence analysis.

In all cases B<0.001, Sn<0.01, Pb<0.01.

Table 2   Weld carbon and oxygen analyses and welding conditions

Weld number	Wire	Gas	Pressure bar	% carbon analysis		% oxygen analysis		Welding conditions, top runs			
				Mean	2s.d.	Mean	2s.d.	A	V	Energy input, kJ/mm	Wire feed speed, m/min
1	B	Ar-5%CO_2	1	0.080 †		0.032	±0.021	305	28.5	1.6	10.4
2	B	Ar-20%CO_2	1	0.091 †		0.042	±0.014	275	30.0	1.5	10.4
3	B	100%CO_2	1	0.093 †		0.073	±0.018	315	33.5	1.9	11.1
4	C	Ar-5%CO_2	1	0.063 †		0.037	±0.008	330	28.5	1.7	13.9
5	C	Ar-20%CO_2	1	0.064 †		0.048	±0.004	320	30.0	1.7	13.5
6	C	100%CO_2	1	0.070 †		0.064	±0.020	300	31.5	1.7	13.8
7	D	Ar-5%CO_2	1	0.103 †		0.033	±0.018	285	25.0	1.3	9.9
8	D	Ar-20%CO_2	1	0.103 †		0.049	±0.004	265	28.0	1.4	10.2
9	D	100%CO_2	1	0.103 †		0.067	±0.010	325	33.0	2.0	13.1
10	B	Ar-1%O_2	1	0.08+		0.020	±0.004	300	30.0	1.6	8.1
11	B	Ar-2%O_2	1	0.08+		0.021	±0.002	300	29.0	1.6	8.9
12	B	Ar-10%O_2	1	0.07+		0.027	±0.016	290	29.0	1.5	8.6
13	C	Ar-1%O_2	1	0.05+		0.023	±0.004	280	27.0	1.4	9.5
14	C	Ar-2%O_2	1	0.06+		0.022	±0.004	245	28.0	1.3	10.9
15	C	Ar-10%O_2	1	0.05+		0.036	±0.020	250	29.0	1.3	8.4
16	D	Ar-1%O_2	1	0.09+		0.014	±0.004	280	30.0	1.5	10.6
17	D	Ar-2%O_2	1	0.09+		0.014	±0.002	290	29.0	1.5	11.8
18	D	Ar-10%O_2	1	0.09+		0.039	±0.028	265	30.0	1.5	10.7
Low carbon tests											
23	E	Ar-5%CO_2	1	0.027 †	±0.006	0.058	±0.016	360	26.0	1.7	3.2
24	E	Ar-20%CO_2	1	0.031 †	±0.004	0.066	±0.006	350	26.0	1.7	3.2
Hyperbaric tests											
20	B	Ar-20%CO_2	5	0.098 †	±0.004	0.054	±0.004	290	32.0	1.7	10.4
22	B	Ar-20%CO_2	10	0.106 †	±0.006	0.052	±0.008	290	32.0	1.7	10.4
19	B	Ar-2%O_2	5	0.083 †	±0.004	0.029	±0.006	300	35.0	1.9	8.9
21	B	Ar-2%O_2	10	0.063 †	±0.016	0.033	±0.010	300	35.0	1.9	8.9
Plate				0.176 †	±0.004	0.006	±0.006				
Wire B				0.08+	±0.02	0.020					
Wire C				0.05+	±0.02	0.009					
Wire D				0.11+	±0.02	0.009					
Wire E				0.010 †	±0.004	0.010	±0.002				

NOTES: † Leco C results
       + Emission spectrograph results

       Welding traverse speed 330mm/min
       Shielding gas flow 24 l/min

Table 3 Comparison of theory with experimental results for carbon-dioxide bearing gases

Test	Wire	Gas	Pressure bar	$[\%C] [\%O]^2$ theoretical	$[\%C] [\%O]^2$ experimental	Effective, α	Effective pressure CO_2 bar
Hyperbaric tests							
2	B	$Ar-20\%CO_2$	1	1.95×10^{-4}	2.47×10^{-4}	60%	0.2
20	B	$Ar-20\%CO_2$	5	(a) 3.63×10^{-4} (b) 2.36×10^{-4}	2.81×10^{-4}	80% 51.2%	1.0
22	B	$Ar-20\%CO_2$	10	2.99×10^{-4}	2.88×10^{-4}	38.8%	2.0

NOTES: (a) Assume $Ar-20\%CO_2$ at 5 atmospheres has the same effective dissociation as $100\%CO_2$ at 1 atmosphere (80%)

(b) Assume $Ar-20\%CO_2$ at 5 atmospheres has a dissociation of only 51.2%

An Evaluation of the Fatigue Behavior in Surface, Habitat, and Underwater Wet Welds

D. K. Matlock*, G. R. Edwards*, D. L. Olson* and S. Ibarra**

*Center for Welding Research, Metallurgical Engineering Department, Colorado School of Mines, Golden, Colorado, USA
**Gulf Science and Technology Company, Pittsburgh, Pennsylvania, USA

ABSTRACT

Fatigue crack propagation in surface, habitat and underwater wet welds was studied. Fatigue crack growth characteristics were found to depend primarily on gas porosity, which varied with both welding technique and supplier.

KEYWORDS

Underwater wet welds; fusion zone fatigue; limit load analysis.

PREFACE

The following study of fatigue propagation in surface, habitat, and underwater wet welds was originally published in the proceedings of the Second International Conference on Offshore Welded Structures, London, November, 1982 (Matlock and others, 1982) and, at the behest of the United States delegation to IIW, has been included here in abbreviated format. The interested reader is referred to the reference cited above for complete details of the work.

INTRODUCTION

Underwater welding has become a topic of interest because of the extensive use of this technique in the repair of many offshore platforms in the Gulf of Mexico. With increasing regulations requiring more assurances of safe offshore structures, it becomes necessary to gain a full understanding of the physical properties of underwater welds. Such information is important because it provides metallurgical data to use as a basis for repair procedure recommendation. There are two main types of underwater welding: "wet" and "dry" habitat. As a rule, underwater "wet" welding is limited to structural members that are not subjected to high stresses. This limitation has occurred because of the poor mechanical properties of the resulting weldments. The reported degree by which the properties are modified varies,

but the average degradation appears to be a 20% reduction in tensile properties and a 50% reduction in ductility (Silva, 1968). Also, the conditions
present during underwater welding are particularly conducive to hydrogen
absorption and cracking, both in the weld metal and the heat affected zone
(HAZ), and limitations on base plate compositions are required (Grubbs and
Seth, 1972; Helburn, 1979). "Dry" habitat welds are normally used on more
highly stressed members because the resulting weldments are capable of pass-
ing surface welding mechanical tests.

To date, limited studies have been performed to systematically compare the
mechanical properties of underwater wet welds with welds produced in dry
habitats or by normal surface welding procedures. In this paper the fatigue
crack propagation and fracture characteristics of the weld metal from a
series of weldments for offshore structures are presented. Specifically,
surface, habitat, and underwater wet welds fabricated by two commercial
offshore welding companies (referred to as supplier L and supplier M) are
compared.

EXPERIMENTAL PROCEDURE

Nine experimental welds representative of standard surface, habitat, and
underwater wet welding processes were fabricated according to commercial
practice by two offshore welding companies. The nine 0.6 m long multiple
pass butt weldments, fabricated by joining two 0.3 x 0.6 m plate sections of
ASTM A-36 (ASTM, 1982) steel, were divided into three groups. Groups I and
II consisted of surface, habitat, and underwater wet welds from suppliers L
and M respectively, while group III consisted of three underwater wet welds,
each fabricated by supplier M with E6013 electrodes having different coatings. All welds used a single-V edge preparation and were made with the
shielded metal arc (SMA) welding process. The approximate welding conditions for all weldments were 28 to 32 volts and 145 amperes. Further
details concerning the experimental weldments are given by Matlock and
others (1982). The welds were evaluated using standard metallographic,
radiographic, chemical, and hardness testing techniques.

The fatigue crack growth characteristics of the welded plates were evaluated
with 12.7 mm thick compact tension specimens machined according to dimensions specified by ASTM (1980). The samples were oriented so that the fatigue crack would propagate along the centerline of the weld bead. The 12.7
mm specimen thickness, which is larger than normally used in fatigue
studies, was chosen to insure that a representative distribution of weld
defects was present in the crack propagation plane. All testing was performed according to ASTM specification E-647-78T. More complete details of
the fatigue testing procedure can be found in Matlock and others (1982).

RESULTS

Specimens for macroscopic, microscopic, and chemical analyses were extracted
from each weld. Photomacrographs representative of the nine weldments are
presented in Fig. 1. The number of passes (~12) was found to be similar for
the surface and habitat welds from both suppliers. The number of passes
used for the wet welds (17 to 43) was significantly greater, and depended on
both supplier and welding rod. Furthermore, the surface and habitat welds
from both suppliers were essentially free of porosity while the wet welds
from supplier M possessed a high density of uniformly distributed pores. In
contrast, the wet weld from supplier L had a much lower density of randomly

distributed voids. Analysis by optical microscopy indicated that the primary microstructural constituents in all the welds were similar and typical of microstructures in A-36 steel weld metal. The average weld metal centerline and heat affected zone (HAZ) hardness values were also determined and found to be HRB 82-92 for all welds, indicating that the strengths of all the welds were similar.

The fatigue crack growth rate data were evaluated to determine generalized crack growth expressions. For constant values of the stress ratio R, da/dN versus ΔK data have been shown to follow the Paris crack growth law (Paris and Erdogan, 1963),

$$\frac{da}{dN} = A(\Delta K)^n \qquad [1]$$

where A and N are constants for a given environment, material, and set of test conditions.

The fatigue crack growth characteristics of all of the experimental welds are summarized in Figs. 2, 3, and 4 for welds in sample groups I, II, and III respectively. In each figure, the data points are omitted for clarity and the appropriate crack growth equations for the high ΔK linear portion of these data are included. Data from the literature (Rolfe and Barsom, 1977a) for wrought ferrite-pearlite steels are also included in each figure.

After the stable fatigue crack growth data were recorded, the samples were overloaded at an approximately constant displacement rate, and the peak loads at failure were recorded as a function of fatigue crack length. Fracture surfaces were then examined by both optical and scanning electron microscopic techniques. Representative optical fractographs of the fractured weldments are presented in Fig. 5.

DISCUSSION

Several important observations are obtained from an analysis of the fatigue crack growth rate data presented in Figs. 2-5. First, in comparison to normal wrought ferrite-pearlite steels, all nine experimental welds exhibited lower growth rates for low values of ΔK. This observation is consistent with previous studies on fatigue in welds (Rolfe and Barsom, 1977b). Also, all of the welds exhibited a higher sensitivity of the fatigue crack growth rate to ΔK, as evidenced by the magnitude of the exponent n in Eq. 1 (n = 3.0 for wrought ferrite-pearlite steel, n = 3.3 to 7.0 for the nine welds).

The data summarized in Figs. 2, 3, and 4 show that the fatigue crack growth characteristics of welds for offshore structures depend on both welding procedure and supplier. Specifically:

(i)   For a given supplier, the fatigue crack growth behavior of surface and habitat welds was similar, as summarized in Figs. 2 and 3.

(ii)   For supplier L, the behavior of the underwater wet weld (n = 4.2) was similar to the corresponding surface (n = 3.8) and habitat (n = 3.3) welds. Also, the absolute growth rates were similar to those for wrought ferrite-pearlite steel (see Fig. 2).

305

(iii)   For supplier M, the behavior of underwater wet weld M3 (n = 6.6)
        was significantly different from that of the corresponding sur-
        face (n = 4.2) and habitat (n = 4.3) welds (see Fig. 3). Also
        the fatigue growth characteristics of underwater wet welds pro-
        duced by supplier M exhibited a higher dependence on ΔK than
        those produced by supplier L (n = 6.6 vs. n = 4.2).

(iv)   The selection of welding rod was shown to significantly affect
        the fatigue crack growth rate behavior of underwater wet welds,
        as indicated by variations in n from 4.3 for weld M6 to 7.0 for
        weld M4 (see Fig. 4).

(v)   A comparison of the data for two underwater wet welds produced
        with the same welding rod (i.e., weld M3 in Fig. 3 and M4 in
        Fig. 4), but by different welders, indicated that the data were
        essentially the same, and that the observed fatigue properties
        were reproducible.

Because the microstructures, hardnesses, and chemical compositions (except
for Mn and O) were similar for all of the experimental welds, the fatigue
properties primarily reflected differences in porosity. As implied in
Fig. 5, the stable fatigue crack growth zones of the surface and habitat
welds were characterized by relatively smooth, flat fracture essentially
free of voids or inclusions. The stable fatigue crack growth zones of the
underwater wet welds from supplier M (shown in Fig. 5) were significantly
different from the zones of the surface and habitat welds, and were characterized as topologically rough, with a high uniform porosity.

The volume fraction of pores was measured using standard point counting
techniques in both the fatigue zone and the overload zone of each weld.
Typical results are shown in Fig. 5. The higher values obtained in the
overload zone reflect the effects of the interconnecting of voids which were
out of the primary fatigue fracture plane.

An evaluation of the fracture surfaces of the welds showed that for low
values of crack extension (and thus low values of ΔK) the pores effectively
pinned the crack front and retarded crack growth. Thus, for low values
of ΔK, the observed crack growth rate decreased with an increase in pore
density. However, for higher values of ΔK, the pores decreased the total
load bearing area, and increased the local stress at the crack tip. Correspondingly, the growth rates at high ΔK were greater in porous welds.

Pores which retard crack growth at low ΔK but enhance crack growth at high
ΔK result in a value of n in Eq. 1 which increases with porosity. This is
verified in the comparison of surface weld M1 (n = 4.2) with underwater wet
weld M3 (n = 6.6) in Fig. 3.

The fatigue crack growth rate data in Figs. 2 and 4 also support the interpretation presented above on the effects of porosity on fatigue. First, all
three welds from supplier L were low porosity welds and correspondingly
exhibited similar fatigue properties as shown in Fig. 2. Second, the fatigue crack growth characteristics of the three underwater wet welds (Sample
Group III) presented in Fig. 4 directly reflect the pore structure, which
was shown to vary with welding rod. In particular, a weld with smaller
pores (weld M6) exhibited fatigue crack growth characteristics similar to
those of the low porosity welds of supplier L and to the surface and habitat
welds of supplier M. Underwater wet welds M4 and M5 with similar pore
structures had similar fatigue crack growth characteristics in which the

pores retarded fatigue crack growth rates at low values of the imposed stress intensity factor range.

The peak fracture loads measured after the completion of stable fatigue crack growth also were directly dependent on the porosity. If it is assumed that the primary effect of the porosity was to decrease the cross sectional area of the weld, then the peak load at failure simply was a measure of the limiting load for ductile rupture in the remaining ligament. By incorporating the effect of porosity into the Paris (1979) limit load analysis, the peak load can be written as:

$$P_L = \frac{0.35 \ \sigma_o \ B \ (1-f_p) \ b^2}{(a + b/2)} \quad\quad [2]$$

where f_p is the volume fraction of porosity, and the term $(1-f_p)$ modifies the specimen thickness B to give an effective thickness.

The peak fracture loads were correlated with crack length and are plotted in Fig. 6. The data indicate the anticipated (Towers and Gerwood, 1979) decrease in fracture load with an increase in crack length. Also the welds which were essentially pore free exhibited higher failure loads for a particular crack length in comparison to the high porosity underwater wet welds of supplier M. Based on hardness data, the tensile properties of the weld metal were shown to be similar for all experimental welds and were estimated to be: yield strength = 427 MPa, ultimate tensile strength = 537 MPa. In Eq. 2, σ_o was taken as 482 MPa and predictions of the limit load for $f_p = 0$ and $f_p = 0.2$ were plotted in Fig. 6. A comparison of the experimental data with the predicted crack length dependence of the limit load in Fig. 6 indicated that the limit load anlaysis for zero porosity directly described the behavior of the pore free surface and habitat welds. Furthermore, the underwater wet weld data were also adequately described by Eq. 2 if a porosity like that observed in the overload zone ($\sim$ 20% - see Fig. 5) was assumed. This fact is shown in Fig. 6.

SUMMARY

The data presented above provided several important conclusions with respect to the general fatigue behavior of weld metal for offshore structures. In addition, specific conclusions were reached concerning the differences in similar welds manufactured by two commercial welding companies.

1. The fatigue crack growth characteristics of welds for offshore structures depended primarily on the bulk defects, particularly gas porosity, which varied with both welding technique and supplier.

2. The underwater wet welds (in this case, particularly those produced by supplier M) contained significantly higher pore densities than the other types of welds; correspondingly, the sensitivity of the fatigue crack growth rates to the imposed stress intensity factor in underwater wet welds was greater.

3. The underwater wet welds produced by supplier M exhibited high and uniform pore densities, with the pore size dependent on the electrode coating. The pore densities in the underwater wet welds were more than an order of magnitude greater than the pore densities in the underwater wet welds produced by supplier L.

4. A plastic yielding limit load analysis indicated that the decreased toughness of underwater wet welds primarily reflects the increased porosity of these welds.

The results of this study indicate that underwater wet welds in ASTM A-36 steel exhibited fatigue behavior which systematically differed from that of pore free surface and habitat welds. For many applications, it may be possible to utilize underwater wet welds if additional weld metal is applied to compensate for the load bearing loss due to porosity. However, this approach must be used with caution. Several factors not included in this study, such as surface geometry of the as deposited weld, are also known to significantly affect fatigue and fracture.

ACKNOWLEDGEMENTS

The authors are expecially grateful to the CSM Center for Welding Research and to Gulf Science and Technology Company for support of this work. Messrs. Duncan L. Hammon, Stephen Liu, Gary J. Coubrough, and W. Scott Gibbs were especially helpful with the experimental testing.

REFERENCES

ASTM (1980). Tentative method for constant-load amplitude fatigue crack growth rates above 10^{-8} m/cycle. ASTM Specification E647-78T, 1980 Annual Book of ASTM Standards, part 10, ASTM, Philadelphia, PA, pp. 749-767.

ASTM (1982). Standard specification for structural steel. ASTM Specification A36-81a, 1982 Annual Book of ASTM Standards, part 4, ASTM, Philadelphia, PA, pp. 137-140.

Grubbs, C. E., and O. W. Seth (1972). Multipass all position wet welding - a new underwater tool. Fourth Annual Offshore Technology Conference, April 30 - May 3, 1972, Houston, Texas.

Helburn, S. (1979). Underwater welders repair drilling rigs. Welding Design and Fabrication, 52, 53-59.

Matlock, D. K., G. R. Edwards, D. L. Olson, and S. Ibarra (1982). An evaluation of the fatigue behaviour in surface, habitat, and underwater wet welds. Second International Conference on Offshore Welded Structures, November 16-18, 1982, London, Paper 15.

Paris, P. C., and F. Erdogan (1963). A critical analysis of crack propagation laws. J. of Basic Eng., 85, 528-534.

Paris, P. C., H. Tanada, A. Zahoor, and H. Ernst (1979). The theory of instability of the tearing mode of elastic-plastic crack growth. Elastic-Plastic Fracture, ASTM STP668, American Society for Testing and Materials, Philadelphia, PA, pp. 5-36.

Rolfe, S. T., and J. M. Barsom (1977a). Fracture and Fatigue Control in Structures, Prentice-Hall, Englewood Cliffs, New Jersey, p. 239.

Rolfe, S. T., and J. M. Barsom (1977b). Fracture and Fatigue Control in Structures, Prentice-Hall, Englewood Cliffs, New Jersey, p. 252.

Silva, E. A. (1968). Welding processes in the deep ocean. Naval Eng. J. 80, No. 4.

Towers, O. L., and S. J. Garwood (1979). The significance of maximum load toughness. The Welding Institute Research Bulletin, Number 10, 292-299.

GROUP Ⅱ: SUPPLIER M

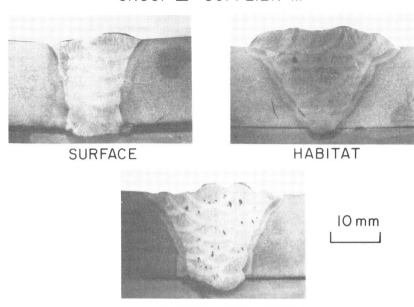

SURFACE   HABITAT

10 mm

UNDERWATER WET

Fig. 1.   Representative macrographs of surface, habitat and underwater
wet welds. Etched in 2%

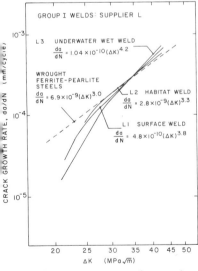

GROUP I WELDS: SUPPLIER L

L3   UNDERWATER WET WELD
$\frac{da}{dN} = 1.04 \times 10^{-10} (\Delta K)^{4.2}$

WROUGHT
FERRITE-PEARLITE
STEELS
$\frac{da}{dN} = 6.9 \times 10^{-9} (\Delta K)^{3.0}$

L2   HABITAT WELD
$\frac{da}{dN} = 2.8 \times 10^{-9} (\Delta K)^{3.3}$

L1   SURFACE WELD
$\frac{da}{dN} = 4.8 \times 10^{-10} (\Delta K)^{3.8}$

CRACK GROWTH RATE, da/dN   (mm/cycle)

ΔK   (MPa√m̄)

Fig. 2.   Fatigue crack growth
characteristics of
surface, habitat, and
underwater wet welds
fabricated by supplier
L.

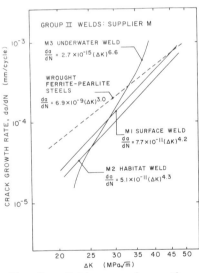

GROUP Ⅱ WELDS: SUPPLIER M

M3 UNDERWATER WELD
$\frac{da}{dN} = 2.7 \times 10^{-15} (\Delta K)^{6.6}$

WROUGHT
FERRITE-PEARLITE
STEELS
$\frac{da}{dN} = 6.9 \times 10^{-9} (\Delta K)^{3.0}$

M1 SURFACE WELD
$\frac{da}{dN} = 7.7 \times 10^{-11} (\Delta K)^{4.2}$

M2 HABITAT WELD
$\frac{da}{dN} = 5.1 \times 10^{-11} (\Delta K)^{4.3}$

CRACK GROWTH RATE, da/dN   (mm/cycle)

ΔK   (MPa√m̄)

Fig. 3.   Fatigue crack growth
characteristics of
surface, habitat, and
underwater wet welds
fabricated by supplier
M.

GROUP II: SUPPLIER M

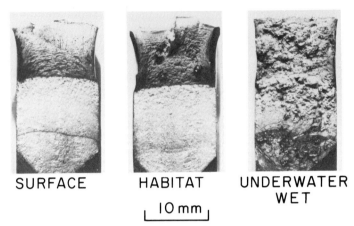

SURFACE   HABITAT   UNDERWATER
                                       WET
                   ⌊ 10 mm ⌋

Fig. 5.   Representative fractographs of fractured compact tension
           fatigue specimens. Pore volume fractions in specimen M3 were
           approximately 0.10 and 0.19 in the fatigue zone and overload
           zone, respectively.

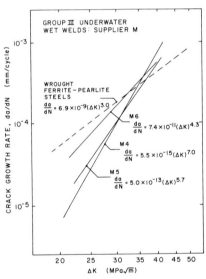

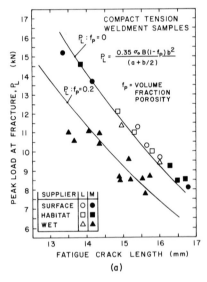

Fig. 4.   Fatigue crack growth
          characteristics of
          three underwater wet
          welds fabricated with
          different welding rods
          by supplier M.

Fig. 6   Comparison of measured
          peak fracture loads and
          the peak loads calcu-
          lated from a plastic
          yielding limit load
          analysis.

The New Method of Mechanized Underwater Welding

A. E. Asnis and I. M. Savich

E. O. Paton Electric Welding Institute of the Academy of Sciences of the Ukrainian SSR, USSR

ABSTRACT

The peculiarities of mechanized wet underwater welding of metal are considered. The effect of the medium (vapour-gas bubble) on the transition of some alloying elements from a filler metal into a weld is investigated. The main advantages of the new wet welding method are described. Certain factors affecting the HAZ crack formation are studied. The data on the equipment for the underwater welding are given.

KEYWORDS

Wet underwater welding, mechanized, flux-cored wire, semi-automatic machine.

A new stage of the underwater welding development began in the second half of the 20 th century, when the off-shore gas and oil production became economically expedient. Besides the works on restoration of oil-producing equipment and service lines, which are indispensable in exploiting the resources of the continental shelf, it has been applied in their construction as well. The latter required the high quality of welded joints and the sharp rise in productivity. A wide range of organizations has become involved in underwater operations with the application of metal welding.

In the postwar years the works on improving the quality of welds and mechanizing the underwater welding process were carried out in our country. Manual electric arc welding, for example, was performed by using tubular electrodes. The outer surface of the tube was coated and argon was supplied inside it. This raised the cost of the process and complicated it, while exerting the negligible effect on the weld quality. The experience of argon underwater welding was described by Avi-

lov I.I. (1965). The attempts to mechanize welding by using
the solid C8-08Γ2C grade wire were made. Here, the molten metal either directly contacted the vapour-gas bubble atmosphere or was shielded by CO_2 supplied from the surface, as was
shown by Madatov N.M. (1967). In both cases the weld quality
was low enough and the welding process was unstable.

In the world practice the problem of underwater welding is
known to be solved by using the dry method. This technology,
together with some advantages, has the drawbacks, the high labour consumption and operation cost being the main of them.

The demand for underwater welding is enormous in our country.
The construction of underwater pipelines, port and other hydraulic engineering structures, the necessity of their repair
stipulated the carrying out of scientific-research and expe-
rimental-design works on the development of the new technology
and equipment for underwater welding of metal.

The E.O.Paton Electric Welding Institute of the Ukrainian Academy of Sciences has developed the mechanized wet underwater
welding technology. To accomplish it a special self-shielding
flux-cored wire was used. The application of this new design
allowed to make welds, whose strength is equal to that of the
base metal. Peculiarities of the new method are described by
Savich (1969).

The wet welding method is quite maneuvrable, versatile and high
efficient capable of producing the quality welds in any spatial positions. In this case no devices for forcing the water
out of the arc zone are needed. Shielding gases are not used
either, since the arc burns in the atmosphere of the vapour-
gas pocket formed from the surrounding water.

As the analysis showed the atmosphere of the vapour-gas bubble
consists mainly of hydrogen. Oxygen, being also the product of
water decomposition, is, as a rule, almost completely spent
for metal oxidation.

We studied the transition of elements, usually used in the
welding engineering for heating and alloying welds from
electrode wire into the molten metal pool. For this purpose
the solid wires of comparatively pure Armco iron, each being
alloyed with one of the following elements: carbon, silicon,
manganese, aluminium, titanium, vanadium, nickel, chromium and
zirconium, were manufactured by the vacuum melting method.

The content of these elements in the deposited metal and the
transition coefficients are presented in Table I. The coefficient of the transition of some elements were calculated on
the basis of the data, obtained by the chemical analysis of
metal of wires and deposits.

As is seen from the Table, silicon, carbon, aluminium and titanium are most actively oxidized in the underwater welding.
Vanadium, nickel and manganese pass into the deposited metal
with less losses. The chrome transition degree according to
the said experiments was not possible to determine since in

building up by wire alloyed with this element the arc is unstable.

TABLE I

	Content of elements, %							
	C	Si	Mn	Ti	V	Ni	Al	Cr
In experimental wires	0.50	1.15	1.0	0.41	1.30	1.80	0.91	1.0
In deposited metal	0.10	0.15	0.55	0.03	0.70	1.20	0.30	traces
Transition coefficient with regard to base metal)	0.3	0.2	0.73	0.07	0.53	0.66	0.33	-

The transition coefficient of zirconium was not calculated either, due to its negligible quantities added to the appropriate experimental wire.

Comparing the degree of affinity of carbon, silicon and manganese for oxygen with the transition coefficients in underwater building up by using the experimental wires, it may be concluded, that the degree of oxidation (burning-out) of these elements is of the same nature as in air welding. However, the process of oxidation under the water proceeds more intensively. This must be accounted for the fact, that water vapours, being decomposed in the arc zone, form a large amount of free oxygen reacting with the molten steel.

The effect of varying the mechanized underwater welding parameters on the transition of such alloying elements as manganese, silicon and carbon from the C8 -08Г2C grade wire into the weld metal was studied.

As Fig.1 shows, with increase of arc voltage manganese and silicon content in weld metal decreases, while carbon content remains practically unchanged. The increase of arc voltage results in the arc length growth, therefore, the time of electrode metal drop dwelling in the region of high temperatures of the arc column, where manganese and silicon react very actively with oxygen, increases.

The mechanized underwater welding is performed by using the power source with a rigid external volt-ampere characteristic. In this case the increase of the welding current value is associated with the increase of the wire feed speed and the

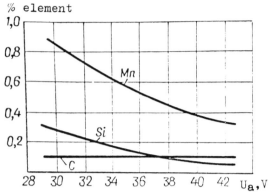

Fig.1. The effect of arc voltage on the transition on Mn, Si and C into weld metal in the mechanized underwater CB -08Г2С wire welding.

rate of wire melting, thus favouring the decrease of time of drop dwelling in the high temperature region and, hence, the slight increase of manganese and silicon content in the weld (Fig.2)

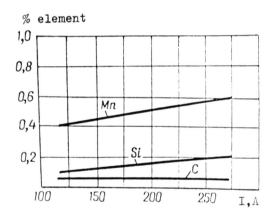

Fig.2. The effect of welding current on the transition of Mn, Si and C into weld metal in mechanized underwater CB-08Г2С wire welding.

When varying the mechanized underwater welding speed within the wide enough range of values, no significant effect of this parameter on the chemical composition of the metal deposited is observed, as is seen from Fig.3.

314

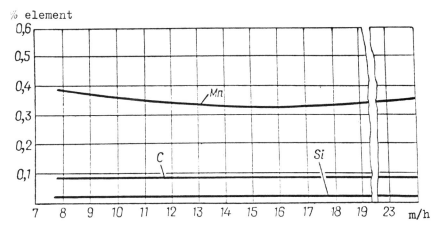

% element

Fig.3. The effect of mechanized underwater
Св -08Г2С wire welding speed on the
transition of Mn, Si and C into weld
metal.

Due to the fact, that the weld metal solidification rate in
underwater welding is high enough and the molten metal pool
is comparatively small and practically has no cold zone of
the so-called "tail", where carbon oxidizes most actively, in
all experiments carried out the variation of conditions parameters does not affect the amount of carbon remaining in the
weld metal.

In decomposition of water vapours in the arc zone hydrogen, in
addition to oxygen,is formed which penetrating into the metal
causes the embrittlement of weld and HAZ metal. To decrease
the negative effect of hydrogen, slag-forming elements are
added to the flux-cored wire. The latter is aimed not only at
deoxidizing and additional alloying the metal deposited, but
at improving the weld formation as well. Besides, the slag
 crust slightly slows down the deposit cooling rate. The
ПпС-АН1 grade self-shielding flux-cored wire, developed at
the E.O.Paton Welding Institute, meets well the said requirements. It is manufactured of 1.2...2.0 mm diameter and widely
used in welding of various metal structures operating under
the water.

The most versatile, as the practice has shown, is 1.6 mm dia.
wire, allowing to make flat, vertical and overhead welds with
the sufficient efficiency. Table 2 presents the mechanical properties of underwater welded joints of ВСт3сп   and 09Г2 grade
steel .

TABLE **2** The values of the mechanical properties
of underwater welded joints

Characteristic of properties	Steel grade	
	BCт3cп	09Г2
Ultimate strength, MPa	406–414 / 409	503–526 / 511
Yield strength, MPa	308–336 / 321	390–408 / 392
Elongation, %	28.7 – 37.3 / 31.5	22.7 – 33.4 / 28.7
Reduction in area, %	57.8 – 68.9 / 66.1	61.4 – 72.1 / 68.5
Impact toughness, J/cm² at +20°C	86 – 100 / 105	87 – 125 / 112
at –40°C	47 – 52 / 50	75 – 125 / 97
Bend angle, degr.	180	180

Note: Numerator – the minimum and maximum values,
denominator – the average of five values.

The welds possess high values of impact toughness both at posi
tive and negative (down to -40°C) temperatures. The data given in Table 2 show that the mechanical properties of underwater welded joints are practically the same as those of the base metal.Weld metal is observed to have the lower content of
carbon and manganese and to be practically free of silicon.
The structure is fine-grained, consisting of acicular ferrite
with the regions of fine-dispersed laminated pearlite and bainite, it furthers to maintain the high enough strength characteristics and resistance to crack initiation and propagation.

To perform underwater welding at depths, accessible for divers
the special equipment reliably operating both in fresh and sea
water is required.

Together with the requirements to the equipment, designed for
welding in the air and under the water, there exist additional
requirements, specified by the environment. When operating at
a depth, it is necessary to have the so-called "immersible
units", including a feeding mechanism with a flexible hose and
a holder, a wire reel, control circuit leads and welding cables. Here,protection of the electric motor and the feeding device reduction gear from the contact with the surrounding wate
is required. Taking into account all the requirements the
E.O.Paton Welding Institute has developed the A1660 type semi-

automatic machine for the wet underwater welding, widely applied in performing different hydraulic engineering operations,
with the regard to these and other requirements.

The semi-automatic machine with an elongated holder hose can
be used in the case of welding at a small depth, e.g. ship
repair afloat. Here the wire feeding device and the reel are
located in the air.

In conclusion it should be noted, that we developed the technology and the equipment for mechanized wet underwater welding. The process is used in construction and repair of different hydraulic engineering structures.

REFERENCES

Avilov G.I. The experience of underwater argon welding application. "Svarochnoje proizvodstvo", No8, (1955)
Madatov N.M. Underwater welding and cutting of metals. "Sudo-
strojenije", Leningrad, (1967)
Savich I.M. Underwater flux-cored wire welding. "Avtomatiches-
kaja Svarka", No10, (1969)

Structural Strength of Underwater Welded Joints

A. E. Asnis, I. M. Savich and A. A. Grishanov

E. O. Paton Welding Institute of the Academy of Sciences of the Ukrainian SSR, USSR

ABSTRACT

The physical-mechanical properties of low-carbon and low-alloy steel welded joints, made by a wet method, were studied to establish the possibilities of using this method in fabrication of special-purpose hydroengineering structures. Mechanical properties and dynamic strength of underwater welded joints were studied. The effect of welded joint metal cooling rate on the crystal lattice change in the HAZ is shown. The effect of negative temperatures on the cold brittleness of underwater welded joints is studied. The cold brittleness of joints, made with flux-cored wire underwater welding, is compared to similar joints made with calcium-fluoride coated electrodes in air. It is established that underwater welded joints are not susceptible to ageing and possess the high cold resistance.

KEYWORDS

Wet underwater welding, cooling rate, strength, ductility, stresses, cyclic toughness, physical properties of welds.

Until now only manual arc underwater welding, developed and reported by Khrenov K.K. (1933, 1946), was used for the underwater engineering jobs. However, this method did not find wide application in repair and construction of special-purpose hydraulic structures because of limited technological possibilities and unsatisfactory quality of the welds.

The E.O.Paton Welding Institute has developed a wet method of underwater semi-automatic flux-cored wire welding of low-carbon and low-alloy steels, a technology of this welding and equipment for its accomplishment. Savich I.M. (1969) presented data on wet method of the mechanized underwater flux-cored wire wel-

ding. The advantages of this method are as follows: good quality of welded joints, sound welds, that can withstand high
pressures, possibility of welding in all spatial positions,
high efficiency of the process.

The present paper is aimed at a comparative analysis of the
mechanical properties of the weld metal, the effect of cooling
rate on imperfections of the HAZ crystal lattice, the cold resistance of welded joints made by the semi-automatic underwater welding and manual welding in the air, using the УОНИ -
-13/45 grade electrodes with a calcium fluoride coating. Crystallization crack formation resistance, sizes and nature of
distribution of non-metallic inclusions and mechanical properties of the welded joints were compared.

Low-carbon and low-alloy steel specimens were welded (Tables
1 and 2).

TABLE 1

Steel	Content, %					
	C	Si	Mn	S	P	Cr
ВСт 3сп	0.21	0.17	0.51	0.0029	0.026	–
09Г2	0.12	0.38	1.8	0.036	0.028	0.29
14ХГС	0.14	1.0	1.0	0.024	0.031	0.70
17ГС	0.17	1.0	1.2	0.310	0.027	0.25

TABLE 2

Steel	σ_y, MPa	$\sigma_и$, MPa	δ, %	ψ, %
ВСт 3сп	262.8	426.5	41.4	67.0
09Г2	304.0	441.3	21.0	57.8
14ХГС	343.2	490.3	22.0	56.6
17ГС	333.4	490.3	23.0	56.0

Welding materials were as follows: tubular flux-cored wire of
1.6 mm diameter, manufactured of 08 кп type steel strip of
0.5x12 mm or 0.4x10 mm for the underwater welding and the 5 mm
dia. electrodes of Э42А-ф type of УОНИ -13/45 grade with calcium fluoride coating for welding in the air. The A-1660 type
semi-automatic machine was used for the underwater welding.

The crystallization crack formation resistance was studied on
T-joint specimens of 25 steel (GOST 1050-74) of 30 mm thickness, containing 0.25% C. Welding was performed in the air
with 5 mm dia. УОНИ -13/45 electrodes at 260...270 A current
and under the water with the flux-cored wire under the conditions, given in Table 3.

TABLE 3

Metal thickness δ, mm	Spatial position of the weld	Joint	Ua,V	I_w,A	$V_w \cdot 10^{-3}$ m/s	Number of weld layers
10...14	Flat	Butt	32...36	250...280	1.66...2.22	2-3
	Vertical	T-joint	26...28	180...200	1.66...2.22	2-3
	Overhead	Overlap	25...27	150...160	1.66...2.22	2-3
6...8	Flat	- " -	28...30	220...240	2.22...2.77	1-2
	Vertical	- " -	26...28	160...180	2.22...2.77	1-2
	Overhead	- " -	24...26	140...150	2.22...2.77	1-2

Note: Electrode stickout 15...20 mm

Five specimens of each series were welded. After cooling the welds were fractured by the impact testing machine. A longitudinal notch was previously made on the weld surface, it determining the weld fracture across the critical section. Cracks were not observed in any case. When welding with УОНИ -13/45 electrodes the tears and cracks of 3...6 mm length were formed in the unwelded craters. Thus, 0.25% C steel joints without cracks and other defects can be performed by the underwater semi-automatic flux-cored wire welding.

The mechanical properties of the weld metal were determined on the five-fold proportional specimens (Table 4).

TABLE 4

Metal	σ_y, MPa	σ_u, MPa	δ, %	ψ, %	J/cm^2 at 293°K	233°K
ВСт3сп , steel weld, УОНИ-13/45 electrodes	475,6	300.1	30.0	66.0	138	67
The same, flux-cored wire	403.0	332.9	31.5	66.1	105	50
09Г2 steel weld,flux-cored wire	501.1	382.6	28.7	59.9	100	45
The same, steel 14ХГС - " -	490.4	380.2	17.1	56.4	92	42
The same, steel 17ГС - " -	470.0	380.2	20.1	57.6	80	40

The content of non-metallic inclusions was studied on the 12mm
thick steel specimens with butt welds made by the flux-cored
wire (Fig.1).

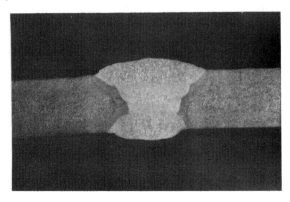

Fig.1. Weld formation in underwater flux-
cored wire welding.

The chemical composition of the weld metal made by the УOHИ -
13/45 electrodes in the air is as follows (%): C - 0.085,
Si - 0.23, Mn - 0.65, S - 0.019, P - 0.020; while the chemical
composition of the weld metal made with the flux-cored wire
under the water is (%): C - 0.065, Si - traces, Mn - 0.20,
S - 0.014, P - 0.016.

When studying the quantity, shape, value, type and nature of
distribution of non-metallic inclusions in the weld metal made with УOHИ - 13/45 electrodes in the air the globular fine-dispersed inclusions FeO were observed, their size not exceeding 1.57 μm, SiO_2, and complex silicates of 3.67 μm size
(Fig.2)

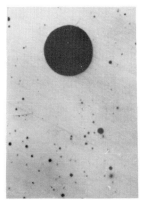 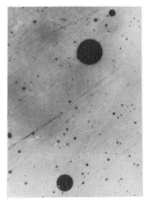

Fig.2. Non-metallic inclusions in welds (x 500)
a-welding with УOHИ -13/45 electrodes in the air;
b-welding with flux-cored wire under the water.

Welds, made by the underwater semi-automatic welding, are contaminated with inclusions of ferrous oxide, which are uniformely distributed accross the weld section and have globular shape and do not exceed, as a rule, 1.58 μm in size; 3.65 μ m dia. inclusions being very seldom observed.

The underwater welding is characterized by a high rate of cooling of the molten and high-temperature heated metal that leads to the structural changes and may cause the drastic changes in its mechanical properties.

The study of weld metal and HAZ of BCт3сп steel showed that in air welding with the above-mentioned electrodes the weld structure is ferritic-perlitic, while in underwater welding an acicular ferrite is formed with regions of fine-dispersed perlite and beinite. Here the cooling effect of environment promotes the metal grain refining (Fig.3).

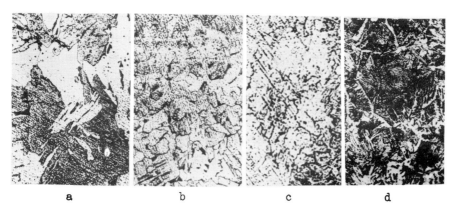

| a | b | c | d |

Fig.3. Microstructure of welded joints (x180)
a,b - welding in the air;
c,d - welding under the water;
a,c - weld; b,d - HAZ.

In both cases the weld metal structure is columnar, being similar to the cast metal structure. The width of crystals in the manual weld amounts to approximately 0.1 mm, while that of the semi-automatic weld amounts to approximately 0.05 mm for the upper layer. In the lower layers the structure is fine-grained, disoriented due to a thermal effect. In the coarse grain region the structure is beinitic with a hypoeutectoid acicular ferrite precipitation at the grain boundary both in underwater and air welding.
The metal hardness in the weld, made with УОНИ -13/45 electrodes, is equal to 122...131 HV, that of the HAZ is 140 HV, the hardness of parent metal being 115...121 HV. The hardness of weld metal made in water with the flux-cored wire amounts to 132...140 HV, that of HAZ is 140...145 HV.

The fatigue tests of BCr 3cn steel specimens cut across the weld of 14 mm thick butt joints were carried out at alternating bending. Two batches of flat and notched cylindrical specimens were manufactured. The latter can give comparative information on weld metal sensitivity to stress concentration. Test base was 5 000 000 cycles. As is seen from Fig.4, the fa-

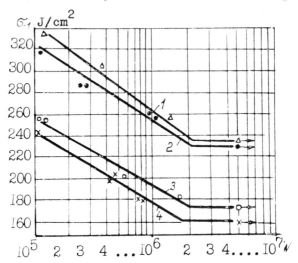

Fig.4. Fatigue strength at alternating bending
1,3-underwater welding with flux-cored wire;
2,4-welding with УОНИ-13/45 electrodes in
the air; 1,2-flat specimens.

tigue strengths in underwater and air welding both of flat and notched specimens were practically similar.

Tensile tests of joints at vibration loads were carried out at the alternating bending in the Afanasjev type machine, until their complete fracture (Fig.5).

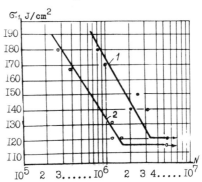

Fig.5. Vibration strength of welded joints. 1-welding under the water; 2-welding in the air.

It was established that the fatigue strength of ВСт3сп steel joints made with ЧОНИ -13/45 electrodes in air and with the flux-cored wire under the water, is practically identical. However, the results of the underwater welding conditions within the whole range are higher.

The steel ВСт3сп joint resistance to impact loads at negative temperatures was determined on the 400x80x60 mm T-beam specimen. The specimens were manufactured of 12 mm thick welded butt joints, made with ЧОНИ -13/45 electrodes in the air and with the flux-cored wire under the water. The tests were performed in the pendulum impact testing machine of the design of the E.O.Paton Welding Institute, the impact energy being 1.5kJ. As is seen from Table 5 all the specimens withstood practically a similar number of impacts at the temperature up to 233°K.

TABLE 5

T_{test}, °K	Number of impacts till fracture of joints made by	
	manual welding with ЧОНИ - 13/45 electrodes in the air.	underwater semi-automatic flux-cored wire welding.
273	$\frac{13...13}{13}$	$\frac{12...13}{12}$
253	$\frac{11...12}{12}$	$\frac{8...13}{11}$
233	$\frac{5...6}{5}$	$\frac{2...4}{3}$

Note: 1. The numerator gives the minimum and maximum quantity of shocks till joint fracture, in denominator is the mean value of testing 3-5 specimens. 2. In air welding the specimen fractured in HAZ while in underwater welding - in parent metal.

It was of interest to determine and compare the cyclic toughness of metal of low-carbon steel welds made with ЧОНИ 13/45 electrodes in the air and with the flux-cored wire under the water. The instrument, the design of which was developed by G.P.Alekseev and described by Asnis (1962), was used for determining the cyclic toughness. The cyclic toughness was determined at 2° twist angle on the cylindrical specimen.

It is established that the relative cyclic toughness of ВС 3с steel welded joints made with ЧОНИ -13/45 electrodes amounts to 8.8%, while that of welds made with the flux-cored wire under the water is equal to 9.2%, it testifies to a higher ability of the metal of underwater weld to absorb the vibrations.

Due to high rates of solidification and plastic deformation considerable distortions of the crystal lattice occur both in the weld metal and HAZ under the effect of the residual stresses of the first kind. Such imperfections of structure increase the strength, decrease the ductility and metal resistance to brittle fracture.

Depending on welding conditions the rate of cooling of the joint metal is considerably changed causing the occurence of imperfections of the crystal lattice.

The experiments were carried out on BG₃ 3сn low-carbon steel 350x180x14 mm specimens.

The mechanized underwater welding was compared to the mechanized welding in the air. In all cases welding was performed with 1.6 mm dia. flux-cored wire under the following conditions: $I_w=240\div260$ A, $U_a=28\div30$V, $V_w=1.9.10^{-3}$ m/s.

Only the cooling rate at the temperatures below the recrystallization point may affect the degree of distortion of crystal lattice of metal that undergoes the allotropic transformation.

The parameters of the fine structure were determined in the diffractometer ДPOH-1 in the cobalt radiation. The lines (110) and (220) of ferrite from the 12 by 12 mm specimen plane parallel to the fusion zone were recorded at the layer by layer removal of the specimen material by the chemical etching. By this procedure the data on distribution of crystal lattice distortions in the whole HAZ volume were obtained.

The measurement of distortions of the crystal lattice a/a(a-crystal lattice parameter) and sizes of sites L were carried out by the method of a harmonic analysis of the intensity curves of X-ray diffraction lines.

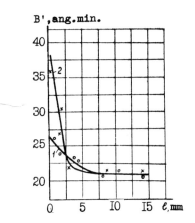

Fig.6. Relationship between the width B of X-ray
line and distance up to the fusion zone
1 - welding in the air;
2 - welding under the water.

Fig.6 shows the effect of distance L from the fusion line on the width B' of the X-ray diffraction line. The width of the X-ray line is a measure of the crystal lattice distortions caused by dislocations, stresses of the second kind and dispersity of the ferritic phase. With the increase of the level of the crystal lattice distortions the width of the X-ray line enlarges.

The crystal lattice distortion is not observed in the parent metal (within the error of measurements).

In the HAZ there are crystal lattice distortions the value of which increases when approaching the fusion line.

The high rate cooling, characteristic of the underwater welding leads to reduction of the metal area width where the crystal lattice distortions are observed and to the increase of values of such distortions. If in the air welded joint the width of the area of distortions amounts to approximately 6 mm, then in those, made under the water, it is halved (approximately 3 mm).

The value of the crystal lattice distortions near the fusion line of the joint, made in the air, reaches $\triangle a/a = 3.0.10^{-4}$, that of the joint, made under the water $-4.8.10^{-4}$.

The sizes of areas of coherent scattering are also different in these cases: in the air $L=20 \times 10^{-5}$ mm, under the water $- 7.10^{-5}$ mm.

When comparing the measurement data, it should be noted that the distortions of the crystal lattice, displaying themselves in increase of the second kind stresses and in the refining of the areas of the coherent scattering is observed only in the HAZ subjected to recrystallization in cooling. Hence, it may be concluded that such recrystallization, proceeding with a high rate, is the main cause of development of the HAZ crystal lattice distortions.

It should be also noted that microdistortions in the bead deposited under the water, are much more numerous than those in the air. The microdistortions characterize the level of distortions of the crystal lattice in metal. In underwater hardsurfacing   more non-equilibrium conditions of the phase transformation are created due to the intensity of the heat removal than during the air hardsurfacing, thus leading to the stronger hardening of crystallites, mostly, due to the formation of structure with a high degree of the crystal lattice imperfections, as shown during the electron microscopic examinations. In underwater hardsurfacing the hydrogen hardening may play a significant role, considered by Karpenko G.V. The second kind stresses can partially be relaxed during annealing of the distorted structure or in weld metal cooling with respect to the welding process. Apparently, the slower heat removal during air welding of the bead leads to a considerable stress relieving due to relaxation. The fact that in air hardsurfacing the structure might be formed with a lower level of the crystal lattice distortion than in underwater hardsurfacing should not

be ignored. By taking into consideration the processes of
structural transformations it is rightful to suppose that the
lower level of the microstresses (second-kind stresses) in the
bead made in the air, is specified to a larger extent by a
stress relaxation in the HAZ metal subjected to the hardening.

The brittle crack initiation near the welds at low stresses
from the external loads can be caused by such two factors:
steel ageing in places of concentration of thermal plastic de-
formations and high residual stresses.

The first factor plays a decisive role. As a result of ageing
deformation the lower threshold of cold brittleness is shifted
into the side of negative temperatures at static loading.

The conditions of welded joints performance under the impact
load are worse than those under static load, since high rates
of applying the loads enable to use the comparatively smaller
volumes of metal. Consequently, the non-uniformity of stress
distribution under impact loads may appear more considerable.
In this case the welded joints are more susceptible to stress
concentration.

The cold brittleness tests were carried out on the specimens
according to the procedure developed at the E.O.Paton Welding
Institute, (Asnis A.E. (1947). The comparison of the cold resistance of underwater joints to those welded in the air, by
manual electric arc welding, for instance, (electrodes УОНИ
13/45) allowed to have a certain idea on the performance of
joints in the underwater welding. When evaluating the cold resistance of welded joints, made in the air and under the water,
the test temperature of specimens and area of section between
notches, occupied by a crack, were taken into consideration.
As is seen from Table 6, the joints welded under the water are
more resistant to brittle crack propagation.

TABLE   6

T,°K	Share of area between notches occupied by crack, %	
	in air welding (УОНИ -13/45)	in underwater welding (flux-cored wire)
213	96–100 / 98	44–49 / 46
233	95–99 / 97	no cracks
253	96–98 / 97	–
273	no cracks	–

Note: The results of testing 3-5 specimens are given.

The high cold resistance of specimens can be explained by the
more narrow HAZ zone in the   underwater hardsurfacing than

in the   air   welding, this reducing the effect of strain ageing of metal. This was described in the work of Babich V.K., Gul Yu.P. (1972).

It is known that the steel susceptibility to ageing is considerably higher with the increase of nitrogen content above 0.008%.

In welding with quality coated metal electrodes the protection of the molten metal from the contact with ambient air nitrogen and oxygen is one of the main problems. However, this problem is not completely solved, bacause some amount of the atmospheric nitrogen is dissolved in the molten steel. Besides, the nitrogen from the parent and electrode metal also transfers into the weld. The air welded joints contain about 0.02% of nitrogen that specifies their susceptibility to ageing.

The high cold-resistance of the welds, made under the water can be explained by the metallurgical features of the welding process. In the conditions of the wet semi-automatic welding the arc atmosphere consists, mainly, of hydrogen and oxygen. It does not contain any nitrogen and moreover contributes to its partial removal from the molten metal. Nitrogen, dissolved in steel, forms NH_3 compound with hydrogen, easily dissolved in the surrounding water. The nitrogen content in the welds, made under the water, does not exceed 0.001 %.

CONCLUSIONS

1. Welds made with the flux-cored wire under the water possess practically the same resistance to the crystallization crack formation as the welds made with the УOHИ 13/45 grade calcium fluoride coated electrodes in the air.
2. High rates of cooling of the molten and heated up to the high temperature metal, characteristic of the underwater welding conditions, provide a fine-grained structure in the weld as compared to the air welding, favourably affecting the results of mechanical tests.
3. The study of microstructure showed that welds made by the underwater welding have the structure of the acicular ferrite with areas of fine-dispersed lamellar perlite, while in the HAZ the beinite structure and hypoeutectic ferrite are present.
4. The strength of joints welded under the water and in the air did not almost differ under fatigue and impact loads.
5. The rise of rate of welded joint metal cooling is accompanied by the increase of the level of distortions of its crystal lattice, i.e. by increase of the second kind stresses.
6. In the underwater welding the width of the area, in which the crystal lattice distortions are observed is considerably less than in the air welding.
7. Welds, made by the underwater semi-automatic welding and having a decreased nitrogen content are slightly susceptible to ageing and that specifies their resistance against brittle crack propagation even at 213°K temperature.
8. The results of comparative tests allow to recommend the underwater semi-automatic flux-cored wire welding for the fabrication of special-purpose hydroengineering structures.

REFERENCES

Asnis A.E. (1962) Dynamic strength of welded joints of low-carbon and low-alloy steels. Mashgiz. M., pp.149-150
Asnis A.E. (1947) Evaluation of steels and welded structures working at low temperatures. "Zavodskaya laboratoria" No.9
Babick V.K., Gul Yu.P., Dolzhenkov I.E. (1972) Strain ageing of steel. Metallurgia, M, pp 7-92.
Khrenov K.K. (1946) Underwater electric welding and cutting of metals. Voenizdat, M.
Khrenov K.K. (1933) Underwater electric welding. Svarshchik, 1,2, pp.23-24.
Karpenko G.V., Katson K.B., Kokotailo I.V., Rudenko B.P. (1977 Low-cyclic fatigue of steel in working media. Naukova dumka, Kiev, pp.11-24.
Savich I.M. (1969) Flux-cored wire underwater welding. Avtomaticheskaya svarka. J.10, 70.

IV

SECTION IV

REPAIR AND OTHER APPLICATIONS / REPARATIONS ET AUTRES APPLICATIONS

Practical Application of Locally Drying Underwater Welding for Steel Pipes

T. Ohmae*, Y. Manabe*, Y. Nagata**, M. Tamura**, N. Ishikawa*** and K. Satoh†

*Hiroshima Technical Institute, Technical Headquarters' Mitsubishi Heavy Industries Ltd., Hiroshima, Japan
**Hiroshima Shipyard & Engine Works, Mitsubishi Heavy Industries Ltd., Hiroshima, Japan
***The Research Institute of Shimizu Construction Co. Ltd., Hiroshima, Japan
†Faculty of Engineering, Osaka University, Osaka, Japan

ABSTRACT

This paper deals with a new technique called "Locally Drying Underwater Welding Process", and can be summarized as follows: The cooling characteristic of welds made by this method was analyzed and the satisfactory mechanical properties of the welds were confirmed. After those fundamental tests, an automatic welding system for this process was fabricated and by means of it, steel pipes were welded in the sea with satisfactory results. Further, the appendix shows methods developed for improving the mechanical properties of underwater welds in high tensile steel.

KEYWORD

Locally drying underwater welding; water jet; mechanical properties; application for steel pipes, retarded cooling method; induction heating.

INTRODUCTION

There are two, wet and dry, underwater welding methods now in practical use. However, reluctance to the application of the former to strength members has been shown due to difficulties in the phases of quality and working efficiency. On the other hand, the latter requires a water discharge chamber to the size and shape of an object to be welded, which gives rise to an economical problem (Misher, Randall, 1970). Under such circumstance, the authors (Nishiguchi, Tamura, 1975) developed an entirely new "locally drying underwater welding process" which could meet the following three key requirement.

1. Welds achieved should be of high quality as if they were made on land.
2. The process should be automated to eliminate dependence upon the skill of diver.
3. The process should ensure safe and efficient operation.

The authors have analyzed the basic phenomena and examined the quality of welds. This paper presents an outline of the new underwater technique so far developed, the automatic underwater welding system practically applied to steel pipes and improvement incorporated in it to achieve welds with more satisfactory mechanical properties.

DEVELOPMENT OF LOCALLY DRYING UNDERWATER WELDING METHOD

Principle

Figure 1 shows the principle of this method. A double-torch having trampet-shaped nozzles at its end is used. By supplying shielding gas from the center nozzle of the torch and delivering a high-speed water jet from its outer nozzle, a stabilized gaseous zone (local cavity) is formed directly under the torch by the various effects of the water jet, where welding can be performed in the same manner as on land.

Test Pieces and Welding Conditions

Test pieces. Table 1 shows the chemical compositions and mechanical proper-ties of the base metals and the weld metal of wire used in the present study. The base metals are 12 mm-thick ship's classified mild steel (A (M.S.)) and 50 kgf/mm^2 high strength steel (B (H.T50)). The welding wire is 1.2 mm in diameter for MIG welding.

Welding condition. In the test, a 100 mm diameter nozzle was used to produce water jet of which flow rate was 70 liters per minute. Then, a gaseous atmosphere was created by discharging an $Ar-CO_2$ (20%) gas mixture at a rate of 150 liters per minute, which served as the shielding gas. Table 2 sum-marizes the welding conditions applied during the present test, and welding was done at flat position.

Test Results

Cooling curves of welds. Figure 2 shows a comparison of the cooling cruves of an air weld and underwater welds made under various conditions by the locally drying method. Comparing the air weld and the underwater weld made under the same welding conditions, the latter showed, in the dry zone (t < tr[*] a very similar cooling characteristic to that of the air weld. After tr, however, the cooling characteristic of the underwater weld greatly deviated from that of the air weld, because the former came in direct contact with water and was rapidly cooled by it. In the meantime, considering the cooling curves of the underwater welds, each of them has a very different cooling characteristic, depending on the welding conditions. Namely, the high-current and high-speed weld (S4) was rapidly cooled from a high temperature at around 800°C, whereas the low-current and low-speed welds (S2 and S3) came in contact with water at low temperatures, and their cooling characteristic was similar to that of the air weld (Tamura, Manabe, 1982).

_____ _ _____ _ _____ _ _____ _ _____

[*1] tr is the theoretical time calculated from a gaseous atmosphere of radius, Ro, and a welding speed, of Vw. tr=Ro/Vw(sec.). However, the gaseous atmosphere radius, Ro, is a little larger than the nozzle radius due to he slant of the water jet. In the present test, therefore, Ro=57 mm.

Maximum hardness of welds. Since the underwater welds show such a unique
cooling characteristic, those made with the same heat input have greatly
different cooling rates, and the maximum hardness of welds cannot be arranged
in relation to either welding conditions or cooling time from 800°C to 500°C
which is normally used to arrange the maximum hardness of air welds. Accordingly, the authors arranged the maximum hardnesses of the welds in relation
to the cooling time from 800°C to 300°C, as shown in Fig. 3. As a result,
good arrangement was obtained, and the calculation curve obtained from
equation (1) (also shown in Fig. 3) agreed well with the experimental results.

$$Hv \; max \; = \; Hvnd(682[C] \; + \; 12.7[Mn] \; - \; 26.2[Si] \; + \; 141) \; + \; 212[C] \; + \; 2.85[Mn]$$
$$+ \; 35.3[Si] \; + \; 141 \tag{1}$$

where, $\Bigg\lbrace$

$$Hvnd \; = \; -1.27 \, \mathcal{T}nd^5 \; + \; 1.20 \, \mathcal{T}nd^4 \; + \; 3.47 \, \mathcal{T}nd^3 \; - \; 4.73 \, \mathcal{T}nd^2 + \; 0.328 \, \mathcal{T}nd$$
$$+ \; 1.00 \qquad\qquad\qquad (0 \leq Hvnd \leq 1) \tag{2}$$

$$\mathcal{T}nd \; = \; \frac{\log \mathcal{T} - \; (5.95[C] \; + \; 1.13[Mn] \; - \; 1.23[Si] \; - \; 1.22)}{-2.68[C] \; - \; 0.347[Mn] \; + \; 2.83[Si] \; + \; 1.98} \tag{3}$$

As shown in Fig. 2 and 3, the welds under low-current, low-speed welding
conditions had long cooling time ($\mathcal{T}$) and presented relatively low hardness.

Mechanical properties of welded joints. Table 3 shows some examples of the
mechanical properties of welded joints and COD test results by this method.
The obtained values are slightly lower than those of the air welds but prove
that the welded joints had satisfactory quality. Based on those results of
the fundamental studies, the authors put this welding method into practical
use, as described below.

APPLICATION FOR STEEL PIPES

Outline of Work

The outline of the underwater welding work is as given below and as shown in
Fig. 4 (Manabe, Tamura, 1980).

(Object pipe)   (Environment)

No. of joint   : Approx. 50   Water depth: 5 - 6 m max.
Diameter of pipe: 700 - 800 mm   Current   : 0.3 - 0.6 knots
Material   : Mild steel   Clearness : Approx. 1.5 m

Welding Procedure and Model Test

Welding procedure. Figure 5 shows the horizontal circular welding procedure
for the work. A suspended-type carriage is mounted on rails provided on the
steel pipe. The torch is installed on the carriage. A water-tight wire
feeder and a water jet pump are installed on the underwater scaffolds in the
vicinity of operation place. A diver will only install the equipment and
adjust its position. After the torch switch is turned on, all welding will
be performed automatically. Welding conditions will be remote controlled
from ship through underwater TV camera.

Model test. Prior to applying this underwater welding method on actual
structures, a model test was carried out in a large water tank as shown in
Fig. 6 and 7, and various problems which were supposed to be encountered in
actual operations (such as the effects of offset of steel pipes and variations

in root gap upon the formation of the gaseous zone and welding procedure were studied. Also, just before the work, the welding procedure test was conducted in site (in the sea), and the obtained welds proved themselves to have satisfactory mechanical properties.

Results of Work

The work was performed according to the following procedure using underwater welding equipment as shown in Fig. 8.

Erection of scaffolds. Special scaffolds were erected to facilitate the underwater work.

Cleaning and grinding. Joints were ground to clean of shells, rust, etc.

Installation of equipment. Circular rails, carriage and wire feeder were installed.

Welding. Three or four passes were normally used under welding conditions almost similar to those on land.

Figure 9 shows an example of the welding work being performed and welded joint. The work was successfully completed with satisfactory results similar to those obtained in the tank test.

CONCLUSION

As shown in this paper, the above-mentioned actual work showed that the newly developed underwater welding process was promising as an underwater joining method for the future.

APPENDIX

Underwater Heat Treatment

Figure 10 shows the post-weld heat treatment method by induction heating (frequency; 60 Hz) which the authors developed for underwater welds in pipe structures as mentioned in this paper. The reason why the induction heating method is adopted is as follows: It needs no gaseous zone formed around it and can provide a required quantity of heat for the treatment. As shown in Fig. 11, the weld heat-treated by this method has greatly reduced hardnesses and has high mechanical properties as if it were made on land (Tamura, Manabe, 1982).

Retarded Cooling Method

A retarded cooling method was also developed as another means to improve the quality of welds made by the locally drying underwater welding technique. In this method, the weld is shielded with insulation materials to prevent quenching and hardening. The test results prove that the retardedly cooled welds have higher mechanical properties. (Tamura, Manabe, 1983).

REFERENCES

Misher, H. W., and M. D. Randal (1970). Underwater Joining and Cutting Present and Future. The Second Offshore Technology Conf., No. OCT-1251, Houston.

Nishiguchi, K., M. Tamura, and A. Matsunawa (1975). Development of Underwater Welding with Local Cavity Formation Method. The Second International Symposium of J.W.S., No. 2-2-(6), Osaka.

Tamura, M., and Y. Manabe (1981). Study of the cooling characteristics and weld hardness in the Locally Drying Underwater Welding of Mild and 50 kgf/mm² H. T. Steel. IIW DOC. XII-B 17-81

Manabe, Y., and M. Tamura (1980). Development and Practical Use of Locally Drying Underwater Welding, I.I.W. Doc. XII-B-289-80, Portogal

Tamura, M., and Y. Manabe, (1982). Study on Mechanical Properties of Locally Drying Underwater Welding Joint and Improvement by Underwater Heat Treatment, J. Ja. W. Soc., 51, 105-111

Tamura, M., and Y. Manabe (1983). Development of Retarted Cooling Method for Locally Drying Underwater Welding, I.I.W. Doc. XII-E-43-83, Norway.

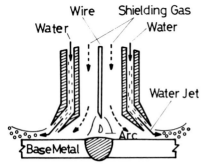

Fig. 1. Principle of locally drying underwater welding process.

Table 1 Chemical Compositions and Mechanical Properties of Base Metals and Weld Metal of Wire used

Material	Mark	Thickness Dia. (mm)	C	Si	Mn	P	S	Y.P (kgf/mm²)	T.S. (kgf/mm²)	El. (%)
Base Metal	A(M.S.)	12	0.12	0.24	1.00	0.016	0.005	32	47	32
	B(H.T50)	12	0.14	0.35	1.23	0.014	0.003	36	52	27
Wire	W1	1.2	0.08	0.53	1.12	0.017	0.015	43	55	34

Table 2 Welding Conditions for Test

Test Condition	Mark	Environment of Welding	Welding Current [A]	Arc Voltage [V]	Welding Velocity [cm/min]	Heat Input [kJ/cm]
Group 1 (Welding Condition Const)	R1	In air	200	26	25.0	12.5
	S1	Underwater				
	R3	In air	200	26	12.4	25
	S3	Underwater				
Group 2 (Heat Input Const)	S2		120	22	6.3	
	S3		200	26	12.3	25
	S4	Underwater	330	32	25.0	
Group 3 (Welding Velocity Const)	S5		120	22		6.3
	S1		200	26	25.0	12.5
	S4		330	32		25

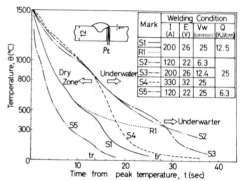

Mark	I (A)	E (V)	Vw (cm/min)	Q (KJ/cm)
S1 —	200	26	25	12.5
R1 ······				
S2 ---	120	22	6.3	
S3 –·–	200	26	12.4	25
S4 ---	330	32	25	
S5 –·–	120	22	25	6.3

Fig. 2. Comparison of cooling curves in various welding conditions.

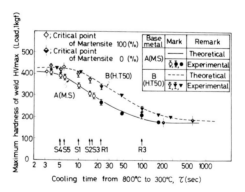

Fig. 3. Relation between maximum hardness of weld and cooling time from 800°C to 300°C.

Table 3 Comparison of Mechanical Properties between Air Weld and Underwater Weld

Base Metal	Environ- ment	Tensile test of weld metal		Charpy impact test				COD test	
		ε (%)	σ_B (kgf/mm²)	vE₋₂₀ (kgf-m) [*1]		vTrs (°C)		$T_{\delta c} = 0.25$ [*2] (°C)	
				W.M.	HAZ	W.M.	HAZ	W.M. [*1]	HAZ [*1]
A(M.S)	Under water	33.6	52.7	5.5	8.2	-10	-15	-85	-100
	In air	37.5	50.2	8.6	11.3	-25	-30	-105	-120
B(HT50)	Under water	32.7	56.5	4.1	6.4	-5	-10	-80	-95
	In air	37.1	52.3	7.5	9.2	-20	-20	-100	-120

(*1) Notch position

(*2) Temperature of $\delta c = 0.25$(mm) in COD test

HAZ WM

Welding condition : S3,R3

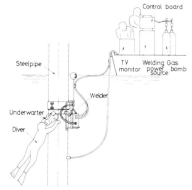

Fig. 4. Underwater horizontal circular welding of steel pipes.

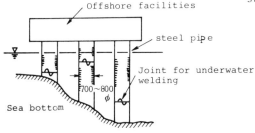

Fig. 5. Outline of underwater welding work.

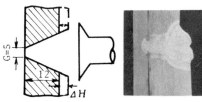

(a) Offset variation ($\Delta H = 4$ mm)   (b) Root gap variation ($\Delta G = 4$ mm)

Fig. 6. Macro structure in various groove shape variation.

(a) View of underwater circular weld.

(b) External view of welding bead and macro structure.

Fig. 7. View of underwater circular welding and test result.

IIW-M

(a) Remote operating panel.

(b) Wire box.

(c) Welding torch.

Fig. 8. Main welding equipment for underwater welding.

(a) View of automatic underwater welding.

(b) View of remote operating and controlling.

(c) External view of welding joint.

Fig. 9. View of automatic circular underwater welding and result.

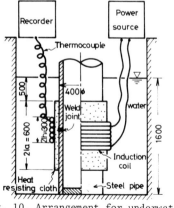

Fig. 10. Arrangement for underwater heat treatment test.

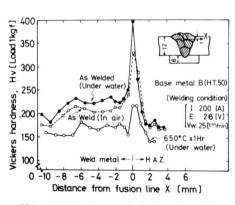

Fig. 11. Hardness distribution of underwater weld after underwater heat treatment.

Deep Hyperbaric Welding.
Mechanization/Automation

H. O. Knagenhjelm

Materials Technology Section, Norsk Hydro Research Centre, Norway

ABSTRACT

The paper is discussing choice of welding process for mechanized/automated pipe-welding in a tie-in situation. The choice between TIG and different MIG systems are presented.

A philosophy on the natural steps in a development towards full automation is presented.

Further a brief discussion is done on some focal points in the development of an orbit system.

Based on earlier and ongoing work it is feasible to use orbit systems for first generation mechanized welding very soon gaining quality and saving welding time.

KEYWORDS

Hyperbaric welding, orbit systems; mechanization; pipelines; Tie-in; Automation.

INTRODUCTION

Based on the work carried out during various research programs on hyperbaric welding since 1974, the DHWP (Deep Water Hyperbaric Welding
Program), inhouse work, the Deep-Ex81 trials and the German/Norwegian
Project on development of a mechanized welding system, we will try to
present a strategy and philosophy for development of hyperbaric welding in
the years to come.

This will mainly consist of the following focal points
- Choice of welding process and welding system.
- Level of automation.
- Root pass welding technology.
- Equipment performance under pressure.
- Power source type and level of electronics.
- Filler pass technology, groove type.

CHOICE OF WELDING PROCESS

There are many topics that need consideration when one want to choose the
main welding process for mechanized/automated welding. The following
topics seems to be the most important:

- Low defect suceptibility of system and of weld.
- Acceptable mechanical properties, ie. CTOD > 0.25 mm,
  Chapy-V $>\sim$ 100 J, HAZ hardness below 245 HV_5.
- Advanced welding system and power supply. Possibilities of
  closed loop control of welding parameters.
- Easy slag removal, preferably no slag at all.
- Reasonably high deposition rates, > 2 kg/arc hour.
- Preferably use of the same process for root and fillers.
- Continuous wire feed.
- Possibility of welding narrow grooves.

This is leaving us with basicly two processes MIG/MIG fluxcored wire and
TIG.

The TIG Process

The recent years of research has demonstrated sufficient arc stability for
practical welding down to $\sim$ 500 m.s.w. Welding with 100% He as
shielding gas has been demonstrated at SINTEF at 320 m.s.w. The area of
fusion is increasing with depth and at 300-400 m the geometrical shape of
the fused area is favourable for welding of heavy root faces. Deposition rates of >3 kg/arc hour could be expected at 300 m.s.w+. The
main items in favour of the TIG process are the following:

- Very advanced orbit pipewelding systems have been in use for
  $\sim$ 10 years in the nuclear industry and can be adapted to
  use in a habitat.
- Mass and heat can be added indepently to manipulate fusion
  characteristics and create tempering effect.
- Very high quality clean weld.
- Low defect suceptibility as no slag is present.
- TIG is established as a reliable and high quality rootpass
  method.
- The fusion characteristics of TIG in superior to MIG and MMA.

- A detailed knowledge of the influence of pressure on TIG exists.

These are the main reasons why Norsk Hydro has continued to develop the process further in solving detailed equipment problems and carrying out the large number of parametric studies necessary to use an automatic or mechanized process.

The MIG Processes

The MIG bare wire process has been used at shallow depths for several years, but at depths > 100 m.s.w the fluxcored process offers much better arc stability and higher deposition rates. The MIG fluxcored process is not suited for rootpass welding which has to be carried out by solid wire MIG or TIG. In the following some of the main topics will be summed up for MIG bare wire and MIG fluxcored wire.

MIG bare wire   This process is almost only used by Comex for rootpass welding and for fillers at shallow depth.

- Lack of fusion problems may occur and tolerance to parametric variations decrease with depth. Has been used by Comex for rootpass welding at 300 m.s.w, but very sensible to parameter choice, power supply characteristics and fit up.

- Very limited penetration capabilities which means that narrow rootgaps leads to lack of penetration, ie. very accurate fit up is required.

- Rather low deposition rates ∿1 kg/arc hour as it is normally used in the dip transfer mode.

- Rather poor mechanical properties.

MIG fluxcored wire   This process has been used by BOC and Comex for fillers.

- Good deposition rates in position ∿2,5 kg/arc hour.
- Rather high heat input capability which reduces HAZ hardness problem encountered with MIG bare wire and MMA.

- The types used so far has a slag that sticks very hard especially at the edges of the bead which are rather uneven, this means that grinding is required as much or more than for MMA. This reduces the net deposition rates and does not save very much time compared to MMA.

- Due to irregular bead and sticking of slag, microslag inclusion problems may occur.

- Some orbit systems are prepared for the MIG process.

- Not very suited for narrow groove welding due to dip transfer mode.

The fluxcored process has been sucesssfully demonstrated to ∿650 m.s.w by Cranfield and Institute de Soudure for filler passes. Even at 300 m.s.w+ the transfer mode is mainly dip transfer. If the process is to be used for mechanized welding the sticking of the slag has to be solved. This probably may require a spray transfer mode which may restrict positional welding.

New developments as synergic MIG and new transistor power supplies for MIG may improve the possibilities for use of bare wire MIG.

Conclusion

Norsk Hydro has chosen TIG because it can be used for all welding passes, can weld narrow grooves and heavy root faces, has no slag and advanced proven orbit welding equipment is available. Using 100% helium as shielding gas has been demonstrated at 320 m.s.w, and this could have significant positive economic and diving effects.

DEVELOPMENT PHILOSOPHY

It is obvious that it is not smart to try and implement a fully automatic system at first. We can see several natural steps in a development towards a fully integrated diver independant system. Below will be given a proposal for the main steps.

1.   Manual TIG root, semiautomatic TIG torch for manual fillers. Conventional V-groove 60-40° or composite grooves. Equipment is ready for use and proven.

2.   Manual TIG root. Orbit TIG mechanized fillers with manipulation from welder by sight or via television. Narrow V-groove 40° or composite grooves. System can be implemented within 1 year.

3.   Full orbit-TIG weld. Closed loop control of root pass by sensing of high/low and root gap. Welder manipulation for small adjustments via television. 40° V-groove or composite V-groove. Implementation within 1-2 years.

4.   Full orbit TIG weld. Welder manupulation for small adjustments via television. Very narrow groove, with 3-6 mm root face, minimum root gap $\backsim$ 0 mm may reduce magnetism problems. 1 bead pr. pass. Implementation 2-3 years.

5.   Fully orbit TIG weld. Automatic controlled from surface. Manipulators for rigging etc. Implementation 5-10 years.

Step 1 and 2 can be implemented as soon as any diving company wants to develop procedures. Step 3 may not be necessary if step 4 is feasible and as promising as we believe. Step 4 will be able to make the weld in 30-50% of the time required today and may also eliminate the magnetism problems. Step 3 involve quite some electronic equipment and sophisticated TV systems if a simple mechanical sensor or sensing via the arc is not possible. Bead on plate testing during Deep-Ex81 has demonstrated that the width and penetration of the beads are suited for narrow groove welding.

FOCAL POINTS FOR ORBITAL TIG WELDING

In the following we will point out some of the points which is critical in this type of welding system development and which needs special attention.

Groove Geometry Considerations

The ordinary groove used in exsisting hyperbaric welding is the V-groove with a 30 or 37.5° bevel (See Figure 1 for parameters). The root pass, which is the most difficult job, is welded by MMA or manual TIG or MIG. Economic benefits are derived from the use of smaller joint volumes. This gives fewer passes and saves time. This can be obtained by introducing compound grooves or very narrow grooves. The cross-sectional area of different grooves are illustrated in Figure 1. It should be noted that the area of a compound groove is about 3/4 of an ordinary 60° groove. The V-groove is relative flexible to rootgap and land size variations and high-low (Offset). By normal pipe fitting an offset of about 5 mm is not uncommon.

Thin material butt welding is ideal for the TIG process. It is normal during TIG-welding to simulate thin material welding using a U-groove (Welding and Material Fabrication, 1972). But this geometry has the drawback that close tolerances are needed, (max 1 mm high-low). An optimalisation study is made by Key, Turner, Hood and Keiser (1980). They report that conventional U-groove joint geometries are much easier to weld than V-grooves.

The narrow groove (U and V) may cause difficulties when using conventional gas nozzles. It is both experienced and reported (Hill, Graham, 1978) that for narrow groove a satisfactory shielding is obtained even with a long electrodes stick out. Electrode wear, may however occur by too high amperage.

The special features of hyperbaric TIG welding could make it possible to use very narrow grooves and heavy root face (∿3-6 mm). It should be feasible to move the pipes axically so that a root gap of ∿ 0 mm could be obtained.

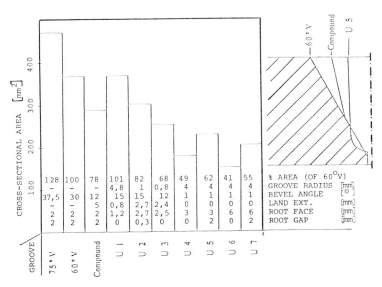

Fig. 1   Groove parameter and cross-sectional area for different groove types. Material thickness: 1" = 25.4 mm.

Root Pass Welding

The main problem is to handle misfit (high/low, root gap variations) and be sure of full fusion and acceptable reinforcement. This is especially difficult in the overhead position. Manual TIG has been able to cope with rather heavy misfits due to fusion characteristics and the independence between heat source and filler wire. In mechanized welding however, misfit is more difficult to handle. Recent experiments have shown that the characteristics of a hyperbaric TIG arc gives wider tolerance to fit up for a given set of welding parameters.

Mechanized welding of V-grooves have demonstrated the feasibility of doing this by varying welding parameters. It is envisaged that parametric studies will give data for a closed loop control of the root pass welding. The root pass is the only really difficult part of the mechanized welding. Only using the orbit system for the fillers will still give great benefit.

Sensor system. If a variable root gap is expected and the tolerances of the welding systems for a given set of parameters to such variations is low, a closed loop system is necessary. We then need to measure the root gap and high/low. This has been done with television systems, but these are bulky and probably unpractical for hyperbaric use. A mechanical sensor or the arc itself is probably more feasible. One will aim at using ∿ 0 root gap and rather heavy root faces (3-6 mm). This may make sensors surplus, ie. one set of welding parameters, or preprogrammed as function of positon around the pipe, could be used. Further development of alignment frames would make it possible to obtain a ∿ 0 mm root gap.

Filler Passes

The following points need attention
- Optimation on deposition rate in position.
- Optimation on weld profile, width to match narrow groove.
- Optimation to obtain 100% recrystallization of weld metal.
- Hot wire addition to increase deposition rate.
- Optimation of pulse technique and weaving to improve positional welding.
- Seam tracking systems based on short measuring cycles using the arc, magnetic arc flickering, pilot arc or televison system.

In general the filler passes are expected to be relatively simple and the points mentioned here are mainly refinement and optimation. Experience has shown that due to the cleanliness of the weld metal the mechanical properties are very good even if the weld metal is 100% non recrystallized. Figure 2 shows a Messer Griesheim Multi-TIG in a dry land habitat test.

Television Systems

The fiberoptic transmission to a color TV camera has shown to give very good pictures of the arc and the pool. Such systems are now under development for hyperbaric use, and is expected to make it feasible to have a remote controlled welding via a pendant and a TV screen. Experiments have shown that welding manipulation of an orbit TIG system via televison is easier than by eye. Experience is gained through remote controlled welding in nuclear power plants.

Fig. 2. Messer Griesheim Multi-TIG system.

Equipment Behaviour

The environment in a habitat at great depth is special, typical features
are: He + 2-3% O_2 atmosphere, 90% RH including salt and temperature of
30-35°C

This environment may cause problems to bearings in electrical motors.
Electronics has normally been encapsulated at ambient pressure to be sure
that no harm is done from decompression of helium which penetrates the
components. However, this philosophy may give impractical solutions.
Experience have shown that some electronic systems are not harmed by this
gas environment, the high relative humidity is probably worse. In the
German/Norwegian Project systematic testing of an orbit welding head is
done down to ∿ 500 m.s.w in a realistic environment. The results of
this work will be presented.

Power Supply and Arc Voltage Control

A key to stable and repeatable hyperbaric TIG welding is a constant arc
voltage. Deep-Ex81 clearly demonstrated that the existing simple analog
types used for atmospheric welding is not suitable. The arc length/arc
voltage gradient is very steep making a constant arc length a necessity.
Analysis of the arc voltage signals during Deep-Ex81 revealed fluctuations
∿20 000 HZ with rather high peak values. This demands a special regulator to be designed and a stage 2 prototype is now ready for testing.
Further development will have to be carried out. The stage 1 prototype
has been able to weld using 100% He as shielding gas at 320 m.s.w with
very interesting results.

A fairly advanced (transistorized) power supply is needed where functions
and parameters can be run by micro processors. A ground pulsing frequenzy
of ∿20 000 HZ has proved beneficial to arc stability and reduction of
arc blowing in roots.

Environmental control

The Deep-Ex experiment revealed Ozone and smoke as the two main pollutants
from hyperbaric TIG welding. Ozone is a local problem, and the background

347

level can be kept low using active coal filter and rapid background gas change. By optical shielding of the arc this problem can be almost eliminated. 20 times the CL value was observed locally during Deep-Ex. Smoke amount as high as 15 mg/m^3, (CL = 5 mg/m^3) were also measured. Smoke removal is more difficult as strong local extraction destroys the gas shield. By remote controlled welding in the habitat these problems can be handled.

CONCLUSION

Mechanization of pipe welding for tie-in operations is feasible for implementation whenever a diving contractor wishes to do so.
Work is now carried out to solve detailed problems and optimize equipment and procedures. Higher levels of mechanization automation is expected within 2-10 years.

Mechanized/automated orbit TIG welding systems seem to be the only feasible method for tie-in operations deeper than 250 m.s.w. To our opinion other methods tested by us give marginal results and have operational difficulties.

ACKNOWLEDGEMENT

The author would like to express his thanks to colleague S. Øvland for valuable help during preparation of the paper.

REFERENCES

Welding Technology Data, (1972), sheet 33 in Welding and Material Fabrication, p. 396.

VV. Rochehin, G.N. Gusakov and YU.M. Frolov, (1968): Gas-Electric Welding Non-rotating Butt Joints in Pipes. Svar. Proig 1968 NaG pp 21-22. English translation: Welding Production, vol 15, No. 6, pp 38-41

J.F. Key, P.W. Turner, D.W. Hood and D.D. Keiser, (1980): Joint Geometry Influence on Mechanized Pipe Welding. Welding Journal, pp 24-32.

R. Hill, M.R. Graham, (1978): Narrow Gap orbital Welding. Advance in Welding Processes Proceedings, 4th International Conference, Harrogate, Yorks pp 351-355

C.J. Allum, (1983): Development of Manual and automatic Methods for Hyperbaric Pipe Welding. Paper Swedish Trade Fair.

H.O. Knagenhjelm et al.,(1982): Deep-Ex81. Deep Weld. Norsk Hydro.

Arc Plasma Underwater Cutting

W. Kalczyński* and A. Klimpel**

*Mining and Metallurgy Academy, Cracow, Poland
**Silesian Technical University, Gliwice, Poland

ABSTRACT

The design and operation of underwater arc plasma cutting torch
and results of cutting steel, copper and aluminium sheets on
the depths to 30m in simulated conditions have been shown.
Optimal underwater arc plasma cutting parameters on the base
of the regression equations have been stated and the influence
of cutting process on the structural changes in severed materials have been tested.

KEYWORDS

Underwater arc plasma cutting torch; automatic control; cutting
parameters; steel, copper, aluminium sheets arc plasma cutting.

INTRODUCTION

The arc plasma cutting process is one of the most efficient
method of underwater metals cutting and from the technical and
economical point of view surpass others underwater cutting methods (Szapiro, 1963; Wodtke, 1976; Stalker, 1977).
The main advantages of the underwater arc plasma cutting process
are: – multiple higher cutting speeds in comparison with others
conventional methods, very important from the point of view of
the divers working real time connected with decompression time
requirements, – lower cutting costs, – high quality of severed
material surface, – one torch can cut many different materials,
–high thickness materials can be cut, for steel up to 40mm,
– continuous torch operation mode without rest time,
– process is easy to automatization and remote control.
The main disadvantages of underwater arc plasma cutting process
are electrical safety aspects because of danger of electric
shock, especially in salt water, considering high open circuit
voltage 180–200 V (Madatov, 1971).

THE DESIGN AND OPERATION OF ARC PLASMA

UNDERWATER CUTTING TORCH

The designed arc plasma underwater cutting torch working in the transferred arc mode has been patented in Polish Patent Office No P-220754, Fig. 1.

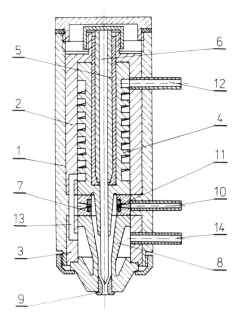

Fig. 1. Scheme of arc plasma underwater
cutting torch.

The torch comprises body 1, made from dialectric material which contains cylindrical metal inserts 2 and 3 and tungsten electrode cooler 4 feeding plasma arc current to the tungsten electrode (cathode) 6. The tungsten electrode 6 is attached by means of clamping sleavs 5 in cooler 4. The front part of cooler 4 is coaxially inserted into swirling assembly 7 placed between head of cooler 4 and gas nozzle 8 containing replaceble plasma nozzle 9. External surface of gas nozzle 8 is covered by heat resisting dialectric material. Plasma gases are fed into the torch through the pipe 10 directly to the swirling assembly 7 and then by nozzles 11 to ionization chamber between the front part of the cooler 4, the tungsten electrode 6 and internal surface of gas nozzle 8. Cooling water and plasma arc current are fed into the back part of the cooler 4 by means of the pipe 12. Cooling water is directed onto external surface of the cooler 4 and flowing by passage 13 to the chamber between the nozzle 8 and the metal insert 3, and is leaving out the torch through the pipe 14 which fed plasma pilot plasma arc current to the gas nozzle 8 during striking of pilot plasma arc in

350

the ionization chamber. Circulation of cooling water is then
in closed circuit under pressure of 0,3 to 0,9 MPa.
The developed plasma arc underwater cutting device include:
cutting torch, power source having an output of 100 kW, control
system, air compressor and plasma gases cylinders, Fig. 2.

Fig. 2. General view of deep water simulator.

The standard arc plasma cutting control system was additionally
equiped with: automatic system supplying air to the plasma torch
when the torch is plunging under water and during rest time,
plasma gases pressure gauge in the conduit feeding plasma gases
to the torch, voltmeter indicating pilot and main plasma arc
voltage, securing sure control of torch-control system arrang-
ment operation and rapid leads breakdown revealing.
During plasma arc cutting torch plunging under water automatic
air flow into the torch and then through the ionization chamber and the plasma nozzle 9 is activeted. Air pressure is suf-
ficently high to prevent water getting into the torch. After
the plasma torch automatic or manned setting-up on the cutting
line the ionization button is pushed on. Then ionization system starts and in the same time by means of the electromagnetic valves argon flow is opened and air flow is closed.
After air pushing out from the supplying conduit and the torch
by argon pressure, time regulator switch on pilot plasma arc
striking voltage and pilot plasma arc is established. When the
torch approaches to the workpiece on the proper distance self-
acting main plasma arc striking and automatic plasma gases mix
flow take place and then plasma arc underwater cutting process
begins. Workpiece is connected to the positive pole of power
source. The described torch and control system assure higher
tungsten electrode and plasma nozzle durability, formation of
slag-metal bridges which flood cutting gap is avoided and low
cooling water consumption.

RESULTS OF CUTTING CONDITIONS TESTS

The underwater arc plasma cutting tests on 0,5m water depth
have been done using atmospheric pressure stand but up to 100m
water depths the deep water simulator, shown in Fig. 2, was used.
On the base of preliminary tests optimal plasma nozzle dimen-

tions (orifice diameter 3mm and heigh 6mm), torch to workpiece distance 7 to 8mm, and plasma gases mixture containing: 50% Ar, 15% N_2, 35% H_2, have been stated.
In underwater arc plasma cutting of low carbon steel sheets tests on 0,5m depth correlations between , from one side the maximum and optimum cutting speeds, and sheet thickness, plasma arc current and plasma gases flow rate on the other have been found. The optimum cutting speed was considered with regard to the quality of severed material surfaces.
In deep water simulated tests correlations between, from one side the maximum and optimum cutting speeds of low carbon steel 12,5mm sheets and torch plung depth, plasma arc current and plasma gases flow rate on the other have been stated.
The ranges of tested parameters variations during experiments are given in Table 1.

TABLE 1   The Ranges of Tested Parameters
Variations

Type of experiment	Sheet Thickness mm	Water depth m	Plasma arc current A	Plasma gases flow rate m^3/s
non-pressure	10 – 40	0,5	210 – 370	$5 - 10 \cdot 10^{-4}$
pressure	12,5	0,1–30	300 –45o	$5,83{-}10 \cdot 10^{-4}$

The experiments results have been treated by means of the regression equations in the form of multinominal using PS/DK-P;α determined static multifactor ortogonal program (Polański, 1978 Analysis of regression equations has shown that principle effect on cutting speed exerts severed material thickness followed by torch plung depth, plasma arc current and at last plasma gases flow rate. Underwater arc plasma cutting parameters of steel sheets and additionaly cut on 30m depth copper and aluminium sheets are given in Table 2.

TABLE 2   Plasma Arc Underwater Cutting
Parameters

Material	Torch plung depth m	Material thickness mm	Plasma arc current A	Plasma gases flow m^3/s	Cutting speed m/min max	optim.
Low-carbon steel	0,5	12,5 25 38	370 370 370	$5 \cdot 10^{-4}$ $5 \cdot 10^{-4}$ $8,33 \cdot 10^{-4}$	2,0 1,0 0,85	1,4 0,9 0,8
Low-carbon steel	10 20 30	12,5 12,5 12,5	450 450 450	$10 \cdot 10^{-4}$ $10 \cdot 10^{-4}$ $10 \cdot 10^{-4}$	1,7 1,6 1,5	0,75 1,5 0,5

Copper	30	12	340	$9{,}38 \cdot 10^{-4}$	–	0,10
Aluminium	30	10	320	$9{,}38 \cdot 10^{-4}$	–	0,75

METALLOGRAPHIC EXAMINATIONS

Metallographic observations revealed that the HAZ of underwater severed steel sheets has widtn below 1,2mm about 40 to 50% narrower then after dry land plasma cutting, Fig 3.

Fig.3. The HAZ of plasma arc underwater cut
steel sheet.

Microhardness of the steel HAZ is about 430HV near severed surface (100HV higher than dry land plasma cut material) but in the distance of 0,6 to 0,9mm from severed surface falls to beneath 240HV. In the case of copper the HAZ is undetectable and without any traces of oxide eutectic, tipical for dry land plasma cut copper, Fig 4.

Fig. 4. The HAZ of plasma arc underwater cut
copper sheet.

Microhardness of the copper HAZ is below 95HV near severed surface but in dry land plasma cut copper about 60HV and in the distance about 0,4mm from underwater severed surface falls beneath 70HV. In the case of the aluminium HAZ structural changes are undetectable but severed surface is covered by thin layer of melted metal containing high level of pores and bubbles shown in Fig. 5.
The HAZ microhardness near the underwater severed surface is about 72HV (in dry land plasma cut aluminium about 88HV) but in the distance 0,4mm from underwater severed surface only 60HV.

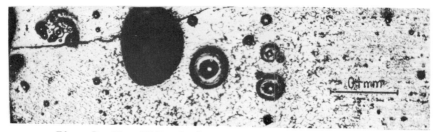

Fig. 5. The HAZ of plasma arc underwater cut
aluminium sheet.

APPLICATION

The economical analysis has shown that plasma arc underwater
cutting costs of 12,5mm low carbon steel sheets are about 8 to
10 times lower in comparison to oxyarc cutting costs.
Plasma arc cutting tests in natural working environments at
sweat water on the depths to 10m has proved unfallible operatio
of the developed plasma torch and control system and good
diver electrical safety.

CONCLUSIONS

1. The developed plasma arc underwater cutting torch and control system make possible effective underwater cutting of
   steel, copper and aluminium sheets in water depths up to 30m

2. The highest influence on cutting speed exert material thickness and the next torch plung depth, plasma arc current and
   plasma gases flow rate.

3. The HAZ is narrower in underwater cut materials then in dry
   land cut materials but microhardness is higher.

4. The costs of plasma arc underwater cutting of 12,5mm low
   carbon steel sheets are 8 to 10 times lower then oxyarc
   underwater cutting costs.

5. The future tests should be conentrated on farther improvements of diver electrical safety and utilizing air as a plasma gas like in dry land plasma cutting.

REFERENCES

Szapiro, J.S.(1963). Weld.Prod., 2, 27-28.
Wodtke, H. (1976). Welding Journal, 1, 15-24.
Stalker, A.W.(1977). A survey of Underwater Cutting Techniques
     The Welding Institute, Abington, England, CB1 6AL, 109-125.
Madatov, N.W.(1971). Weld.Prod., 9, 44-46.
Polański, Z.(1978). Cracow Technical Univ.-Cracow, Poland.

Friction Welding Under Water

E. D. Nicholas

The Welding Institute, Research Laboratory, Abington Hall, Abington, Cambridge CB1 6AL, UK

ABSTRACT

Underwater friction welding of tubes and studs to plates carried out at The Welding Institute are reported. The work demonstrated that the water environment significantly affected the weld microstructure by inducing hard transformation products such as bainite and martensite. Also the severe quenching effect of the water caused surface cracking and modified the metal flow characteristics. By using shroud materials of the foamed plastics type to exclude the water, welds with features similar to 'in air' welds could be achieved. Work when friction welding studs to plate identified that modification to the welding force cycle can reduce the heat affected zone hardnesses in the plate. Similar trends undertaken by some Japanese workers are highlighted. Their results serve to confirm the findings of the work carried out at Abington.

Potential areas for exploitation of friction welding underwater embrace:- pipeline repair; retrofitting of anodes; plate attachments; keying for grouting and fixing of stress measurement devices. Brief reference is made to the development of a submersible friction stud welding facility that is designed to operate at depths approaching 300 metres.

KEYWORDS

Friction welding; underwater; hardness; flash profile; protective shrouding; weld force cycle; equipment.

INTRODUCTION

Friction welding is an autogenous joining process whereby the weld is made in the solid phase. The process is extremely efficient in its utilisation of energy and thus its exploitation for joining a wide range of materials in like and dissimilar combinations combined with short weld times is widely used for many applications throughout a wide spectrum of industries.

For almost all applications the friction welding process is operated in a
normal atmospheric environment. The fact that material flow from the rubbing
surfaces takes place means that the atmosphere is excluded. Clearly therefore, if a gaseous medium can be overcome there is no reason to expect why
a fluid medium cannot also be excluded. This was ably demonstrated in the
late Sixties when by accident, during commissioning of a vertical friction
welding machine, the weld location was flooded with hydraulic oil and still
a bond of reasonable strength was produced. It was at this stage in the eval
uation of the friction welding process that The Welding Institute decided to
investigate the possibility of operating the process in a water environment.
Before reviewing the works of The Welding Institute and other workers in thi
field it is worthwhile to look at the friction welding process in closer
detail.

THE PROCESS

The simplest mechanical arrangement for friction welding, involves two
cylindrical parts held in axial alignment: one of the parts is rotated while
the other is advanced into contact under a pre-selected axial pressure.
Rotation continues for a specific duration sufficient for the frictional
heat to achieve an interfacial temperature at which the metal in the joint
zone is in a plastic state, both at the interface and for some depth behind.
Having achieved this condition the rotating part is arrested while the pressure is either maintained or increased to consolidate the joint. Other
arrangements which may be employed, include rotating a centre piece between
two stationary end pieces, a technique suitable to join wire coils and long
part lengths, rotating two end pieces against a stationary centre piece, as
used for the fabrication of prop shafts and rear axle casings. Friction
welding is sufficiently versatile to be used to join a wide range of part
designs, such as bar to bar, tube to tube and tube or bar to plate.

The two energy variants used in friction welding today, are referred to as
Continuous Drive and Stored Energy. The former is the most widely used
variant in Europe, USSR, and Japan, while the stored energy method is predom
inantly adopted in the USA. Although each variant has its advocates, they
will both provide welded parts of very high integrity and at production rate
which are far better than alternative joining methods such as flash and
resistance welding. Cost savings can also be made because of lower power
consumption, reduced machining and greater conservation of material.

There are four relative motion categories currently available for friction
welding. The most widely used is that of conventional rotation as it is th
simplest system to engineer and the most versatile to apply. The major
limitations of this sytem are that only parts of essentially circular
geometry are suitable, and precise angular orientation between parts cannot
easily be achieved. Angular and linear oscillation together with orbital
motion systems have been successfully developed to overcome these limitatio

At this point it is appropriate to intoduce a method for joining hollow
sections, called Radial Friction Welding. The simplest arrangement is
schematically presented in Fig. 1. To support the pipe walls and prevent
penetration of upset metal formed during the welding sequence a mandrel is
positioned in the bore.

To make a weld the ring is rotated while subjected to uniform radial compre
sion. To terminate welding, ring rotation is arrested while the radial com
ressive load is maintained or increased. Alternatively, by changing the mo

of ring deformation to that developed by expansion, rings can be inserted into hollow cylindrical bodies, or indeed the method can be used for pipe welding where access to the outside is restricted.

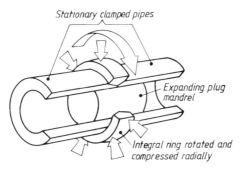

Fig. 1. Schematic arrangement for radial friction welding.

WELDING TRIALS (1969-1973)

The early trials with both mild steel and 1% carbon steel when welding in mediums other than air quickly identified the major problems of increased transmission power, high weld zone hardnesses and irregular metal deformation. A simple example demonstrating the latter two features is presented in Fig. 2, for welding 19mm diameter 1% carbon steel to mild steel plates in air, oil and water. The peak hardnesses in the rod rise from 660 to 740 to 830 HV_{20} when the medium is changed from air to oil to water. Also it is worth while to note that there is a rise in hardness in the heat affected zone of the plate where for the weld in water the value has increased to

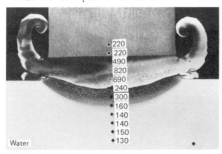

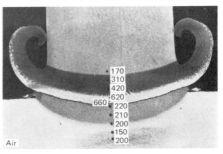

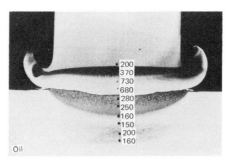

Fig. 2. Macrofeatures and hardnesses of friction welds made in various environments using 1%C steel studs.

357

300HV when compared to the parent plate hardness of approximately 150HV. These results clearly reflect the relative quenching effect of each medium. The very high cooling rate of the water has resulted in circumferential cracks being formed at the rod/flash junction. The photomacrographs further highlight the move away from a smooth regular flash form as the quench rate increases.

Similar trends as described above were found, as expected, when friction welding carbon steel tubular parts (110mm OD by 10mm wall thickness) directly in water at atmospheric pressure. Figure 3 compares the hardness variations for welds produced in air and in water, where the latter welds were made at a high friction pressure of 185N/mm² . In the weld zone, hardness values peaked

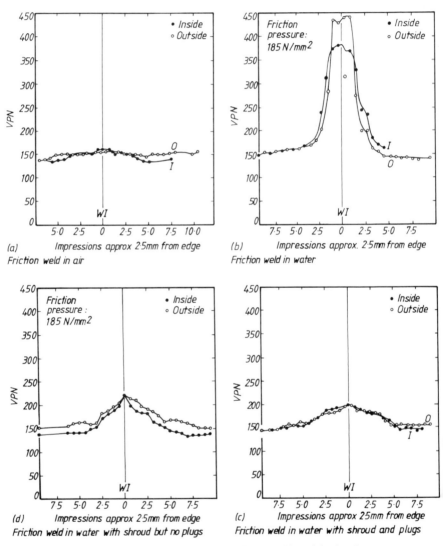

Fig.3. Hardness surveys across tubular steel friction welds made in air and water.

at approximately 450HV, Fig. 3b which is some three times greater than the parent steel hardness of 150HV. These high values reflect the transformation to martensite resulting from the severe quenching effect of the water.

Clearly, it would appear necessary to provide some form of barrier to prevent direct impingement of the quench medium on the weld region. A simple yet very effective solution was obtained by making the welds within a block of polystyrene thus protecting the peripheral regions. The plastic proved most effective in preventing ingress of water while at the same time did not impair the flow of hot flash metal from the rubbing surfaces. The success of this shrouding was clearly evident in the significantly reduced hardness levels monitored across the weld, referred to in Fig. 3c. Further protection of the tubular parts by using a polystyrene plug inserted into the bore served to reduce again the hardness at the weld Fig. 3d, which was only approximately 50HV greater than that measured for the 'in air' weld, Fig. 3a.

If reference is now made to Fig. 4 where sketches of the macrofeatures of the welds referred to above have been produced, it is apparent that the protection offered by the polystyrene shroud and plug has allowed the flash metal to develop in an almost identical manner to that observed for the 'in air' weld. Thus the significant defect of lack of bond at the weld interface observed for the weld produced without shrouding has been eliminated.

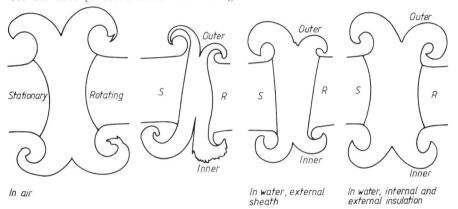

Fig.4. Tubular friction welds made in air and water (weld profiles). R = rotating, S = stationary.

As the introduction of insulation blocks of polystyrene may be difficult at depth, a preliminary appraisal of generating the protection using polyurethane foam was undertaken. The idea was to mix the foaming reagents in a container which surrounded the weld location. Work thus carried out proved successful and again by accident a method was evolved whereby the foamed plastics had strength imparted to it in order to resist pressurisation.

This phenomenon occurred due to the addition of oil to the mould to prevent foam adhesion problems.

Based on the work described a provisional patent was applied for in 1972 and the final specification was published in 1976 (patent No. 1 451 447). Included in this particular patent are references to possible equipment that might be used for underwater repair of pipelines.

It is interesting to note that Japanese workers were quick to follow the path of underwater friction welding. In fact in 1973 they filed a patent application which was granted in May 1977. They decided to protect the weld region by wrapping heat insulating tape around the ends of the parts. With this arrangement they argued that steam is generated which will be trapped between the insulating material and the parts being welded and thus serve to reduce if not eliminate the quench effect of the water. Also reference is made in the patent to the effect of water pressure on weld properties. For a fixed set of welding parameters the effect of increasing the water pressure was to reduce both the amount of upstand (flash collar) and the tensile strength. Details of their work when friction welding 13mm diameter 0.3%C, 0.7%Mn steel bars underwater is given in the Table. The use of various tape materials when welding at a water pressure of 6 bar would appear to afford adequate protection to allow parent metal tensile strengths to be achieved.

TABLE

Water pressure bar	Heat insulating material	Radial height of flash above weld mm	Tensile strength (kg mm^2)	Position of failure
1	None	1.4-11.6	54.8-56.6	Parent metal
4	None	2.0- 6.1	56.0-56.7	Parent metal
6	None	0.5- 5.2	21.3-33.6	Weld zone
6	Plastic tape	1.0- 2.2	54.3-56.5	Parent metal
6	Polyethylene tetrafluoride and glass fibre cloth tape	1.7- 1.9	54.8-56.7	Parent metal

Other workers namely Tazaki and Hirai investigated friction welding of a range of carbon steels in a water environment. Predictably as the carbon content of the steel increases so does the hardness of the weld zone while for a given carbon steel, the upset distance decreases when compared to welds made in air. Their work further confirmed the significant effect of water on flash metal formation and in the case of 0.45%C steel weld highlighted another problem associated with quench cracking.

WELDING TRIALS (1979-1980)

Interest was rekindled in friction welding underwater for the particular application of attaching studs to plates. In this context, one of the major requirements from such a weld, apart from its general strength was that the hardness in the heat affected zone of the submersed structure be kept below a level of 330HV. Trials using the equipment shown in Fig. 5, thus recommenced with a high carbon equivalent grade 50D steel plate (0.5CE) representing the submersed material. The friction welding machine comprises in the main of a welding head and hydraulic power pack. An hydraulic motor provides a maximum spindle speed of 2000 rev/min at 26kW transmission power, while the maximum axial welding force of 60kN is generated from an hydraulic ram operating at 136bars pressure. Some satisfactory welds were produced in a water environment when both copper (shrouded) and pure aluminium (no shroud) were used as the stud materials. In the case of copper the HAZ hardness of the steel rose to approximately 200HV which compared most favourably with the parent steel hardness of 185HV. No change in hardness was observed when the pure aluminium stud was welded. Unfortunately because of rapid heat abstraction through the plate steel, problems persisted.

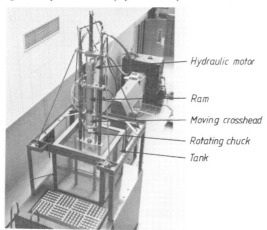

Fig.5. Friction welding machine for underwater welding.

Experiments when welding mild steel studs even in air showed that HAZ hardnesses would rise to values of 440HV as a result of the larger mass of the steel substructure Fig. 6a. Indeed similar hardnesses were obtained with welds made underwater with shroud protection. These values were, however, favourable when compared to unshrouded welds in water where the hardnesses reached around 500HV.

 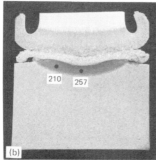 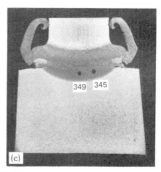

Fig.6. Macrosections of stud friction welds made:

a) in air
b) using a double weld in water
c) with a modified force cycle in water.

The problem was partially solved by making a second friction weld on top of the first stud weld. By choosing the right frictioning characteristics the heat from the second weld served to temper back the HAZ hardness of the first, Fig. 6b. Further consideration of this result raised the possibility that by changing the welding force cycle, i.e. high initial load then reduced to a lower level to promote thermal energy development, then increased to the final forge force, the same effect could be produced. Limited trials using a reprogrammable force as referred to above resulted in a significant degree of success, Fig. 6c. Weld strength was still maintained since the weld withstood a 90° hammer bend test without failure.

APPLICATION AREAS FOR UNDERWATER FRICTION WELDING

1.   Repair of damaged pipelines - In this context a schematic represent
       ation of a possible machine is shown in British Patent No. 1 451 44
       This particular concept allows the attachment of flanges to the
       pipeline ends, which of course can then be bolted together.

       Consideration can also be given to the potential application of the
       radial friction welding method, although it would be necessary to
       develop a splittable rotating/compression welding spindle. However
       an alternative approach can be considered where the spindle is trea
       ed as a throw away item and thus is simply left to decay.

2.   From the theme of the more recent welding trials, the attachment of
       studs is being given serious consideration for the following appli-
       cation areas:- a) retro fitting of anodes (electrical connector
       properties); b) fixing of plates; c) fixing of grouting and
       d) fixing of stress measurement devices.

One particular British Company, namely Thomson Welding and Inspection Limited
based at Inverness in Scotland, has recognised the potential applications for
friction welding underwater and is at present developing a totally submersibl
friction stud welding facility to operate at depths of approximately 300metre
The system will comprise of a welding head where stud rotation is affected
through a hydraulic motor whilst an annular piston will provide the axial
welding force and a separate hydraulic power pack which will deliver oil at
high pressure to the welding head through flexible hydraulic pipes.

CONCLUDING REMARKS

The results contained in this paper are most encouraging as they clearly show
that friction welding can be performed in hostile environments. It is expect
ed that a stud welding facility will be available in the middle of 1983. The
welding head will have been laboratory tested in water pressures up to 21 bar
and the total system site evaluated at water depths of around 150metres. An
important feature of the process is that the weld hardnesses can be kept to
acceptable limits by using shrouding and/or a carefully selected welding forc
pattern.

Finally, it is important to remember that the process can be seriously consid
ed for underwater repair of pipelines although significant investment will be
required to develop such a facility.

SUGGESTED READING

Agency of Industrial Science & Technology (Japan). Method for underwater fri
       tion welding of metallic articles. British Patent 1 473 716. File
       17 Sept. (Japan 48/107132) Publ. 18 May 1977, 2 fig. 6 claims (In
Astrop, A. (1979). Friction welding lines up more jobs. Machinery and Prod-
       tion Engineering, pp 41-43
Nicholas, E.D. and Lilly, R.H. Radial friction welding. Advances in Weldin
       Processes Conference held at Harrogate, UK, 1978, Paper 48.
Tazaki, Y. and Hirai, S. Underwater friction welding. Welding Institute Tr
       lation from Kinzoku Zairyo (Metals Materials) Vol 11, No. 8 pp 91-
The Welding Institute (1979) Exploiting friction welding in production. Inf
       ation package, Series pp 75.

Procedure Development for Wet Welding on Submarine Pipelines

T. J. Reynolds

Sea-Con Services, Inc., New Iberia, Louisiana, USA

ABSTRACT

The feasibility of making wet welded repairs to high pressure, subsea pipelines by using shielded metal arc welding process has been studied.

The pipeline material was API 5LX Grade X52 with 0.344 inch wall thickness.
A full encirclement split sleeve was used to simulate repair of a hole in
the pipeline segment. The repair material was 0.5 inch A36 steel. The
electrode used is a proprietary one developed by Sea-Con Services, Inc.

Main results are summarized as follows:

1. Dye penetrant and solution film tests were made and disclosed no defects
   on the wet welds.

2. A successful wet welded repair can be made without the use of expensive
   habitats.

KEYWORDS

Weld; wet; underwater; subsea; marine; offshore; inspection.

INTRODUCTION

The continuous demand for energy and the dramatic rise in fossel fuel prices
over the last decade has spurred an increase in the exploration and production of gas and oil both on land and offshore. This requires more gathering
and transmission pipelines to provide adequate service to the customers. It
also means that gathering and transmission pipelines used in marginal fields
will be retained in service longer. These remarks are directed towards offshore work.

Concurrently with exploration and production expansion, there has been increasing offshore activity in fishing and pleasure boating, commercial in-

spections of structures, repair work, drilling and new construction. This marine activity increases the chance of pipeline damage by means of an anchor, spud leg, dropping a heavy object overboard or just poor original workmanship. Pipelines that are used well beyond their design life are more likely to experience damage.

Offshore pipeline repairs are very costly. The ability to make repairs to pipelines in the wet would greatly reduce the cost and time. Sea-Con has been researching wet welding to pipelines as an alternate to the current more costly techniques.

REPAIR TECHNIQUES IN USE TODAY

Currently, the most common methods used to make pipeline repairs are:

1. Raising the line and repairing in the dry, utilizing a barge spread. This method has the advantage of making the repair topside with conventional inspection techniques. Disadvantages include availability and cost of a large barge spread, possible damage during raising and lowering operations, and reburial cost.

2. Dry hyperbaric welds utilizing sophisticated habitats. The advantage is the weld is made in a dry environment and can be inspected with conventional techniques. Disadvantages include the availability and cost of specialized habitats, support equipment, and development time.

3. Mechanical repair techniques made in the wet. The advantage is less expensive support vessels and equipment. Also divers can install the mechanical connections. The disadvantages are the availability and cost of the mechanical connection, possible leaks because of sealing problems.

In the event of a pipeline leak, operators have many reasons for expeditiously stopping the leak and returning the pipeline back to service. The reasons may be economic, legal, environmental or contractual. Sea-Con Services believes that wet welding is a fast, cost effective alternative to current repair techniques.

PREVIOUS EXPERIENCE

Repairing underwater pipelines is not new to Sea-Con. Sea-Con made its first underwater pipeline repair in 1972 on 20"Ø and 22"Ø lines. Since then Sea-Con has made dry habitat pipe repairs to lines ranging from 3"Ø to 22"Ø. Wet welded repairs have also been made on pipelines from 20"Ø to 50"Ø and at depths ranging from 20 FSW to 152 FSW.

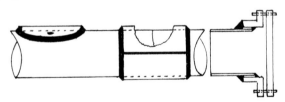

Fig. 1. Horizontal pipe showing patch, split sleeves and slip-on flange repair configurations.

Sea-Con is developing three wet welded techniques consisting of slip-on flanges, a patch, and full encirclement split sleeve - Fig. 1. The customer requirements and severity of damage would determine which repair technique would be used. The slip-on flange technique has been successfully used on a 36"∅, 900 psi., pipeline repair in January of 1981 - Fig. 2.

Fig. 2. Wet welded 36 inch slip-on flange in the Gulf of Suez.

DESIGNING A REPAIR FOR A SIMULATED DAMAGED PIPELINE

Up to the present time, destructive testing of wet weldments has been limited to reduced section tensiles, bends, charpy impact specimens and fatigue analysis specimens. Engineers at Sea-Con decided to design a repair and hydrotest it to failure, for a subsea pipeline that had a hole in it to simulate typical corrosion or puncture damage.

The pipeline material for this test was chosen to be typical of what might be encountered in a repair situation offshore. Therefore, the pipe material selected has the following properties:

Material Specification	API 5LX Grade X52 High Test Line Pipe
Minimum Specified Yield Strength	52,000 psi.
Yield Strength (from certified test report)	60,200 psi.
Ultimate Tensile Strength (from CTR)	86,300 psi.
Carbon Equivalent (C.E.)	0.50
Pipe Outside Diameter	6.625 in.
Wall Thickness	0.344 in.

The repair method used was a full encirclement split sleeve. The sleeve material was rolled from 1/2" thick, A36 plate with an ultimate strength of 75,500 psi. and a C.E. of 0.32. A36 plate was selected because it or similar material can be obtained anywhere in the world. The two split sleeve halves were rolled to an inside diameter of 6.625 inches. The length of the split sleeve was twelve inches.

In developing the test parameters, Sea-Con considered United States Department of Transportation requirements for pipeline safety and specifically the maximum allowable design pressure for pipelines according to the following formula:

$P = (2St/D) \times F \times E \times T$ where:

P = Design pressure in pounds per square inch gauge

S = Maximum specified yield strength in pounds per square inch

D = Nominal outside diameter of the pipe in inches

t = Nominal wall thickness of the pipe in inches

F = Design factor of 0.72 for buried pipeline well away from any structure

E = Longitudinal joint factor which is 1.00 for this test

T = Temperature derating factor which is 1.00 for this test

Using the above formula gives the maximum allowable design or working pressure of 3,888 psi. for this particular pipe. A typical working pressure for this pipe would be 2000 to 2500 psi.

REPAIR PROCEDURE

The X-52 pipe specimen was 38 inches long and had extra strong pipe caps welded onto each end. A $1\frac{1}{2}$"$\emptyset$ hole was burned in the test section of 6 5/8"$\emptyset$ pipe to simulate the leak or damage. One cap had a 6000 lb. coupling welded to it for the subsequent hydrotest. The pipe weldment was fit up, tacked and wet welded in Sea-Con's Training Tank - Figs. 3, 4 & 5. The working depth was approximately 34 Feet Fresh Water. The Training Tank is located at Sea-Con's facility in New Iberia, Louisiana.

The welding source was a conventional rotating type, constant current motor generator set on straight polarity (electrode negative). The electrode used for the test is a proprietary electrode developed by Sea-Con.

The welder/diver was required to fit the two split sleeve halves to the "damaged pipeline" in the typical horizontal position. The two split sleeve halves were fit-up so that two horizontal butt welds, 1/2" in size, could be made using the pipe base metal as the backing material. Finally two circumferential 1/2" fillet welds were made, one on each end of the split sleeves, to close the repair.

Fig. 3. Welder/diver fitting split sleeves to pipe.

Fig. 4 Wet welding split sleeves to
        pipe

Fig. 5 Training Tank
        36'-0 x 22'-0 Ø

TESTING OF THE WET WELDED PIPELINE REPAIR

Underwater nondestructive examination methods were limited to visual examination only. Ultrasonics was not practical due to the difference in microstructure between the base metal and weld metal. Visual examination (VT)
revealed no defects. The pipe weldment was then brought to the surface and
dye penetrant (PT) and solution film (SFT) tested and no defects were disclosed in the wet welds.

The pipe weldment was prepared for a
hydrotest by attaching high pressure
fittings to the preinstalled coupling -
Fig. 6.

According to the following formula the
test pressure was:

Pmax = 2 x UTS x T/D where:

   P = Internal pressure when the
       pipe material is at its
       ultimate tensile strength

UTS = 86,300 psi.

The pipe weldment was calculated to be
at its bursting threshold when the
internal pressure reached 8,962 psi.

Fig. 6 Hydrostatic test set-up

Fig. 7   Bursting pipe specimen

The pipe test specimen was then connected to the hydrotest apparatus equipped with a 0 - 15,000 psi. chart recorder. The pressure was applied in 500 psi. increments with a 3-5 minutes hold at each increment. At 9,200 psi. or 103% of the pressure at the UTS, the test specimen developed a small leak approximately one drop every 20 seconds. The pipe base material had begun to noticeably yield at this point. The test specimen was repaired by wet welding, and re-hydrotested.

The test specimen finally failed at 10,300 psi. - Fig. 7. Subsequent inspection of the test specimen showed considerable yielding had occurred to the pipe whereas the split sleeve had not yielded at all. The failure initiated in the pipe base metal between the pipe cap and wet fillet weld - Fig. 8.

Fig. 8   Close-up of burst
pipe specimen

SUMMARY

The hydrotest demonstrates that the wet welded pipe specimen exceeded the allowable operating pressure by 265%, and the ultimate test strength by 115%.

This test also demonstrates that good quality weldments can be quickly and successfully made by wet welding on subsea pipelines without the use of expensive barge spreads, sophisticated habitats, and complicated mechanical connections. This procedure should be considered by pipeline operators as one of the choices available for submarine pipeline repairs. The wet welded repair can restore a pipeline to its original design pressure and can be stronger than the pipe.

FUTURE TESTS

Sea-Con is encouraged by the results of this test. Continued development of pipeline repairs by wet welding is planned and needed for repairs in different configurations, diameters, materials and depths. Participation by industry is expected and encouraged. Sea-Con has the capability to perform wet and dry welding down to 300 FSW in their Hyperbaric Saturation Welding Facility located in New Iberia, Louisiana.

Data resulting from additional tests will be published.

REFERENCES

U. S. Department of Transportation, Pipeline Safety Regulations, Natural Gas Part 192, Hazardous Liquids Part 195, October 1, 1981 Revision including Amendments thru April 1, 1982.

Mechanized Underwater Welding Application

I. M. Savich and A. A. Ignatushenko

E. O. Paton Electric Welding Institute of the Academy of Sciences of the Ukrainian SSR, USSR

ABSTRACT

The paper presents examples of mechanized underwater welding application in oil and gas pipeline repair, in salvaging operations, etc. Described are certain most typical defects in underwater engineering objects, the elimination of which is expedient with the "wet" mechanized underwater welding. The technology of underwater operations is described. The cost of operations performed with the new method is low, and a high efficiency of welding is achieved.

KEYWORDS

Wet mechanized underwater welding, semi-automatic machine, flux-cored wire, gas pipeline, ship, efficiency.

The gas industry development in our country is characterized by shifting of the main gas production regions to the north. A considerable number of trunk gas pipelines are in operation now, and their extension will be drastically increased in future. The number of underwater passages will grow, accordingly. An important performance index of gas transportation systems is the continuous gas supply to the consumers.

The problem of speeding up the gas pipeline repair, especially in underwater passages, should be considered rather urgent, taking into account the fact that the trunk pipelines transport large quantities of gas daily, and cannot be cut off for a long time through economic reasons.

The need of an effective technique of gas pipeline underwater passage repair is especially strongly felt in the Extreme North, where a severe climate determines the short terms of repair in summer, and makes it difficult to perform any work

IIW-N

during winter. The present repair technique with the pipe hoisting onto the ice is labour consuming and expensive, it takes up a lot of time, and is technically very difficult or quite impossible to perform with large diameter pipes. The installation of plug-in clutches in places of cracks, blow-holes and dents allows to eliminate the gas leakage at low pressure, but the complexity of this repair technique is evident.

The underwater welding is one of the technical means, permitting to markedly simplify and speed up the underwater pipeline repair. Until lately, its application in the oil and gas industry had been held up by the absence of technology, guaranteeing the high quality of welded joints and equal strength of the weld and pipe metal.

The E.O.Paton Electric Welding Institute of the Academy of Sciences of the Ukr.SSR has developed the technology of semiautomatic underwater welding and cutting with special flux-cored wires. The strength of the produced welded joints is equal to that of the base metal, the cuts have sufficiently smooth surfaces and no flash. The laboratory investigations and numerous field tests proved that in low-carbon and low-alloy steel welding the mechanical properties of underwater welds, made by the semiautomatic underwater welding with

ПС-АН1 flux-cored wire, correspond to the mechanical properties of joints welded in air with Э-42 type electrodes (Table 1).

TABLE 1 Mechanical properties of welded joints made by semiautomatic underwater welding with ПС-АН1 flux-cored wire

Welded steel grade	Ultimate strength, MPa	Yield strength, MPa	Relative elongation %	Reduction in area %	Impact strength J/cm^2 +20°	Impact strength J/cm^2 -40°	Welded joint bending angle
ВСт3	406–411 409	308–336 321	28.7–35.3 31.5	57.8–68.9 66.1	86–110 105	47–52 50	180
09 Г2С	503–526 511	390–408 392	22.7–33.4 28.7	61.4–72.1 68.5	87–125 112	75–107 97	180

Note: The average values are the result of three measurements.

Welding is performed by the A-1660 semiautomatic machine, allowing to vary the process conditions within wide ranges, depending upon the weld spatial position. The ПСГ-500 welding convertor, or ВС-600 rectifier or any other one with a flat external characteristics is used as a power source.

An important advantage of the semiautomatic underwater welding is simplicity of arc excitation, good visibility of the welding zone and workpiece edges welded, high efficiency, simplicity of mastering by divers and service personnel, i.e. all those characteristics, which make the process "flexible".

The semiautomatic underwater welding was successfully used in repair of a number of objects.

1.Specialized repair-adjustment department on underwater technical operations and E.O.Paton Electric Welding Institute of the Academy of Sciences of the Ukr.SSR have restored the damaged section of the underwater passage of the gas pipeline across the Enisei river near the town of Dudinka.

The 325 mm dia. siphon (09Г2С steel pipe) failed along the field joint in the place where it comes out of the deepened area near the right bank 100 meters from the water edge, 12 m deep. The type of failure is shown in Fig.1. The welded joint fracture in the HAZ extended for more than 270° around the pipe perimeter. A 150 mm long tearing of the wall along the axis in the base metal was observed, it being, evidently, due to the section rotation under the water current effect. The circular

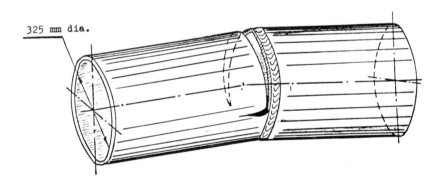

325 mm dia.

Fig.1. The scheme of pipeline field joint fracture.

crack opening in the upper portion of the pipe was 20 mm, the sections in the tearing location were displaced relative to each other by 40-60 mm. The river bed relief in the zone of the siphon laying specified the complex spatial bending of the "river" and "bank" section axes in the region adjacent to the tear. The angles in the vertical and horizontal plane were, approximately, 10° and 7°, respectively, thus requiring a thorough alignment to ensure coaxiality. The repair technology (Fig.2) required the removal of the defective area of the siphon, fabrication and installation of the inner shell, its welding by two double pass inclined welds, weld testing by the working pressure P=30 atm, installation of the external coupling preliminarily mounted on the "river" section, concentric-ally to

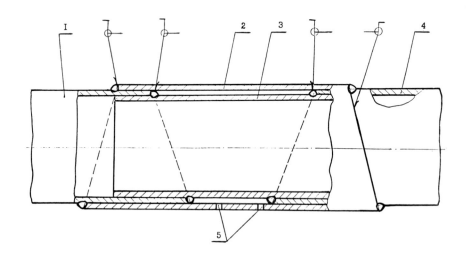

Fig.2. Repair assembly scheme:
1 - pipe end from the fairway side;
2 - external shell;
3 - inner shell;
4 - pipe end from the bank side;
5 - holes for water removal from the circular
    space between the external and inner
    shells.

the inner shell. The external coupling should also be welded under water by inclined welds, the water should be removed from the circular gap between the inner shell and external coupling, and the as-assembled unit should be tested by the working pressure of gas. A threaded hole is provided in the upper part of the external shell the purpose of which is to simplify the deposition of the sealing weld. This hole is necessary, because an intensive, concentrated in a small volume heat source causes an intensive vapour formation and pressure increase in the space between the inner and external shell, and this may hinder the deposition of the last, sealing portions of the circumferential weld. Other versions of the repair assembly were also considered: flange joint design of the inner shell with a bevel cut of end faces, the shell being rotated for 180° around the longitudinal axis after installation, and others. However, they were rejected because of a considerable bending of the axes of the "river" and "bank" sections, limited technical possibilities of installation equipment and short terms of the work performance.

The technical devices used in repair operations consisted of a 290 t barge carrying a self-contained electric power station of 100 kW power, pipe-layer, winches, divers'boat, tug boat and of draft mechanisms, located on the bank, etc. The welding equipment - the semi-automatic A-1660 machine (Fig.3) and the ПСГ -500 power source were installed on the barge.

Fig.3. The submersible assembly of the semiautomatic A-1660 machine (upper cover removed).

A powerful centrifugal pump and the divers'boat fire-extinguishing system pump were used for deepening the river bed under the defective region of the siphon. The usual diver's tool kit was supplemented with jacks, yoke-rests, pulleys.

According to the plan of repairs, the operation sequence was as follows:

After installing the machinery, deepening the river bottom, and welding equipment testing in site, the defective section of the pipe was removed. To simplify the inner shell installation and its welding on, the bevel cuts were made according to a template.

The semi-automatic underwater cutting with ПНР-АН2 flux-cored wire was performed with A-1660 semi-automatic machine at the conditions, ensuring the high-efficiency and good quality of the surfaces, the cutting time amounting to 11 minutes.

The "river" pipe end was lifted by hoists along the vertical approximately by 3/4 of the diameter, the external shell was

moved over it, and an inner shell with a rest mounted on it
was introduced into it (Fig.4,background). After alignment of

Fig.4. The external and inner (background) shells.

the face cross-sections of the pipes, the inner shell was
installed in the place specified in the repair scheme by the
axial force and welded by two-pass circumferential welds in
the conditions, given in Table 2. After performing the root
weld, the slag crust was carefully removed.

TABLE 2 Semi-automatic welding and cutting
conditions.

Conditions				Main parameters			
Operation kind	Steel grade thick- ness, mm	Arc voltage V	Current A	Welding (cut- ting) rate, m/h	Power source, polari- ty	Wire gra- de, diame- ter	Spatial positi- on
Semi-auto- matic wel- ding	09Г2С 10-12	32-34 30-32 27-28	240-260 200-220 160-180	12 8 4-8	ПСГ-500 reverse	ППС-АН1 1.6mm	flat verti- cal overhead
Semiautomatic cutting	09Г2С 10-12	38-39 37-38 37-38	400 380 380	20 16 12	ПСГ-500 straight	ШР-АН2 2.0	flat vertical overhead

It should be noted, that in underwater mechanized welding, the same as in air, the weld deposition in the overhead position requires from the diver-welder a certain amount of training and skill. Here, the welding of the overhead regions of the weld was hindered by the inconvenience of the welder position under the pipe. There was no possibility of preparing the trench of a required depth, since the bottom in site was made up by rocks with pebble-like and silt inclusions. However, despite the inconvenient position and periodical impairment of visibility, due to pileup winds, the high-quality of weld performance was achieved, and the welds were accepted after the first test by the working pressure. The semi-automatic welding efficiency is 3-4 times higher than that of the manual electric-arc underwater welding. The average welding rate is 6-8 m/h. with 10-12 mm weld leg.

After welding on the inner shell, and weld testing under the working pressure by the axial force, created by the on-shore draft mechanism through a system of simple devices, the external shell was moved over to its position specified by the repair scheme. It had been preliminary oriented in such a way, that the large generatrix was located below. The welds, fixing the external shell were also made under conditions, given in Table 2. After performing the inclined welds, the upper threaded hole was plugged, and a weld was made around it, and the holes in the lower part of the shell were used for removing the water from the circular space between the external and internal shells.

In winter the gas supplied to the underwater passage is down to the temperatures of -40°C and lower. This makes higher demands not only of the strength and ductility properties of the welds themselves, but also of the design of the repair assembly as a whole. A rather thick layer of ice is formed on the outer surface of the pipe. An air cushion is provided in the circular space between the shells to prevent the possible defreezing and depressurization of the assembly. The air cushion is created by air supply from the surface under the pressure up to 5 atm through one of the holes, located on the lower generatrix of the external shell, the water having been removed from the space for about 60-70% by volume. After the water removal the holes 5 have been plugged (Fig.2).

The repeated long-term tests of the repair assembly under the working pressure have proved its complete tightness. This work was a completion of restoration of the underwater passage damaged section of the Messoyaha-Norilsk gas pipeline. The underwater passage section has been put into operation after removal of the water having penetrated inside the pipe; defreezing the ice blocks of the permafrost zone in the deepening near the shore and the preparatory operations performed by the operation service.

2. In a number of cases the semi-automatic underwater welding application permitted to considerably reduce labour consumption for the repair jobs and shorten repair terms, the semi-automatic underwater welding technology being used for such operations, which cannot be called underwater welding ones in the

initial meaning of this notion, for these are usually performed in open water resevoirs.

In one of the integrated chemical works of the country an emergency repair was performed of the siphon of the recycling system of water supply for the main production. A 60 mm diameter
hole was formed in 800 mm diameter pipe, made of low-alloy
09Г2С pipe steel, as a result of insufficient cathodic protection and abrasive action of water, which contained a suspen-
sion of hard particles. A large volume of earth-moving was required for an exact location of the defect and its elimination.
The defect was removed with semi-automatic underwater welding.
An operation (technological) access hole was cut out in the
pipe surface in the area of the expected fracture. The diver,
entering the water-filled pipe by the access hole revealed the
defect 20 m from the access hole. A special light-weight small-
sized semi-automatic machine for the underwater welding was
introduced into the pipe, and an oval-shaped patch was welded
on. A noticeable efficiency was achived at the expense of reducing the volume of the preparatory operations and of completely
ly eliminating the earth-moving. The short terms of the work
performance should be especially noted, since the whole work
took only several hours.

3. One more typical example of the semi-automatic underwater
welding application is the welding on of the hull skirting to
the hull of the submerged ship "Mozdok" and sealing of the
cargo hold covers during salvaging (Fig.5,6). The mentioned
ship got a 7 x 14 m breach as a result of collision and submerged to the depth of 30 m in the water area of Odessa port,
creating hazards for normal navigation in this region.

Fig.5. The hull skirting of the "Mozdok" ship, welded on by
semi-automatic underwater welding.

The ship salvaging was performed by the combined methods:
using lifting pontoons and creating the positive floatability
of the hull by pumping polystyrene into the holds. It was necessary to ensure complete tightness of the brought-in patch
and of the ship cargo hold covers to prevent the leakage of
polystyrene pumped into the holds. The semi-automatic underwater welding allowed to perform a large volume of welding
jobs in a short time. The vertical overlap welds, 30-12 m deep,
and the welds in the vertical and flat position 12 m deep, were
made in two passes. The total length of welds was 100 m. There
were no problems of polystyrene leakage after a through sealing, and the ship was lifted by the time fixed.

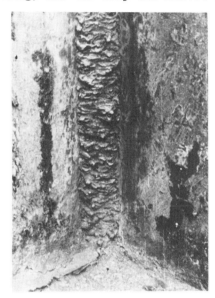 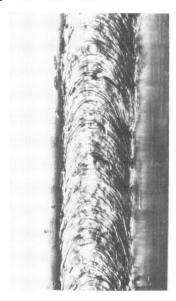

Fig.6. Vertical T-joints on   Fig.6a. Cargo hold cover fi-
the reinforcing elements for   xing weld (flat position, slag
fixing the cargo hold cover.   crust removed).

By now the semi-automatic underwater welding has become widely accepted in the country for performing the repair operations for various-purpose objects. The operations of elimination of the typical defects and accidental damages in oil and
gas pipelines were performed in various climatic zones of the
USSR in all seasons. The experience has been accumulated of
restoration: port facilities, oil tanks, tankers, etc. In all
the cases the semi-automatic underwater welding application
allowed to drastically shorten the terms of the work performance, and to obtain a noticeable economy just for the trunk
pipeline underwater passages about 500000 roubles in 1981.
The duration of repair of the object of medium complexity for
repair is from several hours to 2-3 days, and depends on the
volume of preparatory operations. The semiautomatic underwater
welding application in emergency cases and in case of emergency repair in distant industrial regions gives such important
advantages as mobility and a possibility of a fast mobilizati-

on of the equipment, high speed of the operation performance
due to a high efficiency of the process, moderate expenses
for the welding performance and very high economic efficiency.

Different Underwater Welding Repairs in Practice

W. J. M. Vriens

Vriens Diving Company B.V.,
Smit-Vriens Offshore Diving B.V.,
Van Konijnenburgweg 151, 4612 PL Bergen op Zoom, The Netherlands

ABSTRACT

In the following presentation I will try to portray to you, through several underwater welding case-histories, an idea of how, why and where different methods of underwater welding technics are used. Therefore the title of this presentation is not complete, because it does not sufficiently cover the full contents of the case-histories described in this paper. It gives way to other ideas of repair using construction methods to solve and assist normal every day practical underwater problems. Case-histories are chosen from the fields of inshore construction, offshore construction and the shipping industry.

KEYWORDS

Stick welding; cofferdam welding; habitat welding; stud welding; MAG process; sacrificial anode system; impressed current system.

INTRODUCTION

My first case-history, incidentally was also the very first underwater welding job that Vriens Diving Company B.V. was involved in. During spring of 1959 Vriens Diving Company B.V. was approached by a construction company, who were building a jetty for Shell Petroleum Co. in the Caribean on the Island of Curacao. Our work with the construction company involved assisting with the placing of metal sheeting skirts around mooring dolphins and piers. Welding underwater strengthening bracings between skirt and piers. The idea behind this operation was to try a new corrosion protection between the metal sheeting skirts and the piers by placing oil on the water and that the sea and tidal movement would coat the sheet piling on the dolphin and piers.

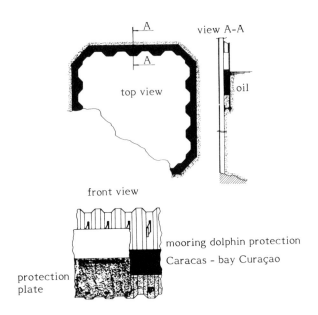

view A-A

top view

oil

front view

mooring dolphin protection
Caracas - bay Curaçao

protection
plate

So why did I choose this project as my opening example, it has a very interesting explanation.
Firstly: In 1959 underwater welding at that time was not as far advanced as
in to-days technical field. The welding rods were standard contact rods dip-
ped into parrafine to protect them from water penetrating into the ceramic
sleeve (coating), also welding to an acceptable standard needed a lot of experience, especially in underwater welding from the diver.
Secondly: This project also demonstrates the urgent need for corrosion protection, which at the time the technique of corrosion protection was very li
ited. Now modern day techniques have countered by the use of two cathodic pr
tection methods.
a) the impressed current system
b) the sacrificial anode system

WELDING OF SELF-SACRIFICING ANODES

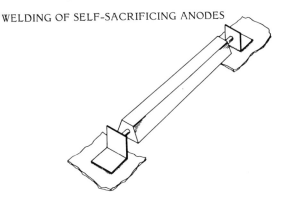

The impressed current system requires, mounting of antennas and ground cables underwater.
The sacrificial anode system requires the placing of numerous anodes positioned on a steel structure. Providing no forseen structured problems are envisaged. The normal underwater stick welding techniques are used.
I mention the next example, of which was a project undertaken in Saudi Arabia, only because of the size and application of this project. In the port of Jubail a jetty requiring 350 tons of anodes to be welded on, giving the theoretical life expectancy to the project of 15 years. A divingteam was able to place 10 anodes per day, each anode weighing 20 kgs. Of course damage to structures is attributed to more than the corrosion factor.

My second case-history relates to a ship, which collided with a cellular type quay, in the port of Freetown, Togo. The bulbuous bow penetrated one of the cells, opening one of the vertical joints of sheetpiling. This situation became critical because through the opening the inside filling started to leak out. The principle of the cellular structure became impaired and the end result was similar to a closed sack full of dry sand with a weight on top of it, when the sack is ruptured, everything collapses. The repair work should therefore primerily restore the horizontal tensile strength of the cell wall. This was achieved by cutting away the damaged joint and using strips of 500 x 100 x 20 mm laid horizontally across the damage, than welded. On completion of the strip welding the hollow spaces inside were backfilled from the topside.

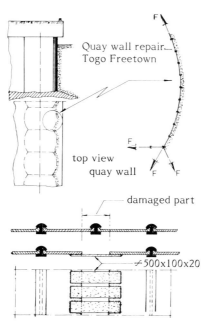

Quay wall repair
Togo Freetown

top view
quay wall

damaged part

500x100x20

This example has been chosen, because the underwater welding repair comes after backfilling under heavy tensile loads and therefore the use of strips enabled the total weld length to be increased to achieve the required strength. In this case the welding process used was of the conventional underwater stick welding. To-days method we would use the MAG-process, enabling, because of the enormous amount of strips involved, a quicker fusion.

My third case-history is a continuation of ships which are involved in colli-
sions as shown in the previous case, in this example two ships collided at
the entrance of the English Channel. One of the vessels sunk and the other,
named "Maroudio", was bought crippled into the French port of Brest. Inspection revealed a large damaged area in the bulbuous bow, measuring 2,5 mtrs.
x 4 mtrs. (8' x 13'). Drydock facilities to accommodated such a large vessel
were not available and the cargo of steelplates were unable to be discharged.
A proposal was tendered and accepted by P & I. This proposal incorporated the
prefabrication of a cofferdam and to install the cofferdam underwater over th
damaged bulbuous nose of the "Maroudio". The cofferdam was welded on using
conventional stick welding to ensure a water tight chamber. Water inside the
cofferdam was pumped out leaving a dry chamber around the damaged area. Stren
thening bracings and fastenings were welded inside the cofferdam. The "Maroudio" was then able to continue her voyage taking 3 months calling at 4 ports
before reaching Japan where she was drydocked and full repairs were made.

BULBUOUS BOW REPAIR "MAROUDIO"

BREST

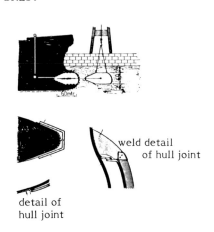

weld detail
    of hull joint

detail of
hull joint

The fourth case-history relates to the repair welding of sheetpiling. This is
indeed a very common application of underwater welding. In most cases during
the driving of the sheetpiling one or two of the sheetpiles become separated
from the guide locking devices. This causes in most cases a vertical triangular opening (with soil leaking through such as a harbour quay wall). The repair procedure for the triangular opening is for a prefabricated patch to be
welded in position. The size of the prefabricated patch is taken from the top
of the sheetpile to ± 50 cm. off the bottom. The prefabricated patch consists
of a relatively thin flexible metal sheet, with a thick edging for the welding process. The flexability of the patch enables it to conform to the shape
of the opening, included on the prefabricated patch at regular intervals are
grout nipples enabling the facility to grout behind the sheetpiling. The grou
ing will give a second line of defence by stabilizing the soil directly behir
the opening and prevent further collapse of terrain directly behind the wall.

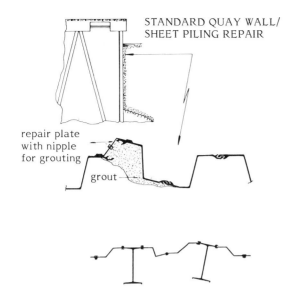

STANDARD QUAY WALL/
SHEET PILING REPAIR

repair plate
with nipple
for grouting

grout

Case-history five takes us a small step from the quay walls and sheetpiling
repairs into an area of construction pits widely used in The Netherlands.
These pits are constructed below ground water level, because in some areas
the drainage is not adequate enough. The construction method works as follows:
Sheetpiles are hammered into the ground and a frame of cross bracings are installed. Then the soil is excavated and the space between the sheetpiles must
now be filled with water. This technique has two functions.
Firstly: to equalise the pressure on the waterside and groundside of the sheetpiles.
Secondly: to prevent the total collapses of the sheetpiling. When construction
depth is attained underwater concrete is poured to make a floor.
After curing of the concrete the water is them pumped out. Prior to the underwater concrete being poured, and while the pit is still full of water, the
concrete must have something to keep it anchored to the sheetpiling, and prevent the concrete from breaking up by the waterpressure beneath the concrete
floor. The welding technique used for this kind of work is called Stud-Welding. Stud-Welding is carried out with a specially designed waterproof weld-
inggun. The advantages of using the Stud-Welding techniques are that the process of Stud-Welding takes less than a second, the process is mechanized enabling a diver with very little welding experience to use it, and that the
quality of the studweld reached is known and easy to control, therefore more
correct based calculation on forces can be made.

STUDWELDING/WELDING OF ANCHORS

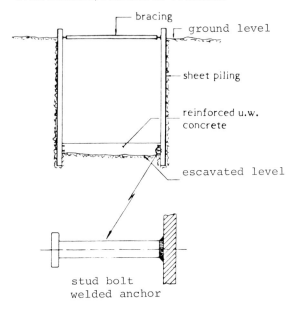

stud bolt
welded anchor

Case history six involves a standard offshore jacket repair using a cofferdam for a nameles oil company in 1982.
In this case the contract required a replacement for a diagonal bracing and
included a thorough inspection of the anode protection.
The depth of the repair was 5 mtr. minus sea level.
Such a job required a cofferdam welding procedure because of the following
advantages:

- dry chamber in which to weld
- working in the splash zone area you are protected from the swell and external eliments including sea conditions upto gale force 6.

Company representatives can supervise the whole operation and surface orien-
tated inspection techniques can be used. Non diving personal can be involved
in this method of work, providing they have attained a level of competence
in emergency escape training. No pressure fluctuation while working in the
splash zone inside the cofferdam.

So with the stickwelding procedure, there is very little difference in the
technique used in the cofferdam as to welding on the surface. For deep water
welding projects a specially designed Habitat is used, which must then involve special diving techniques for a successful underwater welding completion. Welding techniques are such that normal stickwelding is used in the
Habitat, and for the future, trials are now taking place using remote con-
troled welding apparatus.

Cofferdam

Attached to a Production
Platform leg.

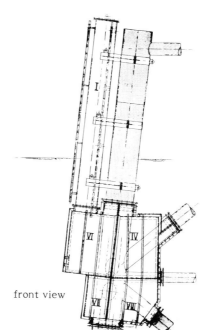

front view

AUTHOR INDEX

SUBJECT INDEX

393

– for safety and quality

Great Achievements often Originate from Simple Ideas.

This also applies to the area of energy. The exploitation of energy is often based on simple and easily comprehensible principles.

For over 3000 years man has obtained energy from running water. The principles originally used are also the basis for present day use of the paddle wheel and turbines in the large water works.

The movement of the waves also contains large, unused energy potential. Present day research utilizes modern principles to exploit this resource. Power from tidewater is exploited on a commercial basis in France. Other countries with large variance between ebb and tide have also planned similar projects. Costs, however, have prevented further development of projects of this kind.

Phillips is greatly concerned about the future energy situation. Several thousand Phillips employees all over the world are involved in research and development projects for the purpose of exploiting existing energy resources better, and developing new ones.

Simple ideas play a large part in these projects too. Phillips has up to now registered 6000 patents in all fields of energy research and the company is actively working to put these ideas into practice.

Energy is more than oil. Phillips is more than an oil company.

Phillips Petroleum Company Norway
Stavanger, Norway

MK STAVANGER RRA 16

AKER GROUP

A|S AKERS MEK. VERKSTED

A/S Norske Shell's olje- og gassfunn på Troll-feltet er gigantisk. Men det representerer samtidig den største tekniske utfordring i Nordsjøen til nå. Både for Norske Shell og for norsk industri forøvrig.

TROLL SKAL TEMMES

Troll kan temmes, er den kjente tittelen på Shake-speare's fornøyelige lystspill.

Med en liten omskrivning gir denne samtidig uttrykk for den innstilling som råder i Norske Shell for tiden: Troll <u>skal</u> temmes.

Aldri har så store gass- og oljeforekomster ligget så vanskelig til som nettopp her ute på Troll-feltet.

Kombinasjonen av 350 meters dyp, en løs havbunn og et grunt reservoar stiller helt spesielle krav til operatøren.

Før feltet kan bygges ut, må derfor eksisterende teknologi videreutvikles.

I denne vanskelige fasen satser Norske Shell sterkt på norsk industri, og dette er helt i tråd med selskapets mål om at den norske andelen av leveranser til utbyggingen av Troll-feltet skal være størst mulig.

Ønsker du ytterligere informasjon om Norske Shell's engasjement på kontinentalsokkelen, kan du henvende deg til A/S Norske Shell, Avdeling for Samfunns-kontakt, Tullinsgt. 2, Oslo 1. Tlf.: (02) 20 02 50.

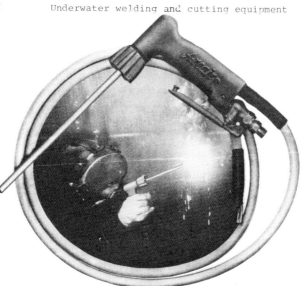

You know something about Smitweld.
You know nothing about Norweld.

The future belongs to those who know.

For instance, that Norweld is Norway's biggest welding supplier. That Smitweld is Holland's – and one of Europe's biggest producers of stainless steel electrodes. And that Norweld and Smitweld merged in 1981 to become one of the biggest welding supplier in Europe.

In 1982 Hilarius, Holland, joined the Norweld/Smitweld group, and we grew even stronger. We already owned AGA/Norweld and Svenska Carbidkontoret in Sweden, and Dansk Terosvejs in Denmark.

We also have subsidiary sales companies in many other countries, including Germany, Belgium and the U.S. And you will find Norweld/Smitweld agents all over the world.

What does all this mean for your business?

It means that in Norweld/Smitweld you will find a supplier with a complete stock of high quality gas- and electric welding consumables like electrodes, filler rods and fluxes as well as other welding equipment.

But aren't there other suppliers who can say the same? Yes, a few maybe, but none with the same position on the Norwegian and Dutch markets. And hardly any with the same expertise in welding with stainless steel electrodes. Nor do we think we're exaggerating when we say that we have one of the world's most advanced welding laboratories.

So what does this mean for you?

It means you can be a part of what's happening at Norweld/Smitweld. You will be able to join in taking the lead in welding development. You will be among the first to adopt new technical innovations and so be able to make better and more competitive products. Faster and at a better price.

Big words?

Maybe, maybe not. But do you want to risk being left out? Mail the coupon instead, and we'll tell you more.

YES, I WOULD LIKE TO JOIN IN LEADING DEVELOPMENT IN WELDING.

☐ Please send me more information about Norweld/Smitweld.

☐ Please send me more information about Norweld/Smitweld's range of welding equipment.

Name:

Position:

Company:

Address:

Telephone:

NORWELD SMITWELD

Mail the coupon to: Norweld A/S
Sandakerveien 64, Oslo 4

Now you know Norweld/Smitweld.
The new welding connection in Europe.

An Ocean
of Welding Technology

THE KVAERNER GROUP

An unusual combination of diversified skills, decentralised organisation and flexible collaboration puts the Kværner Group in a strong position to respond swiftly and competitively to industry needs. Speciali-sing in advanced design and fabrication projects, the group is able to tackle such assignments at all stages from initial planning to final delive-ry. Although preferring to undertake complete installations or systems, Kværner companies can also supply high-quality components — particu-larly where very stringent standards of fabrication expertise are required.

Ranked as the largest mechanical engi-neering group in Norway Kværner's ob-jective has been to become big in a num-ber of specialised markets within heavy engineering, shipbuilding, equipment manufacture and engineering design. Its product spectrum covers a wide field, in-cluding equipment and skills that have found a natural place in the offshore and petrochemicals sectors.

Apart from turbines, heavy plate struc-tures, submersibles, pumps, fire-fighting gear and anti-pollution equipment, this ca-pability also embraces advanced gas tan-ker and refrigeration technology. Altho-ugh all owned by Kværner Industrier A/S, the holding company, the group members are organised as independent profit cent-res with responsibility for developing and applying their own specialities. But they can readily combine to pool mutual ex-pertise within special project organisa-tions set up to link two or more group firms for specific assignments. These can either be contracts beyond the capacity of a single company or deliveries where it is necessary to draw on the experience and technical know-how available in several firms.

Such collaboration ventures may also in-clude outside partners when additional skills or resources are required. At the same time, more informal or limited forms of cooperation are easy to arrange within the group structure — allowing people or facilities to be drawn on for particular aspects of a delivery either as sub-contractors or advisers, for instance.

This high degree of organisational flexi-bility allows the group to cope with pro-jects of virtually any size or complexity and to meet rapidly-changing customer requirements with maximum effective-ness. At the same time, Kværner's ability to respond to the need for technical inno-vation is backed by a concentration on tailor-made products that frequently invol-ve large dimensions, a high degree of ac-curacy and skilled metal-working.

None of the member companies are in-volved in mass production, nor do they have any output worth mentioning of stan-dard lines in large series.

Kværner Brug A/S P.O.Box 3610 Gb., Oslo 1 Telephone + 47 2 67 69 70 and 67 65 50 Telex 71650 kb n Telegrams "Kvaerner" **Div. Eigersund** 4370 Egersund Telephone + 47 44 90 088	**Hydro-power plants:** Turbines, regula-tors, valves, dam hatches, penstocks. **Heat engineering:** Steam and gas tur-bines. **Marine equipment:** Hatch covers, motorcar decks, ro/ro equipment, self-unloading equipment for dry bulk car-go, cargo tanks for liquefied gases and chemicals, controllable pitch prop-ellers, outboard elevators.	**Mining and cement industry:** Grinding mills. **Offshore and other industries:** Steel constructions, offshore structures, loading platforms, process equipment, trenching system for burying pipelines. **Environment protection:** Recirculation plants for solid urban waste.
Kvaerner Engineering A/S P.O.Box 222, 1324 Lysaker Telephone + 47 2 59 50 50 Telex 16996 kveng n Telegrams "Kvaerncon"	Consulting and engineering firm carrying out integrated projects in the oil/gas and industry/energy fields.	
Moss Rosenberg Verft A/S **Moss Yard** P.O.Box 1053 Jeløy, 1501 Moss Telephone + 47 32 52 221 Telex 16334 mvd n **Rosenberg Yard** P.O.Box 139, 4001 Stavanger Telephone + 47 4 54 50 00 Telex 33029 mrv n	**Newbuilding** up to about 160 000 tdw. Specialities: LNG, LPG, ethylene and chemical tankers. **Ship repairs** **Marine equipment:** Inert gas systems, gas detection systems, fairleads, boilers.	**Industry:** Boilers, smoke cleaning, heat recovery. **Oil and gas:** Construction, building and outfitting of offshore platforms, re-pairs, floating gas liquefaction/storage plants.

Norsk Hydro pioneer in the North Sea

Norsk Hydro is the largest industrial undertaking in Norway, and also the largest electrochemical group in Scandinavia. The company makes a large range of products in the fields fertilizers, oil and gas, aluminium, magnesium, petrochemicals and industrial chemicals including welding gases. Hydro and its subsidiaries have 10 500 employees in Norway and 7 200 abroad.
Altogether the Hydro Group currently comprises about 50 wholly owned companies in Norway and abroad, and in addition has holdings in a further 60 companies.

Oil and gas

Norsk Hydro's oil activities started as early as 1963, when the company formed the Petronord group together with the French oil companies Elf, Aquitaine and Total.
Hydro now has interests varying from 5 to 34,6 per cent in more than 40 blocks in the Norwegian sector of the North Sea. This makes Hydro one of the most active companies on the Norwegian shelf.

Hyperbaric Welding

Norsk Hydro has been working on development of hyperbaric welding procedures since 1975.

1975-1978 Frigg-Karmøy pipeline study (DHWP). Development of 2 welding procedures for the Petronord group and Statoil. One TIG + MMA and one full TIG weld. Extensive testing in simulator at Sintef and Institute de Soudure.

1978-1981. Inhouse and Sintef research on specific problems regarding arc stability of TIG and different aspects of MMA. Study on state of the art of welding processes for hyperbaric welding.

1981. Deep-Ex. 81. Deep-Weld.
Norsk Hydro operated welding experiments to 500 m.s.w. including occupational hygiene aspects.

1982. Norsk Hydro operator for German/Norwegian project on mechanized/automated orbit-TIG welding system.

Future: Norsk Hydro will continue to play a leading role in the development of welding technology for deep water applications (250-400 m.s.w.) both in North Sea and off Northern Norway.